PRAISE FOR *THE ORGANIC SEED GROWER*

"John Navazio has written a fantastic guide for organic seed breeders and producers. He has taken organic seed production to a higher level with extensive information on selection, genetic integrity, isolation distances, and seedborne diseases. Although his focus is on plant breeders and commercial growers, much of the information is also applicable to small-scale farms producing seed for on-site use."

—**SUZANNE ASHWORTH**, author of *Seed to Seed*

"John Navazio has made a keystone contribution to the future of the grassroots organic seed movement. *The Organic Seed Grower* is a fundamental resource for the preservation and improvement of agricultural biodiversity. It is an essential guide to high-quality, organic seed production: well grounded in fundamental principles, brimming with practical techniques, thorough in coverage, and remarkably well organized, accessible, and readable."

—**JEFF MCCORMACK**, PhD, founder of Southern Exposure Seed Exchange

"With *The Organic Seed Grower,* well-respected plant breeder and seed expert Dr. John Navazio has written the definitive book on organic vegetable seed production. Encyclopedic, yet well written and approachable, this seminal work deserves a place in every grower's library. From the organic farmer seeking a comprehensive reference, to the family farmer who wants to learn how to guarantee access to a favorite variety, to the progressive seed saver committed to success—all will find this book an indispensable guide and Navazio a trusted partner in organic seed improvement."

—**JIM GERRITSEN**, president, Organic Seed Growers and Trade Association (OSGATA)

"There is nothing more important right now than growing and saving seeds—that most essential aspect of life. While we may have all done this once upon a time, we have mostly lost these skills to private industry or urbanization. Until now. John Navazio reveals all the techniques and tricks, some simple and some complex, that he's learned only after decades of careful observation and practice. Incredible photos help tell the story of life that seeds represent. *The Organic Seed Grower* is what we need to take back community control of seeds from those who have taken it from us."

—**TOM STEARNS**, president, High Mowing Organic Seeds

The Northeast Sustainable Agriculture Research and Education (SARE) program is a USDA competitive grants program, and grant funds have supported the writing, editing, and production of this book. Northeast SARE invests in projects like this one to improve the understanding and adoption of sustainable farming techniques. SARE defines sustainable farming as profitable, environmentally sound, and good for communities.

There are four SARE regions that serve the United States and the Island Protectorates. The Northeast SARE region is made up of Connecticut, Delaware, Maine, Maryland, Massachusetts, New Hampshire, New Jersey, New York, Pennsylvania, Rhode Island, West Virginia, Vermont, and Washington, D.C.

To learn more about SARE and the grant offerings in each of the four SARE regions, go to www.sare.org.

The
ORGANIC
SEED
GROWER

The
ORGANIC
SEED
GROWER

JOHN NAVAZIO

A Farmer's Guide to Vegetable Seed Production

Chelsea Green Publishing

White River Junction, Vermont

Project Manager: Patricia Stone
Developmental Editor: Ben Watson
Copy Editor: Laura Jorstad
Proofreader: Eileen M. Clawson
Indexer: Margaret Holloway
Designer: Melissa Jacobson

Printed in the United States of America
First printing November, 2012
10 9 8 7 6 5 4 3 2 1   12 13 14 15 16

Our Commitment to Green Publishing
Chelsea Green sees publishing as a tool for cultural change and ecological stewardship. We strive to align our book manufacturing practices with our editorial mission and to reduce the impact of our business enterprise in the environment. We print our books and catalogs on chlorine-free recycled paper, using vegetable-based inks whenever possible. This book may cost slightly more because it was printed on paper that contains recycled fiber, and we hope you'll agree that it's worth it. Chelsea Green is a member of the Green Press Initiative (www.greenpressinitiative.org), a nonprofit coalition of publishers, manufacturers, and authors working to protect the world's endangered forests and conserve natural resources. *The Organic Seed Grower* was printed on paper supplied by QuadGraphics that contains at least 10% post-consumer recycled fiber.

Library of Congress Cataloging-in-Publication Data
Navazio, John.
  The organic seed grower : a farmer's guide to vegetable seed production / John Navazio.
     p. cm.
  Includes bibliographical references and index.
  ISBN 978-1-933392-77-6 (hardcover)—ISBN 978-1-60358-452-4 (ebook)
1. Vegetables—Seeds. 2. Vegetables—Propagation. 3. Vegetables—Organic farming.
4. Organic farming. 5. Seed technology. I. Title.

  SB324.75.N38 2013
  635—dc23
                               2012034338

Chelsea Green Publishing
85 North Main Street, Suite 120
White River Junction, VT 05001
(802) 295-6300
www.chelseagreen.com

CONTENTS

Over the past decade it has become apparent that there's a real need for a comprehensive guide to growing organic vegetable seed. For some time now it has been obvious to dedicated practitioners of organic farming that, to be in harmony with the philosophical underpinnings of organic agriculture, it's important to use seed produced using organic practices. This same dedication to organic principles is integral to the mind-set of a number of worldwide organic certification agencies' governing bodies, all of which have added an organically grown seed requirement since 2002. Of course such regulations can't be implemented overnight, but the very existence of the rules has dramatically increased the market for, and use of, certified organically grown seed. Several other important factors are also creating the increased interest in growing seed among organic farmers, though, and these are based on more than just the economics of supply and demand.

WHAT THE GROWERS NEED

For instance, there is a growing awareness among many organic vegetable farmers that we need a reliable supply of high-quality organic seed that's truly adapted to the challenges found on organic farms. What's more, astute farmers—those with years of experience under their belts—are increasingly realizing that the number of vegetable varieties suited to their operations is diminishing.

Growers are seeing a real narrowing of the vegetable varieties that are commercially available, due in large part to the consolidation that has taken place within the seed industry. A series of corporate mergers among the most important seed companies has been ongoing since the 1970s, and the trend has only accelerated in the past decade. Whole market classes of vegetable varieties are being lost as an inevitable result of this. Many varieties that have a certain specific climatic or cultural adaptation, or perhaps have specific market traits that are considered limited in their sales potential for the new corporate owners, are cut from a company's sales list and replaced with varieties that have more universal appeal. In almost all cases, the varieties that remain are those exceptionally well suited to high-input production systems and geographic areas with ideal climates. The varieties that are dropped are the ones less well suited to large-scale centralized agricultural areas. It goes without saying that a large percentage of diversified organic growers are producing vegetables regionally, across the many and varied climates of North America, and not under the ideal cropping conditions nor with the high external inputs that are taken as a given by large-scale conventional farmers.

Amid this climate of consolidation and diminishing choices, there is also the fact that varieties in many crops have been reduced almost exclusively to F_1 hybrids. While it's true that most commercial organic vegetable growers have been using a good number of

F_1 hybrid varieties for their market production, the standard open-pollinated (OP) varieties that have been around for years have also played an important role in many of the planting slots that make organic cropping successful. Many of these successful OP vegetable varieties were bred during a prolific era of plant breeding that extended from the late 1940s through the 1970s. And many of the best varieties from this era were carefully maintained by seed companies, becoming reliable workhorses for organic farmers—notable for their ability to produce good crops in less-than-ideal situations. Unfortunately, a major trend in the seed industry since the 1980s has been a gradual abandonment of these varieties, leaving us with a syndrome I call "hybriditis," where virtually every variety available in certain crop types is a hybrid. The common refrain repeated by large seed companies has been, "Hybrids are much better than open-pollinated varieties." This has certainly become a self-fulfilling prophecy, as many of the OP varieties haven't been adequately maintained through selection and proper varietal upkeep for many years.

With all these factors contributing to the loss of crop diversity and crop choices, the idea of organic farmers producing vegetable seed began to gain traction in the 1990s. For some growers it was purely an act of necessity: Seed of an OP variety important for their production was no longer commercially available. For others, growing for a nascent organic seed market seemed a potential moneymaking opportunity. Still others simply felt that the seed industry was going to hell in a handbag, and it was time to take back the seed.

Many of these vegetable farmers had never considered producing their own seed before. They might have saved some of their own pea or bean seed, yes, and many were already saving seed from certain heirloom tomatoes or peppers for their markets, but growing seed of the more difficult dry-seeded crops was a different kettle of fish. Much of this process seemed to depend on technical know-how and fancy threshing and cleaning equipment, all of which made the prospect of seed growing seem completely out of reach. But necessity once again proved the mother of invention, and a number of pioneering farmers started to find ways to grow, harvest, thresh, and clean seed.

Until very recently the growing of seed was an integral part of all agricultural practice, in all agricultural societies. Growing, harvesting, cleaning, and storing seed was simply something farmers did to ensure that they could plant the same crop the following year. Keeping an eye on a crop's performance and selecting seed from the best plants was a vital part of the process. Indeed, a farmer's ability to maintain a good seedstock was—and still is in many parts of the world—one of the key elements in determining his or her prosperity. Maintaining a good seedstock has many parallels with maintaining good breeding stock in livestock, and both have always been good indicators of the overall health and well-being of a farm.

The model of vegetable seed companies being the exclusive purveyors of seed has really only existed for roughly the past 50 to 100 years. This model developed in the global north due to many of the same forces that were at play in the industrialization of agriculture. Growing vegetable seed commercially has become an increasingly specialized skill—one most farmers have little knowledge of—handled by large specialized companies that both do the research involved in breeding vegetable varieties and produce large quantities of seed. This seed is then disseminated through smaller distribution and retail companies that generally have little or no involvement with seed growing. As the production of vegetable seed has become concentrated into the hands of these very few, large "production research" companies, as they are commonly known, it's left fewer and fewer people in agriculture who possess the skills to produce high-quality vegetable seed. In a very real sense we have lost the diversity of *people* who know how to perform all the steps in this process, which isn't about just growing the seed but also maintaining the variety, keeping it free of seedborne diseases, and harvesting and milling it to the point where it's suitable for commercial use. Also, because seed companies often only breed and produce any given seed crop in one or two of its most ideal climates, few new vegetable crop varieties are adapted to a wide array of conditions—something that was once an important part of the picture when there were many more regional seed companies distributed around the world.

WHY A BOOK?

All the factors that led to this modern renaissance in diversified seed production have prompted the need for a book that is written for the organic farmer who wants to learn the basic techniques and methodology to successfully produce a quality vegetable seed crop. In my own journey to learn the craft and the art of seed growing, I have been amazed at how few good texts there are explaining the basic cultural methods used in growing seed. Indeed, many aspects of the seed-growing craft distinguish it from growing the same vegetables as produce for the marketplace. The timing of the crops through the seasons, the spacing of the crops in the field, and their irrigation and fertilization needs can all be radically different when you're growing vegetable seed. And factors such as the temperature and relative humidity at the time of flowering, the presence of pollinating insects, and the isolation distance that is required between two crops of the same species usually aren't a major concern to the vegetable producer but can make or

break the seed grower. Then there are issues like overwintering biennial crops, roguing and selecting a crop to type, and monitoring the crop for seedborne diseases—all outside the realm of what most vegetable growers ever think about.

This book covers all the major vegetable crops that are normally propagated with "true seed," which is produced through sexual union during the flowering process. Crops such as potato, asparagus, and globe artichoke are not included, because they're routinely propagated asexually with clonal material when grown as vegetables. Most of these clonally propagated vegetables do not come true to type when reproduced via true seed. For these clonally propagated crops, true seed is normally only used during the breeding process.

In this text I also do not describe the intricacies of producing F_1 hybrid seed. There are several reasons for this, both practical and philosophical. First of all, the techniques used to produce hybrid seed (where one parent provides pollen and the other is either pollen-less or does not receive its own pollen) are often quite high-tech and laborious and require additional seasons and isolated fields to produce parental seedstocks in preparation for the actual production of the F_1 hybrid seed. While most hybrid vegetable seed production uses techniques that are acceptable under all current organic standards, a number of the more high-tech methods fall outside the scope, both economically and practically, of most small regional seed companies and the organic farmers interested in producing seed. Information on the standard production methods used to produce hybrid seed is available in a number of currently available vegetable seed production texts that cover many of the techniques used by larger seed companies.

The purpose of this book is to make seed growing accessible to average organic farmers. My goal is to supply the practical techniques and theory that will enable you to successfully grow high-quality seed crops. I have had the good fortune of working with and getting to know some of the great organic seed pioneers of North America over the last dozen years, and my aim is to impart many of the same basic strategies that they use, as well as techniques that have been handed down through the generations, to reestablish vegetable seed growing as a viable part of all agriculture. My hope is that I have been able to deliver this information in a simple enough fashion to be easily understood and used, while at the same time remaining scientifically based and giving you the opportunity to understand the theory behind the principles and practices that I am describing. I'll know that I've been successful in this when I see a dog-eared, dusty, and smudged copy of this book on the front seat of a seed farmer's pickup truck.

ACKNOWLEDGMENTS

The knowledge captured within this text would not have been possible without the good graces of a number of farmers, seed growers, plant breeders, and scientists, who shared their knowledge of the craft and science of growing vegetable seed.

The farmers and seed growers: Frank Morton, Nash Huber, Fred Brossy, Marko Colby, Sebastian Aguilar, Beth Rasgorshek, David Podoll, Rich Pecoraro, Martin and Atina Diffley, and Scott Chichester, as well as a number of other great stewards of the Earth, have been instrumental in my quest to learn how to grow seeds organically. They are the reason that I get up each morning excited to learn just a little more in our quest to put farmers back in control of humanity's inherited wealth, the seed.

I am also very grateful for the advice and support that I received from the plant breeders and scientists who freely gave of their time and energy to carefully read and suggest improvements to this text. They include: Phil Simon, Laura Merrick, Jim Myers, Bill Tracy, Irwin Goldman, Lisa Stephens, Bruce Carle, Jen Kingfisher, and Lindsey du Toit. They have greatly enhanced the accuracy and content of this book. Thanks, too, to the late Stan Peloquin, who awakened in me a true love and appreciation for the reproductive biology of flowering plants.

I owe a huge debt of gratitude for the love and support of my dear friends and colleagues Matthew Dillon and Micaela Colley, who have made my life with seeds possible and the Organic Seed Alliance a reality. My gratitude also extends to Jared Zystro and the rest of the team at OSA for all their tireless work in spreading the word on all aspects of seed sovereignty. I am also grateful for the opportunity to work with the Washington State University Extension in furthering my ability to educate farmers and students in the fundamentals of organic seed production.

This book would not have gotten off the ground without a generous grant from the USDA Sustainable Agriculture Research and Education Program. David Holm and Helen Husher of Northeast SARE had the patience and belief in this project to see it through to the end. A special word of thanks goes to Ben Watson for all of his varied editorial skills and his saint-like patience. Thanks also to Patricia Stone and the rest of the staff at Chelsea Green for guiding me through this process.

Very importantly, I want to extend my deepest thanks for the unyielding love and steadying presence of my daughters, Emilia and Zea. Without their respect for my work and their patience for all of the time I have devoted to preparing this book, this project would not have been possible.

Lastly, I would like to dedicate this book to the memory of my dear aunt Millie, who first instilled in me a deep appreciation for the wild and opened my eyes to the wonders of nature, as my path in life has led directly from my earliest days tromping through the woods of Virginia with her.

PART I

Historical and Botanical Information

A SHORT HISTORY OF AGRICULTURAL SEED

Seed is the vital link to our agricultural past. All of the changes that have occurred in the domestication of our crops are a reflection of the ingenuity and perseverance of our agricultural ancestors. This is even more wondrous when we realize that all of the plants we rely on for our very existence are derived from wild plants that were not usually succulent; in fact, they were often bitter or sour, far beyond what our modern palates would tolerate. The modern crops that grace our tables daily are all derived from a much larger number of plants that were found to be edible and were used extensively by our hunter-gatherer forebears. While natural selection had already shaped these wild plants, the selection pressure of the people who first grew them under cultivation further changed them to better suit their needs. Certainly these early farmer-breeders were interested in plants that were easier to harvest, were more productive, stored better, had better texture or flavor, and had larger fruit, seed, shoots, buds, or leaves, depending on which part of the plant they harvested as food. But natural selection was still the dominant selective agent. These early crops were grown in very rustic agricultural conditions with few or no inputs. The environmental challenges of drought, predation, low-nutrient stress, and excessive heat or cold were ever-present and played a central role in the evolution of our crop plants for thousands of years in the development of agriculture.

Through many generations of cultivation, crops changed based on both the selection pressure of the environment and the selection of humans to meet their cropping methods, culinary needs, and cultural preferences. While some of the early crop domesticates flourished under agriculture, other potential crops were abandoned. The most promising crop plants spread across the countryside via trade and human migration. This dissemination placed these budding crops into many diverse environments under the stewardship of people of different cultures, who often had quite different goals in their selection criteria. What is amazing to ethnobotanists (or anyone who studies the spread of cultivated plants around the globe) is just how genetically diverse, elastic, and adaptive all of the widely cultivated crop species became in a relatively short period of time. The variation that unfolds as diverse challenges of natural and human selection are applied seems to yield an unlimited array of genetic possibilities. Certainly the evidence presented by the fantastic spectrum of shapes, colors, flavors, and sizes, as well as the range of climatic adaptation of our crops when compared with their wild ancestors, is proof positive of this.

Over thousands of years of practice, agriculture became increasingly sophisticated. The cultural methods used by many agricultural societies moved from hand tools to implements

◀ **Cecilia Joaquin, a Pomo woman, using a seed beater to gather seed into a burden basket, 1924.** Edward S. Curtis/Library of Congress Prints & Photographs Division, Curtis Collection

drawn by animals; out of necessity a number of farming communities devised sophisticated irrigation systems; many learned the value of nutrient cycling; and all improved the genetics of the crops they used. Most agricultural societies developed deep traditions and ritualistic methods of gathering seed that were carefully handed down from one generation to the next. In many cases there was an emphasis on how to select for important, desirable traits from the correct number of plants without too much emphasis on any particular type of plant. But despite the increased sophistication in agricultural production and accumulated wisdom in maintaining and improving seedstocks, the forces of natural selection remained the dominant factor in shaping the crops that they grew right up until the industrialization of agriculture in the 20th century. The best farmers knew this instinctively and worked in concert with natural selection. If the crops they relied on weren't suited to the local environment and growing conditions, then the farmers would have to abandon them for something else that might work.

▼ Selecting seed corn in the field allows for selection of both plant and ear traits. Bain News Service/Library of Congress Prints & Photographs Division, Bain Collection

SEED AS TRADE

Trade in seed has been a key component in the spread of crops since very early in the domestication process. Seed of a promising crop has probably been a large enticement in many bartering situations. The commercial sale of seed, however, didn't become commonplace in North America and Europe until the second half of the 19th century, and the seed trade of that era consisted almost entirely of small packets of vegetable, flower, or herb seed that served as starter packets for future seed-saving activity. The growing and saving of farm and garden seed was as much a part of any agricultural pursuit as was working the soil and planting in the spring. With the advent of commercial seed companies, farmers and gardeners bought seed of specialty and hard-to-find items. Sometimes they purchased seed of a particular species that was difficult to produce in their climate. And sometimes people purchased seed packets to try something new, exciting, and full of promise.

Starting in the 1880s and continuing into the 1920s there was a steady increase in the number of start-up seed companies in North America. In addition to many new companies entering the vegetable and flower seed market, there were also companies specializing in seed corn and other agronomic staples on a scale that hadn't been seen before. It is impossible to get a comprehensive list or an accurate count of all of these companies—some were quite small, and some only lasted a few years—but by the 1920s literally hundreds of them had sprung up in every agricultural region of North America. These companies were all regionally based, and they concentrated on supplying seed of crop varieties that were regional standards as well as finding unusual or unique varieties that would give their customers something new or exotic to grow. Most of the seed was grown within the company's geographic region (a small amount was imported from Europe and elsewhere), and a company would rise and fall on the quality of its seed. Quality had two components:

1. The seed had to be relatively clean, free of debris, and undamaged, as well as offer a decent germination rate.
2. The seed had to have reasonably good varietal integrity, being true to type, largely uniform, and able to perform in the field with some semblance of what the catalog claimed.

Standards for seed quality and varietal integrity were not nearly what they are today, and standards for uniformity were much more forgiving, but farmers did notice differences and knew good seed from bad.

Concurrent with this growth in commercial seed was a series of innovations in cultural practices, science, and engineering. These set the stage for a revolution in agriculture that would change the fundamental landscape of farming in the 20th century. Toward the end of the 19th century the first commercial pesticides became available, with Bordeaux mixture being used for fungal diseases of fruit, and various forms of arsenic and pyrethrum used to control insects in a number of high-value crops. At the beginning of the 20th century came the rediscovery of Mendel's principles of heredity, which laid the foundation for genetics and modern plant breeding. By the 1910s several American manufacturers were making gasoline-powered tractors; the first synthetic nitrogen fertilizer was being produced in

Germany by harnessing atmospheric nitrogen to make ammonia using the Haber-Bosch process. Throughout this entire period hundreds of smaller hydroelectric dams were being built across North America, laying the groundwork for the large irrigation projects yet to come.

These changes didn't "take" with farmers overnight. First of all, many of these inputs were expensive, and most farmers were not operating on a cash-intensive system—they produced all or most of their own fertility, feed, and seed for their farms. Pesticides, nitrogen fertilizer, and even tractors wouldn't become commonplace on North American or European farms until after World War II, and even later in other parts of the world. The main source of fuel on the farm was the grain and hay produced on-farm for horses. It's hard to believe now that only 100 years ago, even in countries that were rapidly industrializing, most of the population lived on farms that were largely self-sufficient, breeding their own animals and growing their crops from seed they had grown.

THE CHANGE

Many social, cultural, and political events occurred between the 1920s and the end of World War II that had a profound effect on agricultural practices in many industrial societies. Modern plant breeding came of age during the 1920s, and the development of hybrid corn received most of the attention. Much of the early breeding work in corn concentrated on improving its stalk strength and ability to produce under challenging conditions. This proved invaluable in many areas of the Corn Belt when several cycles of drought, extreme heat, and high winds conspired with poor farming practices to create the Dust Bowl crisis in the American Great Plains during the 1930s. The Dust Bowl, coupled with a worldwide economic depression and a near collapse in world markets for wheat, corn, and cotton, shrank many farmers' profit margins and forced many people off the land. Further displacement of people from farms occurred throughout many industrialized societies with the onset of World War II as young men left to fight, people moved to cities to work in factories for the war effort, and whole nation-states were occupied by invading forces. This set the stage for increased agricultural mechanization when wartime demand for agricultural production became crucial to the Allies as the conflict in Europe became a world war.

The war pushed industrial and scientific development to the hilt in dozens of areas. Many American factories were built solely to produce enough ammonia for gunpowder and bombs used by Allied forces worldwide. US agricultural production played a pivotal role supplying troops and allies with staple foods. Production was bolstered through federal programs such as the Lend-Lease Act, which promoted a more industrial farming model geared toward maximizing yields and raising quality standards of important agricultural commodities. American farmers increased production as much as possible with limited resources throughout the war. Seed production for a number of crops boomed on this side of the Atlantic, with vegetable seed for war-torn Europe in high demand. Some cabbage seed growers in the Puget Sound region of Washington State were said to have paid off their mortgages with three good crop years during the war. By the end of the war tractors outnumbered horses on American farms for the first time. Agriculture in industrial societies was changing.

THE POSTWAR INDUSTRIAL AGRICULTURE MODEL

In the aftermath of World War II there was a gap to fill in the wartime industrial output,

Efforts to increase production during World War I are ▶ *reflected in this famous 1918 poster of Liberty sowing seeds.* James Montgomery Flagg/Library of Congress Prints & Photographs Division

and agriculture came to the fore. Factories that had produced ammonium for explosives switched their production to nitrogen fertilizer. Organophosphate insecticides and herbicides that were newly developed before the war became inexpensive and widely used after it. Hydroelectric dams that were started as Works Progress Administration projects during the 1930s realized their full agricultural potential as western states expanded acreage, developing irrigation districts and building canals and ditches to adequately utilize the water.

Perhaps the most profound change in this postwar era was the mind-set of the new generation of agricultural innovators. In America the returning soldiers were given generous benefits, including the opportunity to go to college with their tuition and most of their living expenses paid for by the GI Bill. Many soldiers who took advantage of this had been raised on farms, and a fair share of these farm boys chose studies in agriculture. Among those who excelled in school, a good number went on to graduate school in agriculture. With great increases in federal funding for agricultural research at universities and the USDA, as well as rapid expansion in all areas of agricultural business, many new positions were being created at this time. These led to leadership roles in both public and private agricultural policy and research in the postwar era for this new generation of young agronomists and scientists that had returned from wartime service. Among this generation there was a real sense that all of the problems in agriculture could be solved with innovations created through scientific understanding, technology, and modern industrial muscle, if coupled with the same "can-do" spirit that had helped the Allies to beat the Axis powers during the war. It was an innocent time, with blind

SAVE SEED CORN NOW!

An alarming shortage exists!

The quality is poor

The situation is everywhere the same—

Very little SEED can be purchased.

1. *Save every good ear.*
2. *Test each ear.*
3. *Cure in a dry place.*
4. *Plant seed grown near home.*
5. *Save your surplus.*

If you buy seed, insist on a statement showing county and state where grown, and percentage of germination

Unless small supplies can be collected from numerous farmers, many counties in the Northwest will have nothing to plant in the spring. *Report any surplus to your bank*, which will arrange for its sale at a good price, or write to the

FEDERAL RESERVE BANK OF MINNEAPOLIS.

◀ This 1917 poster presents saving seed corn for replanting as a patriotic duty, with Uncle Sam advising and supporting the American farmer. Scott Printing Company/Library of Congress Prints & Photographs Division

faith in scientific solutions to vexing problems. The skeptical questioning of technological fixes in agriculture that has become the norm, especially amongst the organic farming movement since the 1970s, was virtually nonexistent at this time.

In the postwar era the Marshall Plan, a massive US foreign aid program to help rebuild the economies of war-torn Europe, spread this industrialized model of agriculture across Europe and beyond. Conceived as a way to boost reconstruction of all aspects of each European country's economy, industrial capacity, and infrastructure, this plan had a strong agricultural emphasis. Hunger was the immediate problem at war's end, and flour, fuel, and cotton were among the first shipments to Europe from America. The next shipments were of manufacturing and mining equipment, farm machinery, fertilizer, and seed. A priority of the Marshall Plan was to stimulate agricultural production to offset the balance-of-trade deficit that existed in these countries. American technical advisers were sent to Europe to assist farmers in the use of the modern agricultural equipment. These advisers also served as ambassadors for modern farming methods and the use of external inputs such as fertilizer, pesticides, and the seed of modern crop varieties that were being developed during this era. (Also remember that inputs like nitrogen fertilizer and DDT were relatively cheap—they were based on petroleum, and oil was extremely cheap in the postwar boom years!)

SEED AT THE CROSSROADS

As productivity increased with the new industrial model, many aspects of agriculture became more specialized. Seed was no exception. With the rapid success of hybrid corn before the war, the stage was set in this postwar era for seed companies to offer improvements in other crops that would convert more and more farmers into purchasing seed every season. It should be remembered that, even at this point in the transition from traditional farming to what we now refer to as modern agriculture, farmers still grew most of the seed to meet their own needs. This was as true of farmers in industrialized societies as it was in true agrarian societies in the mid-20th century. For farmers, producing their own seed was clearly an integral part of their operation. It was a part of every farm. The ability to maintain good seedstock, which is fundamental to producing a good crop, was a testament to their true ability as farmers, much as the breeding and upkeep of their livestock. It was woven into the fabric of the farm. The seed was part of their farm and their farm was part of the seed. Each variety that was selected over time to meet the environmental conditions and the farmer's needs became part of the whole system used on the farm. That was the way it had always worked.

The process of replacing farm-grown seed with seed developed somewhere else was gradual and only occurred when the new seed had a quality that made it hard to resist. This process would still take another 30 or 40 years in even the most industrialized agricultural areas, with seed of some crops such as small grains and cover crops still often produced on-farm today, long after growers have become accustomed to buying vegetable, corn, or pulse seed.

SEED IN THE POSTWAR ERA

The American seed companies that grew considerably during the war were keenly aware of the benefits that had been realized during the intensive scientific breeding efforts applied to hybrid corn by both university breeding programs and corn seed companies in the 1920s and 1930s. With the exception of corn, almost all modern plant breeding in the prewar era was done by a relatively small number of genetic researchers at universities. Most seed companies at that time sold seed of established, public-domain crop varieties, with new varieties only occasionally developed by astute stockseed personnel at the companies or by farmer-breeders. Before the war most seed companies did not have trained plant breeders. After it, given the sharp increase in the number of plant-breeding programs at universities, there was soon a great increase in the number of young plant breeders being trained. This coincided with a demand for professional plant breeders at many of the larger seed companies, which had prospered and wanted to develop new varieties in order to successfully compete. The zeitgeist of modern agriculture at this moment in time was placing a strong emphasis on seed and the promise of plant breeding to solve problems.

Modern plant breeding in the mid-20th century used an assortment of methods based on genetic concepts that had been developing ever since the rediscovery of Mendel's principles of inheritance in 1900. Plant breeding certainly provided a way to incorporate any number of important traits such as disease resistance, climatic adaptation, nutritional quality, and yield potential into a single crop variety in relatively few generations. It was also possible to produce crops with a greater degree of uniformity than was the norm for that time. Many farmers didn't strive for this level of uniformity in their own seed selection, as it wasn't that important in their own crops. In vegetable crops much of the production of this time was done using diversified cropping systems with multiple harvests for local consumption. Very little mechanization was used in these systems, and standards of varietal uniformity for

Professor George F. Sprague was a prominent figure in plant breeding research at Iowa State College and trained many of the best postwar plant breeders during his career. Ames, Iowa, 1942. Jack Delano/Library of Congress Prints & Photographs Division, FSA/OWI Collection

commerce were not as strident as they would later become. The uniformity that came with the new wave of breeding, in both hybrids and the new open-pollinated varieties that were being created, certainly fit the bill with many of the other new industrial elements that were quickly coming into agriculture.

While seed was beginning to be shipped far and wide during this era, with American seed routinely going to Europe due to the postwar rebuilding efforts there, seed companies were still predominantly regional in their focus. Most crops were still produced and used regionally, and thus a regional emphasis on offering seed of varieties with climatic adaptation, ones that satisfied market preferences for a particular region, was a solid business strategy. Especially in vegetable crops regional production was still the name of the game, and postwar prosperity across North America only increased the regional demand for vegetable seed. In Northern Europe and Japan the story was much the same, with vegetable seed companies that had prominence before the war emerging from the ashes, becoming regional powerhouses, and eventually starting up breeding programs during this postwar period.

THE GOLDEN AGE OF PLANT BREEDING

With the boom in seed sales both during and after World War II, the most progressive corn and vegetable seed companies wasted little time building a whole new infrastructure with improved stockseed departments, technicians accustomed to working with plant breeders, and highly skilled plant breeders, who were emerging from the handful of top agricultural schools with a wide-eyed enthusiasm for their craft and its potential to change the face of agriculture.

This new generation of seed-company plant breeders was concerned first and foremost with supplying the growers with varieties that worked for the soil, climate, cultural practices, and markets in the region where each company was based. Most of these breeders had been raised on farms, and they had a real sense of the needs of the average farmer and understood how the variability of the land and the weather across seasons demanded genetic resilience to adapt to all farmers' needs in any one region. They knew that most of the farmers they were serving didn't have the best growing conditions for their crops. Perhaps they didn't have ideal soil, their fertility was uneven, or the soil moisture was erratic, as most growers didn't have irrigation at that time. The best of these breeders didn't treat their breeding materials with kid gloves—instead they did their breeding under average field conditions with minimal external inputs, testing and evaluating across several environments in their region. The resultant varieties were usually genetically elastic and often proved to be workhorse varieties that adapted to the environmental variability that could be expected within the region. Yet even when these field-tough varieties were sold outside their area of intended use, they frequently performed well due to their built-in adaptive capabilities. Despite the specialization that was occurring in seed and the steady increase in seed sales outside their specific regions over the next 30 years, the production-research seed companies that emerged in the post–World War II era remained largely regional until the 1970s.

THE OWNERSHIP CONUNDRUM: SEED AS BIG BUSINESS

Selling agricultural seed had become a good business in the middle of the 20th century. It wasn't exactly fast money, since profits were made based on the quality and volume of seed sold. Yet if a seed company offered good varieties that catered to the farmer's needs and sold seed of varieties that met high quality standards, then it was sure to have a place at the table when farmers laid down their hard-earned cash. Seed companies were run by seedsmen, as they were then called, and indeed they were a special breed, with a connection to seed that far surpassed their business interests.

From the late 1940s to the 1970s there was unmitigated growth in all matter of agricultural goods globally, and seed was no exception. The spread of modern agricultural techniques such as mechanization and the use of inputs like synthetic fertilizer and insecticides spread to many of the major commercial agricultural centers around the world during this period. Seed of modern crop varieties bred in the United States and Europe wasn't far behind. Plant breeders had learned that when crops were selected and evaluated across environments, they could produce varieties with wide adaptation. Occasionally these varieties would have such good adaptation across environments and geographic regions that they would perform well in numerous locations around the world. Indeed, the breeders involved in the Green Revolution designed their breeding programs around this model; a number of their rice and wheat varieties spread across several continents.

Unfortunately, there was an emphasis in much of this breeding work on coupling the use of modern farming techniques and increased external inputs with these widely adapted varieties. As varieties were selected and bred under high-input regimens, they tended to perform best with the cash-intensive, purchased inputs. But as these modern methods of farming spread, so did these seed, and the market for seed became big business.

Since the inception of seed companies in the 1800s, many companies that offered seed of unique varieties had been interested in finding ways to own or "protect" these varieties. Until the advent of hybrids there was no way to prevent your competition from buying seed of your newest variety, then turning around and growing it for seed and producing their own version of it to compete with you.

The earliest breeders of F_1 hybrid varieties of corn were simply looking for the fastest way to isolate favorable traits, combine multiple traits, and get them into a uniformly true breeding strain that would benefit the farmer. It was the shrewd business owners of several seed companies who saw that hybrid varieties could give them de facto proprietary ownership as the hybrids did not produce seed that is a true breeding version of the F_1 variety. Hybrids are usually produced from two inbred parental lines (which are unique extracted versions of normal crop varieties that have been purified by repeated inbreeding) that are not available to anyone outside the company of origin. Therefore, above and beyond the sales pitch that hybrids had superior uniformity and vigor, there was a powerful incentive for breeding hybrid varieties based on their instant proprietary status by virtue of their genetic makeup.

The uniform F_1 progeny hold a heterozygous set of uniquely different genes from these two different inbred parents. Because of this each plant in the next generation (F_2) is usually quite diverse, with many different combinations of traits due to genetic recombination.

This also explains the other great business advantage for seed companies in selling F_1 hybrids, which seed companies realized right from the beginning: Farmers can't save seed of a hybrid and expect to get the same variety in the next generation that is true breeding for all of the same traits. So farmers have to repurchase seed of the hybrid variety each season. Best of all for seed companies is that producing hybrids in and of itself is a form of proprietary ownership with no need for formal legal protection or the cost of lawyers. As long as a seed company has plant breeders capable of producing good F_1 hybrids from their own elite inbred parental lines that are kept secretive and under their control, it is possible to have unique proprietary varieties that no one else can produce.

The combination of a steadily expanding world seed market and this new form of proprietary ownership of seed began to be noticed by large agricultural companies in the 1970s. To them it really made sense to include seed as part of their marketing of everything from fertilizer, insecticides, and herbicides to machinery or in some cases processed foods or commodities. Adding seed companies, especially those with a research-and-development emphasis, was an essential component of vertical integration in the quest to control agricultural markets across regions and around the world. Seed was truly in the process of becoming just another external input like fertilizer and pesticides, one that all farms would need to purchase on a continued, seasonal basis.

In 1970 the Plant Variety Protection Act (PVPA) was passed in the United States as a form of intellectual property protection for non-hybrid, seed-propagated crop plants. The PVPA allowed farmers to save seed of protected varieties for their own use and allowed breeders to use these varieties as parents in their breeding programs. While these exemptions are unusually lenient by today's standards, the PVPA did pave the way for the extremely restrictive use of utility patents for protecting gene sequences in crops and even specific crop traits such as heat or drought resistance.

In the 1980s major oil, agrochemical, and pharmaceutical companies started to buy many of the major seed companies worldwide, having recognized that seed was an untapped intellectual property resource of major proportion for the future. Seed company mergers became the order of the day, and by the mid-1980s Royal Dutch Shell, through a series of mergers and acquisitions, became the largest seed company in the world.

Decisions on the direction of research or the types of services the seed companies would direct toward different market sectors were now made by corporate managers or "bean counters" and were no longer under the purview of agronomists. Many minor agricultural regions, market sectors, and crop types were increasingly ignored as the bean counters from corporate headquarters eliminated the breeding programs and crop varieties that served the least lucrative markets. Increasingly, the commercial development of agricultural seed was directed at the highest-profile, highest-profit market: farms on prime agricultural land in large-scale, centralized production areas that are favored by agribusiness. This trend resulted in a narrowing of the breeding focus by all of the seed companies. The deathwatch for regionalism in commercial seed was beginning.

By the 1980s breeding new crop varieties for larger farms in climates favorable for the respective crop became the order of the day. As these farms began to rely on more and more external inputs and a greater degree of mechanization, the new crop varieties were shaped to fit these systems. The new varieties that were bred and selected in these systems became adapted to the ample quantities of fertilizer and water required in these high-input systems. While they were frequently bred to have resistance to one or two of the major diseases facing a particular crop, they were often overly protected from all other pathological maladies that could conceivably attack, from the seedling stage to full maturity.

Breeding for resistance to insects, however, has been largely ignored in modern crop breeding; as a result, insecticides were applied in almost all selection nurseries to ensure the absence of confounding variables during the selection process. This style of plant breeding has created crop varieties that are adapted to these external inputs and are often not field-toughened like many of the varieties bred during the 1940s to 1960s, when breeders were developing seed for more diverse, lower-input systems. These modern varieties have frequently been called "prima donnas" by organic growers, who are frustrated by their limitations.

At the same moment the seed industry was undergoing consolidation and increasingly catering to an industrial agricultural model in the 1970s and 1980s, the organic farming revolution was catching on in earnest in North America and Western Europe. Regional organic vegetable production led the charge with a rebirth of farmers' markets and ultimately a demand for local agricultural production of all stripes—a demand that is only getting stronger each year. Many of the practices found in organic farming at the end of the 20th century were in sharp contrast to the industrial farming model that was becoming the norm in the global North. Organic farmers were producing vegetables across a much greater range of terrains, soil types, and fertility regimens than what had become the norm in the much more homogeneous, high-input conventional systems. Those organic farmers that survived and thrived were a hardy lot who devised new cultural tricks such as flame weeding and relearned old tricks such as using stale seedbeds to control weeds.

One thing that wasn't very different between these two systems in these nascent early days of organic production was the choice of crop varieties used in the global North. Organic farmers at this time didn't usually question growing crop varieties that were in common use in their region. In fact, the source of the seeds wasn't questioned at all, and the common assumption was that whatever crop varieties were coming from the premier seed companies must be the best for any style of farming. The specialization of producing high-quality seed was far outside our frame of reference and we shared a blind faith that seed was being produced with our best interests in mind. We questioned many of the assumptions of farming in those heady times, but it was a rare person who questioned the goals of plant breeding. Fortunately, many of the modern varieties did have useful traits and the best organic farmers would find the best available varieties for each crop type and would cater their cropping practices to making those varieties work on their farms.

After many organic farmers had been in the trenches for 15 or 20 years they increasingly developed sophisticated, sound farming practices. They learned how to build soil fertility, structure, and health through on-farm nutrient cycling and lessened their trips through the fields applying steel to the soil. Through this attentiveness to the land and the natural systems they realized that many newer crop varieties didn't fulfill their needs.

In the introduction to this book I mention a number of the choices that the best organic growers are now making concerning seed. Since necessity is the mother of invention, many of the strands aimed at recreating decentralized seed systems are now being woven together and such local and regional systems are being called for by farmers of many cultures worldwide who seek seed sovereignty for their agricultural communities. In fact, a number of people and organizations worldwide are working to keep crop genetic resources available to all farmers through the creation of a collective commons using the Open Source model, which is becoming increasingly important in this age of restrictive ownership of many important resources that are the common heritage of all of the peoples of the Earth.

In sharp contrast to this movement toward local control of seed resources is the use of genetic modification (GM), where the order of the day is to have corporations legally control the varieties they transform with foreign genetic trait sequences that are often from completely different species, then control the varieties through utility patents. This is the last step in making seed just one more commodity where all modern innovation is tightly held as intellectual property and controlled in large part by the three largest seed companies in the world: Monsanto, DuPont, and Syngenta.

Early in the 21st century we are at a juncture in the history of agriculture where the people of all agricultural societies have a choice as to which of these paradigms they are going to follow. The seed is a reflection of the farming system as it is grown, cultivated, selected, and fully incorporated into that system. Are the crop varieties and the crop genetic resources of our ancestors going to keep adapting to fit the needs of organic agriculture at the hands and through the innovation of farmers and regional seed companies that have a relationship with farmers? Or will we allow the power to go to a corporate elite that shape our agricultural future based on shareholder profits? The road ahead for agriculture will be determined in large part by those who shape and ultimately control the seed.

2

REPRODUCTIVE BIOLOGY OF CROP PLANTS

Learning the basic morphology and mechanics of a plant's reproductive system is very important in understanding how to maximize the yield and quality of the seed you produce. The inherent biological processes of the flower and its sexual parts are often the first place a seed grower looks to when a particular seed crop is not performing adequately—that is, when the quality and quantity of seed are below expectations. Seed growers who view learning the ins and outs of reproductive biology as mysterious or "too complicated" are selling themselves short. The biology behind these processes is elegant, yet very utilitarian to the job at hand, and is easily learned by anyone interested in the wonders of nature and in understanding how a seed is formed.

FLOWER MORPHOLOGY

The flower is a modified shoot that contains the reproductive parts of the angiosperms. Most flowers have four distinct floral parts—sepals, petals, stamens, and carpels—which form four whorls that make up the individual flower. There is a common pattern in the organization of these floral appendages, although there may be variations in the details of their arrangement based on the method of pollination and seed dispersal for specific families, genera, or individual species.

The sepals and petals are leaf-like and usually form the two outer whorls of the flower.

The **sepals** are often green and thicker than the petals. Collectively the sepals are known as the **calyx,** and they constitute the outermost structure of most flowers. They generally serve as protection to the developing flower bud before it opens. The **petals** form the next whorl, which is known as the **corolla.** Petals are often brightly colored, are scented, and have nectaries at their base, all of which lure insects and other animals to the task of efficiently pollinating flowers. In some wind-pollinated species like beets and spinach, which don't require insects for pollination, the petals can be small, dull, and inconspicuous.

The two inner whorls of most flowers comprise the stamens and carpels. These are the fertile parts of the flower, with the **stamens** being the male, pollen-bearing appendages, collectively known as the **androecium.** Each stamen consists of a stalk or stem, called a **filament,** upon which the anther is borne. The **anther** contains two pairs of pollen sacs in which pollen is formed and develops. The female corollary, the **carpels,** are the ovule-bearing appendages that are collectively known as the **gynoecium.** Carpels can occur individually, but are often several are fused together to form the **gynoecium.** The pistil has three parts: the ovary, the style, and the stigma. The **ovary** contains the ovules, which become seed after fertilization; the **style,** a tube connecting the ovary to the **stigma;** and the stigma, a nutrient-rich surface where the pollen grains germinate on their way to the ovary.

THE JOURNEY OF THE MALE GAMETES

The male gametes or sex cells are produced in large numbers in the pollen sacs of the anthers. Meiotic divisions result in single-celled microspores or pollen grains so small, individual grains are only discernible by the human eye through the use of magnification. At maturity the pollen is released, either through pores in the anther walls or by a splitting or tearing of the anther (referred to as **dehiscence** of the anther). In species that are insect-pollinated, the pollen often has barbs or is sticky. Wind-pollinated plants have light, smooth pollen that is designed to be airborne. When pollen is released, a specific set of circumstances must exist for the pollen to successfully complete its journey, ultimately fertilizing the ovule in an ovary of the same species.

Pollination: The first step in what can be considered a journey frequently fraught with obstacles is **pollination**, the movement of pollen from the anthers to the stigma. Upon release the environmental conditions must be such that the pollen grain remains viable and reaches the stigma of a flower of the same species. If ambient conditions are too hot, the pollen may be denatured; too dry, and the pollen may desiccate and lose viability before reaching a receptive stigma. If it is too cool or rainy the activity of pollinating insects can be greatly impeded: Honeybees are especially sensitive to these conditions and will not fly when it is too cool or wet. Everyone who knows someone with an appreciable number of fruit trees in cooler climates has heard the hard-luck stories of poor "fruit set" in a year when there was a prolonged cool, wet period during the relatively short period that fruit trees flower. Rainy conditions can also impede the movement of pollen in wind-pollinated species by soaking pollen as anthers open and washing much of it to the ground.

Another condition that can impede pollination for most self-pollinated and all wind-pollinated species is to have little or no airflow at the time of pollen maturation and release. In selfers the anthers are always borne in close proximity to the stigma within the **cleistogamous** or closed flowers common to all selfers. In some cases the anthers are so close to the stigma that just the act of dehiscence (splitting open) will cause the pollen to cascade onto the stigmatic surface with little or no prompting. However, in many cases this short journey requires some type of external movement to literally shake the pollen from the anthers onto the stigma. In many selfing species (tomatoes, peppers, common beans, peas) this is easily accomplished by wind moving and jostling the plant. In some selfers (runner beans and fava beans are good examples) insect visits actually increase seed set by the movement they cause when landing on the flower in an attempt to get pollen, *even when they don't get pollen*! Any grower who has ever grown tomatoes through to fruiting in a greenhouse knows that you have to shake them daily during flowering to ensure good fruit set in a greenhouse when they are shielded from the wind. While wind-pollinated crossers like corn, spinach, or beets are grown in sheltered locations, away from the prevailing wind and airflow during flowering, there are times when several days of unusually still air can hinder full seed set (or random mating across the plants in a population that increases the genetic mixing that is essential for the health of crossers) due to low pollen flow in the air.

Pollen Germination: The next step in the journey of the male gametes is the germination of the pollen on the stigma. When the pollen comes in contact with the stigma, a biochemical signal from the moist, rich nutrient medium of the stigma surface promotes hydration and subsequent germination of the pollen grain. Problems can arise that hinder pollen germination if the ambient temperatures are too high or if the stigmatic surface is too dry from extremely low relative humidity. Low relative humidity has been found to be the culprit in cases of poor seed set in several instances of vegetable seed production in the western United States, where either the pollen or the stigma was desiccating before the pollen could properly germinate. Cold temperatures can also hinder pollen germination at different thresholds for different crops.

Pollen Tube Growth: As each pollen grain germinates, it forms a pollen tube that begins to grow down into the porous stigma. A pollen grain contains a vegetative nucleus and a generative nucleus. The vegetative nucleus migrates to the tip of the pollen tube, where it controls the growth and development of the tube, which is rapidly growing through the stigma and into the style. Meanwhile the generative nucleus divides, producing two sperm within the pollen tube. Each growing pollen tube eventually grows through the style with the purpose of delivering the two sperm cells to an individual ovule in the ovary.

This process of the pollen tube germinating and growing through the stigma and style is actually an entire alternate generation of the plant (the gametophyte generation) from the generation of the plant that we are most familiar with (the sporophyte generation). The pollen tube is essentially a free-living plant! For most of the crop plants we grow, the pollen tube life cycle lasts for only 18 to 36 hours, depending on the species. It must complete its journey, delivering the sperm cells to an ovule, in that period of time or it will run out of stored energy and die. In the time that it is alive the pollen tube also has environmental parameters in which it can grow much like a plant does. If the ambient temperature becomes too hot or too cold while the pollen tube is growing, then the tube can stop growing; in most species it will not start to grow again. The pollen tube's growth parameters are much akin to the plant in which it's found. In a heat-loving crop like watermelon, which grows prolifically at temperatures above 80°F (27°C), watermelon pollen tubes grow prolifically as well, and both are able to maintain good, steady growth with temperatures upward of 95°F (35°C). By contrast, in a cool-weather crop such as spinach, which produces its most luxuriant growth at temperatures between 60 and 65°F (16 to 18°C), the pollen tube also expresses optimal growth. But when temperatures rise above 75°F (24°C), pollen tube growth will stop and spinach seed crops can suffer severe drops in yield.

THE DEVELOPMENT OF THE EMBRYO AND ENDOSPERM

The embryo is formed when the pollen tube finally delivers the two sperm nuclei to the mature embryo sac of an ovule. This last critical step in the forming of the seed is known as fertilization. The fertilized ovule will become the seed. In the angiosperms this step is a **double fertilization,** with one of the sperm entering into the egg cell in the embryo sac and fusing with the egg nucleus to form a **zygote** and the other sperm traveling to the center of the embryo sac, where it fuses with two **polar** nuclei to form the primary **endosperm** nucleus, which has three sets of chromosomes. These two fertilization events happen concurrently and must occur for the seed development to progress normally. The endosperm starts to divide before the first mitotic division of the zygote; it divides and grows at a more rapid pace than the embryo to develop the storehouse of food for the seed in the form of endosperm starches, lipids, fats, and proteins.

This steady, rapid growth of the endosperm is important for two reasons:

1. The normal, healthy growth of this tissue is crucial to the development of the embryo, which can abort if anything interferes with the development of the endosperm.
2. There is only a short developmental window for the endosperm to be formed, and the more endosperm nutritive tissue is formed during this period, the better equipped the seed will be, with ample food reserves if its initial growth occurs under adverse conditions.

As with pollen tube growth, this endosperm growth is most readily done when the climatic conditions are most favorable for the particular plant's growth and development. Heat or cold during this period can arrest growth or slow it to the point where the resultant seed is much smaller due to the lower amount of endosperm that was formed. This can lead to lower germination rates in the seed crop and less vigor in the seedlings even if it does germinate. This is why it is so important to grow seed crops in suitable climates!

3

UNDERSTANDING BIENNIAL SEED CROPS

The vegetable crops with a biennial life cycle have a special set of circumstances that must be met to successfully produce a seed crop. A **biennial** is defined as a plant that takes 2 years to complete its life cycle. Biennials grow vegetatively during the first growing season, storing carbohydrates in storage tissue. In the second season this stored food is used to quickly establish a robust plant that has the potential of producing prodigious amounts of seed before the end of the season. The storage organ of a biennial is usually a root (for example, carrot or beet) or a bulb (onion), but can also be the stem (celery), leaves (cabbage or kale), swollen stem (kohlrabi), or the unusual enlarged fleshy hypocotyl and upper root tissue that makes up the root crop of turnips, radishes, and rutabagas. After storing food in their varied storage organs at the end of the first season, all biennial plants must go through a period of relatively cold temperatures in order to promote flower initiation the following season. This cold treatment or **vernalization** requirement (see "Vernalization," below) is the reason that seed of these crops is almost always grown in temperate climates and is not easily produced in tropical climates. Unfortunately, there are relatively few temperate climates where the winters supply enough cold to fully vernalize the crop without destroying it if the crop is overwintered in the field.

THE BIOLOGY OF THE BIENNIAL LIFE CYCLE

The evolutionary strategy of the biennial life cycle is simple. In the first season the plant grows a substantial amount of foliage and stores much of the photosynthates that it gathers in the form of carbohydrates in storage tissue for the next season. The plant then goes into a resting phase for the winter at a low level of **respiration** (the oxidative breakdown of stored food within the cells to maintain cellular processes), using as little of the stored carbohydrates as possible. During this period the plant is naturally exposed to a prolonged period of cold temperatures, and this is a critical factor that contributes to the induction of flowering in the plant's second growing season.

Vernalization

The term *vernalization* essentially refers to a particular length of time at or below a certain temperature that each biennial crop requires for flowering in its second season.

This cold treatment or thermal induction is essential in the initiation of the floral stem in the plant that will become the inflorescence soon after the plant resumes growth in spring of the second season. Without the proper vernalization period the crop will not flower.

The length of the vernalization period varies from one biennial crop to another and is usually measured in weeks by seed growers. It differs from species to species, from crop type to crop type within a species, and even from variety to variety within a specific crop type.

Carrots are a good example of a crop species that has a wide variation in the time required for vernalization. For example, carrots adapted to the subtropics usually require 2 to 4 weeks of low temperatures to initiate flowering, whereas most varieties of the western type require at least 8 weeks of vernalization. There are also differences between carrot varieties within these types: Among the western, temperate-adapted carrots some of the older true Nantes varieties are completely vernalized after 6 weeks, whereas some of the Northern European Flakkee storage varieties need a full 10 weeks (and sometimes more) of cold treatment to be fully vernalized.

The temperature range at which vernalization occurs will also vary from one crop to another. The vernalization temperature requirement for most biennials is somewhere between 32 and 60°F (0 to 16°C). There is a considerable range in the optimum vernalization temperature for different crop species. For example, sugar beets have the quickest flowering response when vernalized at 54°F (12°C), while a number of onion varieties from northern Russia have an optimum near 37 to 39°F (3 to 4°C). In general, an optimum temperature range for most biennial crops is between 41 and 50°F (5 to 10°C).

In practical terms most experienced biennial seed growers know that when most biennial crops are exposed to temperatures that are consistently below 50°F (10°C) for at least 8 to 10 weeks after the first season of growth, they will reliably flower the following season.

Floral Initiation

Plant growth regulators that stimulate flowering increase as a response to vernalization, long before any outward manifestations of floral initiation become evident. The first morphological change that is visually apparent (without cutting into the growing point of the crown) is the formation and elongation of the flower stalk. In most biennials, the flower stalk initiated by the bolting response is first evident at or near the crown, or apical growing point, of the plant. When flower initiation has occurred, the physiology of the storage tissue will change, directing nutrients to flower production. Once this biochemical cascade has started there is nothing you can do to reverse the flowering process and force the plant to return to the vegetative stage. In other words, the misguided practice of cutting off the flower stalk of any biennial will not stop the biochemical processes once they have been initiated. Even though flowering may be prevented when the inflorescence is cut from the plant, the re-initiation of vegetative growth necessary for a commercial vegetable crop will not occur at this point.

All biennials have a period of growth, early in their development, before they are responsive to the effects of low temperatures that induce vernalization. This period is known as **juvenility**. This period is different for each biennial crop species and for different varieties within each species. While chronological age and size are not the sole determinants of juvenility, most biennial crops are in fact quite small, often only 4 to 6 weeks old when they are physiologically mature enough to begin the vernalization process, if temperatures are low enough. Information on when this point of maturity occurs is not easy to determine for most crops; experience in growing the crop often serves as the best tool to judge the proper stage at which a biennial plant is ready to begin receiving the vernalization treatment. As an example, more is known about carrots and their approximate size for vernalization than for most other biennials. Carrots must have at least 8 to 12 leaves and a storage root with an approximate diameter between 0.15 and 0.3 in (4 to 8 mm) to start vernalizing (the common diameter of a pencil is 6 mm). While this may seem quite small, it is important to realize that carrots must usually have between 6 and 10 weeks of steady, initial growth to achieve this kind of size. Other biennial crops probably need a comparable period of growth before the vernalization process can commence; anything less than this in the first season of growth, especially when producing a seed crop using the seed-to-seed method, could lead to less than complete flowering of the crop during the second season.

OVERWINTERING THE CROP

There are two options when it comes to having a biennial crop survive from the first season of growth to the second season. The first option involves two techniques of keeping the crop in the field under ambient environmental conditions during the winter. In this first option the crop is either left standing in situ or is placed in pits and covered by soil. In these two field options the environment cannot be too cold, as there will be a risk of frost damage, or too mild, as the crop may grow excessively and not receive a full vernalization treatment. The second option involves bringing the **propagules,** the parts of the biennial plant with the stored carbohydrates that are used to produce seed, out of the field and storing them in a controlled environment. In this second option the propagules of the crop are placed in either a cooler or some form of root cellar. In these controlled environments you must make sure that the artificial storage conditions maintain a sufficient temperature and humidity to properly store these plant parts till spring. You must also ensure that the storage environment is sufficiently cold (see "Overwintering

the Crop in a Coldroom" on page 20) and that the duration of storage is of an adequate length of time to fully vernalize these plants. Remember that the ideal temperature for storing the propagules for an extended period of time may be different from the required range of temperature required for vernalization, depending on the type of crop stored.

Overwintering the Crop in the Field

The challenge is successfully overwintering the plants after the first (vegetative) season through to the second (reproductive) season. Most of the common biennial vegetable crops originally come from mild temperate regions with mild winters. They are frost-tolerant, but depending on their size, stage of growth, previous exposure to cold, and duration of cold, many of these crops can be damaged at temperatures below 29°F (–2°C). Conversely, a number of biennial vegetables can tolerate temperatures much lower than this if they are at an optimum size for overwintering and have been slowly hardened off to the cold. Under these circumstances there are some biennials that can consistently tolerate temperatures between 14 and 18°F (–10 to –8°C), including some varieties of Brussels sprouts, European kale, escarole, and leafy chicories. In fact, many parsnip varieties are famously cold-tolerant to temperatures of at least 10°F (–12°C) or below.

There are a number of factors to consider concerning both the plant and the environment that determine whether you can successfully overwinter the crop outdoors. Many temperate regions are excessively cold during peak winter weather, and it will not be possible to overwinter biennial crops in the field. An alternative to this is storing the crop by "pitting," or by storing it in a cooler or root cellar. Pitting—long used in many temperate regions—involves storing root vegetables in a pit that is dug below the frost line and is back-filled with soil. In seed production of certain root vegetable crops, like beets and turnips, the roots are dug, topped, and windrowed; then enough soil is mounded over them to prevent freeze damage for that specific climate. While this is still called pitting, it could more accurately be described as "mounding," and it involves the use of soil as insulation (see Garden Beet in chapter 5 on page 51).

Overwintering the Crop in a Coldroom

If the seed crop is of a great enough value you may choose to store the roots or plants at the end of the first season in a cooler or root cellar. The advantage of storing the plants in some type of artificial environment is the ability to avoid extremes of cold in the field for many temperate climates and also to protect the plants from potential damage from other environmental factors such as diseases, insects, or animals. Of course the size of the crop and availability of storage space are limiting factors in many instances.

Storage Temperature: The storage facility must be cooled to at or near the appropriate storage temperature when the crop is first put into the room. This should be the same temperature that the crop will be stored at for the duration of its time in storage. This is easy if you're using a refrigerated cooler with temperature control. Using a root cellar can be much more challenging, as the ambient temperatures of these earthen structures are often not cool enough at the time of the fall harvest to be acceptable, thereby limiting their use to certain geographic districts where the onset of cold weather is early enough in the fall to coincide with harvest.

In general, for long-term storage, most biennial propagules that are stored for replanting in the spring require a temperature as close to freezing as possible without the risk of damaging the plant's tissue by freezing at any time during the storage period. For this reason most biennials are stored at temperatures from 34 to 37°F (1 to 3°C) to avoid the chance of the temperature dipping below 32°F (0°C) with the normal temperature fluctuation of the cooler. Some growers use makeshift coldrooms—say, insulated outbuildings—that don't maintain an optimum temperature range, fluctuating with ambient conditions. Coldrooms where temperatures drop below freezing cannot be tolerated, but there are instances when coldrooms that occasionally warm up into the 37 to 45°F (3 to 7°C) range have successfully been used for storing roots for some period of time. The problem with storing roots at temperatures above 38°F (3°C) is that they will respire at a higher rate, using more of their carbohydrate reserve, and will prematurely put on both root and apical shoot growth, often long before the appropriate spring weather for replanting in the field has arrived.

Storage Relative Humidity: In addition to being at the right temperature, the other primary concern when storing any biennial propagules in any type of coldroom is maintaining the relative humidity at a level that is appropriate for long-term storage of the crop. Most biennial propagules, both the root crops and the biennials that store energy in leaves, stems, or petioles require a very high relative humidity of at least 95% to remain in optimum health for replanting at the end of the storage period.*

The biennials that are not propagated by their roots are not well suited to storage and are rarely stored in coldrooms over winter,

---

* Biennial bulbing crops of the Alliaceae family (e.g., onions and shallots) are a notable exception requiring temperatures between 34 and 38°F (1 to 3°C). Onions and shallot bulbs also have dry outer scales, which require a drier relative humidity (RH) in the 65 to 70% range for long-term storage.

especially when they are fully developed at the end of their vegetative cycle. Most of the non-root biennials are large, bulky plants such as cabbage, kale, broccoli, or cauliflower, and they are best stored with their roots "healed-in" into healthy soil. Some of these non-root biennials have lots of vegetative tissue that can easily rot during long-term storage (the petioles or stems of Swiss chard and celery, the curd of cauliflower, the leaves of cabbage or kale), and so the commercial production of seed in almost all instances for these crops is done in climates suited to successfully overwintering these species in the field. Specifics on the best practices for all non-root-propagated biennial crops can be found in the individual chapters devoted to each crop species (see part 2).

Storage Relative Humidity for Root Crops: Achieving and maintaining a relative humidity (RH) of 95% or greater is essentially done in two ways for root crops. You can place the roots directly into either totes in a high-humidity environment, bins with a storage medium for root cellars, or specially prepared plastic bags that can be placed into any cold-room that may not supply ambient RH. In all cases you must consider the extent to which the roots or "stecklings" must be prepared for each different storage option. (See "Preparing Roots for Storage" on page 22.)

If you're using a cooler that is equipped with a reliable humidifier capable of producing 95% RH, then the roots can be stored directly into open wooden or plastic totes with small openings between the slotted planks on the sides to allow the free flow of humidity to reach the roots. Covering the totes with a 2 to 3 in (5 to 8 cm) layer of *clean* wood shavings (not sawdust, as this will cake, preventing air circulation) will keep the roots free of standing water that can accumulate on the surface of the uppermost roots under high-humidity conditions. This helps eliminate a source of potential rot. Cedar wood shavings may be superior for this purpose as cedar is reported to have a higher level of antimicrobial factors than most other woods.

Root cellars or other earthen storage can supply a high level of RH naturally, though it can be quite variable from cellar to cellar and may also vary across seasons. If you are unsure of the ability of the storage environment to supply humidity constantly at this level, then it is wise to place the roots in some type of medium or container that helps keep the roots at a constant RH. In root cellars, roots are traditionally stored in wooden bins with layers of clean, slightly damp sand or clean, undecayed deciduous leaves. (In New England maple leaves, which are low in tannins, are sometimes used.) The roots are laid carefully between layers of this material so as not to touch, thus slowing the spread of rot through the lot of roots should any decay start. (Remember the old adage that states "one rotten apple can ruin the whole barrel.") These strategies may require some tinkering depending on the materials you have available and the often variable temperature and humidity conditions of each root cellar.

When a high-RH environment isn't available or economically feasible, then biennial roots can be placed in plastic bags and stored with great success in coldrooms with a lower-than-optimum RH. A high-RH environment can be maintained within a plastic bag for a long period of time; however, there is a risk of the roots rotting if you don't make provisions to allow moisture to escape and prevent the formation of standing water. This can be accomplished through several simple and inexpensive techniques that a number of seed workers have perfected over the years. Most seed growers use relatively large plastic bags as described here when holding enough roots for a sizable seed crop. Any grower or breeder wanting to store smaller numbers of roots will need to extrapolate down from these sizes.

Large plastic commercial produce bags (25 lb/11 kg) should have at least three to four rows of small, circular, penny-sized holes across the middle section of the bag. These holes allow some moisture to escape the bags over the course of the storage period. This is necessary because the roots are respiring and water is released in the process. While much of the moisture can escape the bags through these holes, there is always some that condenses and remains in the bag, potentially promoting rot. To trap this free moisture it is recommended that you spread three large handfuls of wood shavings evenly among the vegetable roots in each bag. As in the humidified cooler, cedar wood shavings may prove superior for this purpose: Cedar is reported to have a higher level of antimicrobial factors than most other woods. Care should always be taken not to place warm roots from the field into any of these various cold storage situations directly, as this will cause excessive condensation that may promote rot. These plastic bags should be checked every 6 to 8 weeks during the storage period to identify any rot and remove it from the bags before it spreads. The bags should also be turned and given a shake at this time to redistribute the shavings.

Preparing Roots for Storage: For long-term storage in a cooler or root cellar, the roots need to be properly prepared. Roots that are stored for replanting to produce a seed crop are called **stecklings**. Stecklings are much more likely to rot in these simulated storage environments than they are when stored in pits in the field. Therefore, several methods are used in preparing the stecklings for storage in a controlled environment. First, the roots should be cleaned of most of the soil that is still attached to them, but without the use of water, since getting the roots wet before storing them only encourages decay organisms to grow. In fact, some researchers believe that the balanced microbial community that exists in healthy soil may act as a deterrent to many destructive pathogens that might otherwise grow in storage. Soil that is not easily removed by hand or shaking should not be removed by vigorous rubbing; this can damage the root, making it more prone to disease in storage.

The next step is to remove any portion of the taproot that is longer than 3 to 4.5 in (8 to 13 cm) and also remove most of the petioles (leaf stems), as both of these tissues are the most susceptible portions of the root to rot. Remember that both the taproot and the apical shoot enclosed in the deepest layers of the petioles are where the active growth will emanate in the spring. The taproot, or the lowest portion of the bulb, is where the roots will emerge during the second year; hence the need to retain the 3 to 4.5 in (8 to 13 cm) taproot. The petioles will slowly rot away in storage, and because they enclose the apical bud for next year's shoot growth, this rotting, if excessive, can infect and destroy the bud before the root is replanted. A preventive measure is to trim away as much of the petiole as possible at the beginning of the storage process without damaging the apical bud. This requires deft handwork, a sharp knife, and knowledge of where the apical bud resides. Essentially you will cut much of the petioles off, starting very close to the crown of the root, near the base of the petioles, and cut *upward*, so that the center of the petiole mass is about 0.5 in (1.3 cm) above the base of the crown (see chapter 6, Carrot, "Root-to-Seed Method," page 87). In essence you are eliminating as much of the petiole as possible without damaging the apical bud.

PREPARING STECKLINGS FOR WINTER STORAGE: (A) Make ▶ an initial cut to trim back the leaf tops; (B) and (C) Using a sharp knife, carefully trim the petiole mass farther, to about 0.5 in (1.3 cm) of the base of the crown, but leaving the apical bud intact; (D) The final appearance of the trimmed carrot tops; (E) The last step in this process for carrot stecklings is to remove, with a clean cut, the bottom portion of any taproot longer than 3 to 4.5 in (8 to 13 cm). It is best to do this after winter storage in spring. Photos courtesy of Micaela Colley

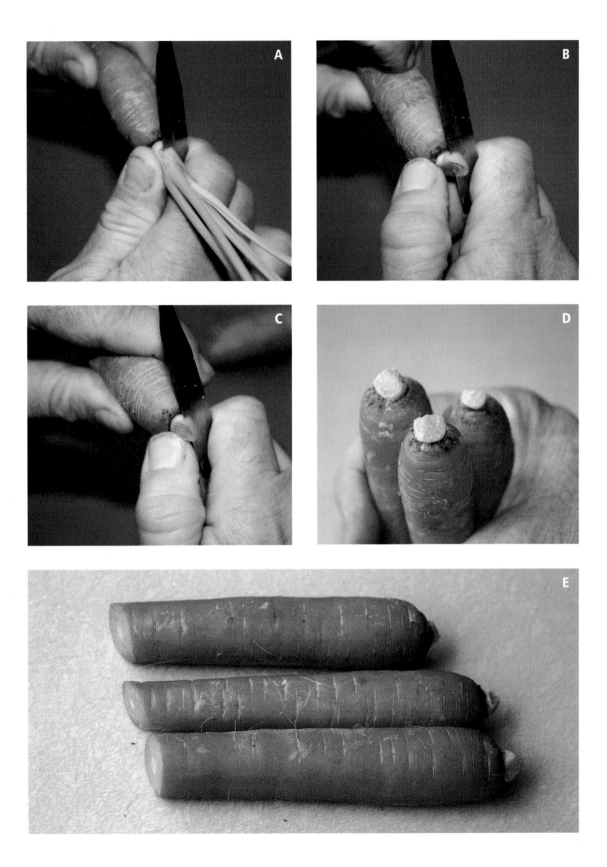

PART II

The Families of Vegetables

– 4 –
ALLIACEAE

The Alliaceae is a family of plants that have historically been lumped into either the Liliaceae or Amaryllidaceae families by many taxonomists. They are now recognized as a separate family in a subclass of the angiosperms, known as **monocots,** which are characterized by having only one cotyledon or seed leaf. All of the important cultivated species in this family belong to the genus *Allium,* which contains about 500 species of plants. The most widely recognized characteristic of this genus is their unmistakable "oniony" smell and flavor. There are more than 15 vegetable crops from this genus grown around the world, with at least five that are significant across multiple cultures. These five widely grown vegetables are from four *Allium* species: (1) common onion and shallots (*Allium cepa*); (2) leek (*A. ampeloprasum*); (3) garlic (*A. sativum*); and (4) Japanese bunching onion (*A. fistulosum*). Two of these economically important crops, garlic and shallots, as well as a number of other lesser-known alliums like Egyptian onions, potato onions, and multiplier onions, are propagated primarily by replanting asexually produced cloves, bulbs, or bulbils.

Allium crops are among the most ubiquitous agricultural plants in the world; almost all culinary traditions use one or more of these crops for practically every meal that is served. Onion, which is the most widely grown allium,

is produced in climates from the tropics to the sub-Arctic. The crop plants of this family, many of them from the middle latitudes in the steppes and mountainous regions of Central Asia, have slowly been adapted to produce both bulbs and seed across many different latitudes in their slow, methodical march around the world. Onions and garlic spread as far as northern Africa over 5,000 years ago, with evidence of their common use in depictions found in ancient Egyptian temples and tablets. Indeed, the migration of onions occurred so early in agricultural history that secondary centers of diversity exist across many parts of

Garlic is an Alliaceae crop that is propagated asexually, most commonly by planting individual cloves from disease-free bulbs. ▶

Eurasia, with much genetic diversity among the traditional varieties that predate the modern crop. Unfortunately, modern hybrid varieties of several of the seed-propagated alliums are becoming so ubiquitous in commercial agriculture that the number of people growing and maintaining seed of many genetically diverse, adapted open-pollinated varieties is decreasing across many cultures.

High-quality seed of the alliums is grown in regions with warm, dry summers with low relative humidity and a reliable source of irrigation throughout the season. Precipitation should reliably be at a minimum during the latter stages of seed maturation and harvest. Since Alliaceae flowers are often not as desirable to pollinating insects as the flowers of plants from other families, it is important to grow allium seed crops in an area with a rich, diverse habitat for pollinating insects.

A number of the common allium crops, including garlic, shallots, and Egyptian onions, are usually propagated asexually by replanting bulbs, cloves, or the small bulbils that occur in the inflorescence. While both shallots and garlic were generally thought by growers a generation ago never to form true flowers, they are both now propagated to some degree by true seed. However, as this germplasm and the techniques used to produce true seed of these two crops are controlled exclusively by a small number of commercial businesses, these crops are not available to farmers as seed crops and will not be covered in this manual.

FAMILY CHARACTERISTICS

Reproductive Biology
All of the common allium crops have flowers that are borne on a simple oval umbel at the top of a seed stalk or scape that arises from the apical growing point. The number of flowers that arise can vary considerably across *Allium* species and even across varieties within a particular species, with fewer than 50 to as many as 2,000 in some types. All of the cultivated alliums have perfect flowers with six stamens and a compound pistil with three chambers. Nectaries are often present at the base of the ovary across the various species. Despite the presence of nectaries, onion seed growers can have problems getting adequate pollination on their crop due to a reluctance of honeybees to visit most allium species flowers, especially when another, more attractive flowering crop is present. This is another case where the best strategy is to have diverse populations of wild pollinators, many of which readily visit onion flowers, living in close proximity to fields to ensure maximum pollinating activity in a seed crop.

Life Cycle
It is likely that all of the most commonly cultivated alliums are biennials requiring a vernalization period that consists of a combination of chilling hours at or below 50°F (10°C) and increasing daylength in order to initiate flowering and produce seed in the second season of growth (see "Vernalization" in chapter 3, Understanding Biennial Seed Crops, on page 17). When grown as a seed crop the cultivated alliums are routinely direct-seeded sometime during midsummer of the first year to produce the first year's storage bulb. The goal is to produce a moderate-sized bulb* the first year that can then be either stored in the ground or stored in a root cellar or cooler for replanting in the spring. Often the size of this storage tissue need not be very big: According to Joel

---

* The term *bulb* will be used generically to describe the storage tissue of the cultivated biennial alliums, despite the fact that several of these crop plants do not have true bulbs but store energy in cloves like garlic, or in foliage leaf bases like Japanese bunching onions.

Reiten of Seeds of Change, many onion varieties need only three or four well-established true leaves to go through the vernalization process in the field for the seed-to-seed method (see Onion, "Growing the Seed Crop" on page 40). These plants will then gain some more bulk in the following spring before flowering. Alternatively, when growing onions for the bulb-to-seed method, the bulbs only need to be 1 to 2 in (2.5 to 5 cm) in diameter to store properly. The key is to have a bulb or plant that is able to go through the winter with a healthy, intact growing point that survives the long storage period and resumes vigorous growth in spring.

There are generally three methods used for overwintering biennial Alliaceae crops:

1. Plant the crop in the field of intended seed production. Overwinter the crop in the field with flowering and seed production in situ in Year 2.
2. Grow the crop in a nursery plot in the first year, allowing it to overwinter in situ; lift the bulbs or plants the following spring, transplanting them to a seed production field.
3. Grow the crop in a nursery plot in the first year; lift the bulbs or plants in the fall, transplanting them to overwinter in next year's seed production field.
4. Grow the crop in a nursery, lift it in the fall, and store bulbs in a cool, dark room until replanting them in spring of Year 2 for seed production.

The advantage of the first method is that it doesn't involve the time and labor of transplanting; it does require more land in Year 1, however, as the crop is planted and thinned to the wider spacing required for seed production in Year 2. The thinning can be done in stages as a selection tool during Year 1, but if there is danger of winter losses the crop numbers should be larger than the final count to accommodate any losses. For biennial crops this method is known as the seed-to-seed method. The great disadvantage of using the seed-to-seed method with a bulbing crop like onions is that you don't get an opportunity to select for most of the bulb characters important to the particular variety, as the overwintered crop rarely is grown to full bulb maturity. However, a crop like leeks—where the crop can be grown to maturity and all of the important commercial characters can be viewed without lifting the crop—is easily selected using this method.

The second method also overwinters the crop in place but then pulls the bulbs or plants in the spring to transplant into the seed production plot. This method doesn't require thinning like the first method but does put the plants through a winter selection depending on the climate.

The third and fourth methods for overwintering biennials use a nursery plot to raise the first-year plants before storing them or transplanting them directly for seed production in the second season. In general the nursery plot will only consume between a tenth and a fifth of the space needed in the seed production field, where the plants must be given a wider spacing to accommodate the much larger flowering plants in the second season. Both of these methods when used with allium crops are considered the bulb-to-seed method even when they are used with alliums that don't actually make bulbs. The bulbs or plants are lifted, selected, and stored before replanting for flowering and seed production.

Climatic Adaptation

All of the alliums that produce seed require cool weather during the spring with ample soil moisture for early foliar growth and bulb development. Climatically, the crops of the Alliaceae require a moderate climate with extended dry weather to produce high-quality seed. The ideal climate for seed production for most of these crops is a dry, Mediterranean climate during summer with low precipitation amounts from late spring until early fall; such conditions significantly lower the chances of disease. For most of these crops a relatively hot summer with temperatures that routinely reach 86°F (30°C) or slightly above during active flowering and early seed growth is advantageous to producing high-quality seed. It is also very important to have a reliable source of irrigation water applied with either a drip or a furrow system, as overhead irrigation may promote disease in the seedheads. Irrigation should be applied regularly until at least 80 to 90% of the crop is exhibiting the early signs of seed maturation.

Allium seed crops can vary considerably in their response to daylength and flowering. Therefore, it is very important to understand the needs of both the specific crop you are growing (onion, leek, bunching onion, and so on) and the specific variety or crop type within each of these crops. Certainly, the daylength requirements for flowering in different types of onions vary widely, from types adapted to the low latitudes of the subtropics to long-day types that produce seed at high latitudes. Alternatively, J. L. Brewster of the Wellesbourne Station in England has reviewed the literature and found that flower initiation in leeks can occur at different daylengths. Minor allium crops like chives (*A. schoenoprasum*) and Chinese chives (*A. tuberosum*) are little studied, and little is known about their response to daylength. As in many other instances with vegetable seed crops, there is always a need for some level of experimentation and observation with any seed crop after you have found out as much as possible about how and where it is commercially produced.

ISOLATION DISTANCES

Isolation for all alliums follows the general rule for cross-pollinated crops that are primarily insect-pollinated. Two crops within a given species of the same crop type, the same form, the same daylength class (as in onions), and the same color need only be isolated by a minimum of 1 mi (1.6 km) if grown in open terrain or 0.5 mi (0.8 km) if there are substantial physical barriers between the crops (see chapter 13, Isolation Distances for Maintaining Varietal Integrity, on page 333). An example of two seed crops of the same crop type that could use these isolation recommendations would be two yellow, long-day round storage onions. The isolation distance would definitely need to be increased to 2 mi (3.2 km) in open terrain if the second onion seed crop is of a different color (white, red, or brown), different shape (bottle or flattened globe), or of a different daylength response. This increased isolation requirement between two crops of the same species but different types can be cut to 1 mi (1.6 km) if there are substantial physical barriers between the two crops. An example of different crop types in leeks would be leek varieties with different-colored leaves (green versus blue) or a taller summer–fall variety versus the older stout, short 'Giant Winter' winter-hardy type. These would also require the greater isolation distance, whereas two similar summer–fall harvest types with the same leaf color could be grown with the less stringent 1 mi (1.6 km) in open terrain and 0.5 mi (0.8 km) with physical barriers on the landscape.

LEEK

There are four different cultivated crops that share the species *Allium ampeloprasum*. The four crops or horticultural groups in this species includes leek, kurrat, pearl onion, and great-headed garlic, also known as elephant garlic. They are all probably descended from the wild *A. ampeloprasum* that grows from Southern Europe across the Mediterranean eastward toward modern-day Iran.

The leek plant is essentially a sheath of leaves; the base of these leaves forms a long edible "pseudostem" that is the primary part of the crop used for cooking. It is grown to harvest maturity as a vegetable in the first year of the biennial cycle and often overwintered in areas where winter temperatures don't usually go below 15 to 25°F (–9 to –4°C), depending on how cold-hardy the variety is.

The kurrat is quite similar in appearance to the leek, although it has been selected over the course of its cultivation for its foliage and has a very short pseudostem. The crop is harvested on a regular basis for at least 18 months for its flavorful, tender leaves, which are used in both raw and fresh dishes in Egypt and parts of the eastern Mediterranean. It is produced from seed, much of it grown regionally in the area of use.

SEED PRODUCTION PARAMETERS: LEEK

Common names: leek, kurrat, pearl onion
Crop species: *Allium ampeloprasum* L.
Life cycle: biennial
Mode of pollination: largely cross-pollinated with some self-pollination
Pollination mechanism: insect
Favorable temperature range for pollination/seed formation: 65–90°F (18–32°C)
Seasonal reproductive cycle: spring to summer of Year 1 through to fall of Year 2 (~14 months)
Within-row spacing: 8–12 in (20–30 cm) if planting to a final spacing
Between-row spacing: 28–36 in (71–91 cm) or two rows/bed at 14–18 in (36–46 cm) apart
Species that will readily cross with the crop: The true pearl onion and kurrat are fully fertile with leeks. This species also includes great-headed garlic, which is asexually reproduced and has never been found to have fertile flowers.
Isolation distance between seed crops: 0.5–2 mi (0.8–3.2 km), depending on crop type and barriers that may be present on the landscape

◄ 'Scotland' leeks.

Pearl onion is primarily cultivated as a garden vegetable in several parts of Western Europe. The plants look much like a miniature leek and form several small edible bulbs at the base of the plant that are winter-hardy and can be used as planting stock. Pearl onions produce fertile flowers and can cross with leeks.

Leek is the only crop in this species that is widely grown around the world. Leeks have long been cultivated in Europe and across the Mediterranean basin, but have become much more prominent across the globe in the past 20 to 30 years. Their utility as a vegetable crop that is harvested throughout the winter in milder temperate regions has made them popular with many people looking for regional food options in temperate climates with limited winter cropping options. Leeks are very versatile in their ability to be marketed during the winter months, especially in milder temperate zones. They can be held in the field at a harvestable stage for several months during winter if they have reached a harvestable stage before their growth slows in late fall. They are also very versatile in being able to be held in a coldroom, under high-humidity conditions, and remaining in a sellable state for at least 8 to 10 weeks.

Great-headed garlic is often called elephant garlic in North America. It produces a plant that is indeed very similar to common garlic (*A. sativum* L.). The bulbs of great-headed garlic usually produce six large cloves and sometimes small, hard yellowish bulbils near the base of the plant. The flavor of this close relative of leek is similar to garlic, but considerably milder. It forms a single umbel, similar to leek, with white or purple flowers, but it doesn't produce viable seed. The crop is propagated using the large cloves, though the small, hard bulbils can be used to propagate the crop when the bulb gets infected with disease.

CROP CHARACTERISTICS

Reproductive Biology

Leek is a biennial crop and produces a more vigorous, heavier plant than an onion in the first year of the cycle, one that is 18 to 30 in (46 to 76 cm) in height. The leek plant doesn't usually form a bulb, but some degree of bulbing can occur when the plant is produced in cooler climates. In the second year of the biennial cycle the plant develops a floral stem that is often from 30 to 48 in (76 to 122 cm) tall at full flowering. The plant then produces a single scape, which has a large single compound umbel (2.5 to 6 in/6 to 15 cm in diameter) with many white to pinkish or lavender flowers. The flowers are perfect and contain six petals, six stamens, and a three-chambered compound pistil with two ovules in each, much like onions. Leeks are self-compatible and can sometimes produce bulbils or topsets in the umbel.

The environmental factors responsible for the initiation of flowering in leeks are less well understood than the factors that promote flowering in onions. An accumulation of cool temperatures (vernalization) and increasing daylength seem to be important in their flower initiation, but there have been experiments where the crop has been grown at temperatures above 68°F (20°C) and the plants still bolted under long daylengths. This said, there are a number of cases where young leek plants that are exposed to appreciable amounts of cold weather during the spring of the first year are much more apt to flower when they are subsequently exposed to longer daylengths later that first year than when the same variety experiences milder spring weather. Hence, it does appear that at least 6 to 8 weeks of temperatures at or below 55°F (13°C) during short daylengths, followed by the longer daylengths of

temperate summers, ensures complete flower initiation in most leek varieties. However, as there are still unanswered questions about the environmental parameters of leek flowering, you should experiment and observe specific varieties and their bolting response closely in your climate before attempting to produce large-scale commercial lots of leek seed.

The leek flower is a large single umbel. ▶

Climatic and Geographic Suitability

The production of high-quality leek seed requires a moderate climate with extended dry weather during seed production that is very similar to the environmental parameters for onion seed production (see Onion, "Climatic and Geographic Suitability" on page 39). Importantly, leeks also require cool weather with ample soil moisture for early stages of development. However, as mentioned in the previous section, leeks do not appear to be as daylength-sensitive as onions are across different latitudes.

As with all dry-seeded vegetable crops a dry, Mediterranean climate during summer with low precipitation amounts from late spring until early fall is best, as it significantly lowers the chances of any diseases forming on the seedheads as they mature. To realize a seed crop with favorable yields, however, it is very important to have a reliable source of irrigation water applied with either a drip or furrow system, as overhead irrigation may promote disease in the seedheads.

Unlike onions, where the bulbs are often stored during winter for replanting, leek plants grown for the seed crop are almost always overwintered in the field between the two biennial seasons required to produce seed. Therefore, it is important to determine how cold-hardy a particular leek variety is and whether it can tolerate the winter low temperatures in your region before growing a sizable crop. As with most biennial crops, the portion of the crop that is storing food for the second season—in this case the leaf tissue (especially the pseudostem)—is more cold-hardy when it is smaller. Leek seed crops are often produced with plants that are somewhat smaller than the plants that are marketed as a vegetable for this reason. There are some leek varieties at a smaller stage of growth that are able to tolerate temperatures as low as 10 to 15°F (–12 to –9°C).

SEED PRODUCTION PRACTICES

Soil and Fertility Requirements

The soil and fertility requirements for onion seed production are equally well adapted to the leek crop (see Onion, "Soil and Fertility Requirements" on page 40). However, leeks have a larger, more prolific root system than onions, which seems to make the crop better adapted to varied soils for seed production. Some growers believe that leeks need more fertility than the onion seed crop because of the heavier, stockier plant and the longer amount of time that the seed needs to mature for a leek crop.

Growing the Seed Crop

Because leek is a biennial crop, producing leek seed is almost always done using a modified root-to-seed method, transplanting seedlings during the first season, overwintering the crop in the field, and producing seed in the second year. It is a root-to-seed method in that the plants are transplanted at some point in the life cycle, but unlike the root-to-seed method used for biennial root crops this method doesn't use nearly mature plants for transplanting. It is generally called the root-to-seed method and not the bulb-to-seed method as with onions, because leeks do not make bulbs. In fact, transplanting or storing leeks for later transplanting is more akin to working with a biennial stem crop such as celery (see chapter 6, Celery & Celeriac, "Root-to-Seed Method" on page 100). It is also possible to use the seed-to-seed method for growing the seed crop, realizing that this method does not allow as much of an opportunity to select the crop as when the plants are transplanted using the root-to-seed method.

Seed-to-Seed Method: The seed-to-seed method, where seed is sown directly into rows from late spring to midsummer during the first year of the biennial cycle, is not often used for leek seed crops. If sown directly in spring or summer the grower must be certain to keep ahead of the weeds during the early months of seedling growth. Rows are spaced at 28 to 36 in (71 to 91 cm) or two rows on a bed 18 in (46 cm) apart. Plants can be thinned initially to approximately 4 in (10 cm) apart within the rows, using this opportunity to select the plants to type over the course of their vegetative growth until they are 8 to 12 in (20 to 30 cm) apart and before they have initiated flowering. If the crop is grown in an area where cold winter temperatures can inflict some damage, then it may be best to save the final thinning until after the plants have gone through winter, as there is almost always genetic variation for cold hardiness in any hardy crop, which you can select for. Also, remember that with other biennials it is possible to produce seed from plants that are less than what their full size is at vegetable maturity (see chapter 6, Carrot, "Seed-to-Seed Method" on page 86).

Root-to-Seed Method: Leek seed is most commonly produced using a modified root-to-seed method, which involves using a seedling nursery or greenhouse to raise seedlings at the start of the first season, growing the crop to some degree of maturity in the first season, then overwintering the crop in the field and allowing it to produce seed in situ. In essence, during the first year the crop is grown in much the same way that leeks are grown for vegetable production. The major difference is that the vegetable crop is often planted into the greenhouse very early in the year in order to produce large leek plants for harvest at the end of the first season. When producing leek seed using the root-to-seed method the crop doesn't need to be sown as early in the first season, as the goal is not to get as large a plant as is normally grown. The goal is to get plants large enough to be able to effectively select to type, while being mindful that the largest leek plants aren't usually as cold-hardy as a moderate-sized plant.

Seed can be sown in mid- to late spring into rows in greenhouse flats or into beds in an outdoor seedling nursery. In nursery beds sow into several rows 4 to 6 in (10 to 15 cm) apart, thinning to 1 or 2 in (2.5 to 5 cm) within the rows after all of the seedlings emerge. Supply even moisture and good fertility to produce vigorous seedlings. In 8 to 12 weeks, when plants are the diameter of a pencil, lift the plants and transplant into the field. There are several traits that the plants can be selected for at this stage (see "Genetic Maintenance" on page 36). Transplants can be planted into row centers of 28 to 36 in (71 to 91 cm) or two rows to a bed, spaced 14 to 18 in (36 to 46 cm) apart.

Within the rows the plants can be spaced to 8 to 12 in (20 to 30 cm) if you're planting to a final spacing. If you feel that another round of plant selection is necessary, then the transplants can be put in at two or three times the density, 4 to 6 in (10 to 15 cm) apart in the row, and thinned to their final spacing after they have achieved a larger size or after they've gone through winter.

Seed Harvest

Seed harvest for leeks is very similar to the procedures given for onion (see Onion, "Seed Harvest" on page 43). The only practical difference is that leek seed is usually later maturing and often needs to be consciously harvested ahead of inclement weather and dried artificially, with supplemental heat, in many temperate regions. This is why high-quality leek seed is often from regions with extended dry weather into mid-fall.

Seed Cleaning

Seed cleaning for leeks is very similar to the procedures given for onion (see Onion, "Seed Cleaning" on page 44). The only practical difference is that leek seed may be somewhat tougher than onion seed when it comes to damage during threshing and cleaning. However, it is still fragile seed and needs to be harvested and cleaned gently, as this is a crucial stage in ensuring that an allium seed crop retains its quality.

GENETIC MAINTENANCE

Genetic selection for most of the important traits is simpler for leeks than it is for many of the other biennial crops. This is because all of the important traits of the vegetable crop, except for a small portion of the pseudostem at the base of the plant (which is largely dependent on the extent of hilling), are evident in the aboveground parts of the plant. The most important question is whether you have to grow the plant to full size in order to evaluate and select for its vegetative characteristics. One response to this is that it is certainly important to know what the variety looks like when it is grown to full size; however, many growers agree that, once you have a good sense of a variety's characteristics, then it is possible to select for those traits in a proportionately smaller leek plant. This is important when overwintering the crop in situ in the colder reaches of the possible seed production zone for leeks, as smaller plants are usually more cold-hardy than fully grown plants.

The traits that are important to monitor when selecting the crop are the length and relative diameter of the pseudostem. The uniformity and relative degree of whiteness of the pseudostem are also important traits when you are attempting to maintain the norm for the variety. The leaf width and length will also vary, as can the color of the leaves, which can be green, gray-green, or blue-green. The leaves can also exhibit a waxy coating that can vary in its intensity. The relative height and stature of the plant, and whether it has erect leaf blades pointing skyward or a more horizontal leaf stature, are also variants that determine the phenotype of a variety.

ISOLATION DISTANCES

Isolation for leek seed crops follows the general rule for cross-pollinated crops that are primarily insect-pollinated. Two crops within the same crop type, the same form, and the same color need only be isolated by a minimum of 1 mi (1.6 km) if grown in open terrain or 0.5 mi (0.8 km) if there are substantial physical barriers between the crops (see chapter 13, Isolation Distances for Maintaining Varietal Integrity).

Examples of different crop types in leeks would be leek varieties with different-colored leaves, different height, different stature, or a marked difference in the length, diameter, or degree of white coloration in the pseudostem. A prime example would be a taller summer–fall leek variety versus the older stout, short 'Giant Winter' winter-hardy type. Any two leek varieties with any of these major phenotypic differences will require a greater minimum isolation distance and should be grown at least 2 mi (3.2 km) apart in open terrain, though when there are notable physical barriers on the landscape they can be grown at a less strict 1 mi (1.6 km) apart and maintain a high degree of purity.

ONION

The common or bulb onion is one of the most widely cultivated crop plants in the world and usually ranks as one of the five most important vegetable crops by any measure across all agricultural societies. Its use as a vegetable goes back more than 5,000 years; it was probably first cultivated in the mountain valleys of the Pamir highlands in present-day Afghanistan, Tajikistan, and Pakistan. The glacial peaks of the Pamir Altai were known as the Onion Mountains when Marco Polo traversed the region, en route to China, because of an abundance of both cultivated forms and wild relatives of this crop in the area. Still, modern ethnobotanists have been unable to identify a definitive progenitor of our modern crop, which is often an indication of the great length of time that has passed since the crop was first domesticated. The use of the versatile onion undoubtedly spread via the Silk Road, and the ancient Egyptians immortalized it in hieroglyphics. An inscription on the Great Pyramid at Giza records the total cost, in silver, to procure the onions, garlic, and radishes needed to feed the workers for the 20-plus years it took to complete the project.

As the cultivation of the onion spread, so did its ability to adapt to different photoperiods, varied seasonal temperatures, and the length of storage time that is necessary between seasons in its biennial life cycle. Farmer-breeders growing this crop through the ages also captured the natural genetic variation for color, flavor, moisture content,

SEED PRODUCTION PARAMETERS: ONION

Common names: onion, shallot, multiplier onions, some bunching onion varieties
Crop species: *Allium cepa* L.
Life cycle: biennial
Mode of pollination: largely cross-pollinated with some self-pollination
Pollination mechanism: insect
Favorable temperature range for pollination/seed formation: 65–90°F (18–35°C)
Seasonal reproductive cycle: spring of Year 1 through to fall of Year 2 (~14 months)
Within-row spacing: 8–12 in (20–30 cm)
Between-row spacing: 28–36 in (71–91 cm) or two rows/bed at 18 in (46 cm)
Species that will readily cross with the crop: All of the above crops are fully sexually compatible, though most shallots and multiplier types do not usually flower.
Isolation distance between seed crops: 0.5–2 mi (0.8–3.2 km), depending on crop type and barriers that may be present on the landscape

and shape of the bulbs, producing many forms before the advent of modern plant breeding. Onions are grown in almost all temperate and subtropical agricultural areas and are limited in their distribution in the wettest tropical areas, including much of Southeast Asia, due to unfavorable climatic conditions.

The initiation of bulb formation in onions is in large part a response to the length of day during the growing season. It is very common for people in the North American seed trade to classify onions as short-, intermediate-, or long-daylength types. For instance, an onion variety that is best suited for the southern reaches of the United States or Mexico will to be described as a short-day onion. This tells you that this variety will initiate bulb formation

in the lower latitudes, where the daylength during the growing season doesn't exceed 11 or 12 hours. Likewise, intermediate-day types only require a minimum of 13 to 14 hours, and long-day types need at least 16 hours to begin bulbing. While other parts of the world may not use this language, the knowledge that using regionally adapted onions is a key to a successful crop has become common among onion growers.

To cater to optimum growing seasons, onions are planted at different times of the year across regions, and also to fit the particular daylength requirements of the variety. Short-day varieties are grown at lower latitudes, where the daylength is close to 12 hours year-round. Intermediate types are adapted to the middle latitudes and are often planted in late summer to establish a robust vegetative frame in order to begin bulbing in spring and to produce a mature bulb before the heat of summer of the following year. Long-day onions are generally grown at latitudes above 38 to 40 degrees, north or south, and are sown in early spring.

Commercial onion seed production is done in suitable latitudes where the summer through mid-fall season is sufficiently dry to mature high-quality seed. High-volume commercial seed production areas include Italy, Spain, southern France, South Africa, and Australia. In the United States the Columbia Basin of Washington, the Treasure Valley of Idaho, and the Imperial Valley of California, as well as Arizona are all important production areas. However, as with all widely grown crops, onion seed is also produced in many less-than-ideal climates around the world by farmers maintaining locally adapted varieties.

◄ 'Rossa di Milano' is an Italian intermediate- to long-daylength onion that stores well, an essential trait when using the bulb-to-seed method for seed production.

CROP CHARACTERISTICS

Reproductive Biology

The onion is a biennial and requires two growing seasons to produce seed. During the first year of growth the crop produces a bulb, which begins with the swelling of the leaf sheaths near their attachment to the stem disc at the base of the aboveground part of the plant. As bulbing continues the newer leaves no longer form blades, but develop into swollen bladeless leaves called scales that essentially form the many overlapping layers of the onion bulb. During the second season onions develop a floral stem that is 3 to 5 ft (0.9 to 1.5 m) tall with extensive branching and can cover a diameter of more than 3 ft (0.9 m) on the ground.

A simple oval umbel is borne at the apex of each scape (floral stem). Each umbel will usually have between 200 and 600 individual florets. The flowers are perfect and contain six off-white petals, six stamens, and a three-chambered compound pistil with two ovules in each. Onion flowers are **protandrous,** with anthers from a single flower opening and shedding all of their pollen within 24 to 36 hours, while the stigma doesn't become receptive for at least 24 to 48 hours after pollen shed. This promotes outcrossing between different onion plants within the population. Onions are almost exclusively insect-pollinated, and many species are attracted to both the pollen and the nectar of the flower. Open-pollinated seed crops are usually better adapted to producing seed than hybrid crops, where male-sterile female parental lines can have reduced visitation from some insect pollinators due to their lack of pollen.

Onions can form from 1 to 20 floral stems or scapes, sometimes more, based upon their genetic background, the storage history of the bulb, and the environmental conditions at time of planting. Onion pollination is increased in clear, warm weather, which tends to increase pollen and nectar production as well as insect activity. Warm temperatures also increase the percentage of pollen tubes that successfully fertilize the ovules. The endosperm of the seed will reach the "dough stage" in 3 to 4 weeks after fertilization, and the seed coat will begin turning black. However, the crop will still require at least 3 more weeks of favorable conditions to produce mature seed.

Climatic and Geographic Suitability

The production of high-quality onion seed requires a moderate climate with extended dry weather during seed production that gets enough heat to reliably mature seed, but not extreme heat. All onions require cool weather with ample soil moisture for early foliar growth and for the early stages of bulb development. The ideal climate for onion seed production includes a mild, cool spring and a dry summer with temperatures that routinely reach 86°F (30°C) or slightly above during active flowering and early seed growth, but generally don't reach above 95°F (35°C) until late in the summer, after the seed is set and maturing.

As with all dry-seeded vegetable crops, a dry, Mediterranean climate during the summer with little precipitation from late spring until early fall is best, as it significantly lowers the chances of any diseases forming on the seedheads as they mature. To realize a seed crop with favorable yields, however, it is very important to have a reliable source of irrigation water applied with either a drip or furrow system, as overhead irrigation may promote disease in the seedheads.

Onion seed production is based largely upon which daylength type is being produced and whether the climate allows for the

An onion's umbel is typically composed of between 200 and 600 individual flowers.

overwintering of the crop. As with most of the biennial crops, the portion of the crop that is storing food for the second season—in this case the onion bulb—is more cold-hardy when it is smaller. In some cases onion seed crops are overwintered without significant bulbing. Some varieties at this stage of growth are able to tolerate temperatures as low as 10 to 15°F (–12 to –9°C).

SEED PRODUCTION PRACTICES

Soil and Fertility Requirements

Onions are best grown on friable soils that are relatively light, preferably sandy or silty loams or peat soils. As the onion's root system is relatively small, without a taproot, and relatively shallow, it is very important for the soil to be fertile and have a high water-holding capacity. The soil can be slightly acidic (5.9 to 6.5 pH) if it has high levels of organic matter, and can be more basic (above 6.5 pH) with mineral soils. Nitrogen should be readily available but not excessive for seed production. Organic practices that emphasize the use of compost and the subsequent decomposition of humus are ideal for onions and other alliums. A slow and steady release of nitrogen and phosphorus throughout the season from humus is preferable, both for growing well-formed bulbs and for good seed yields.

Growing the Seed Crop

As a biennial crop, producing onion seed is done using either an overwintering seed-to-seed method or a bulb-to-seed method very similar to the root-to-seed method used for both carrot and beet, where a "mature" storage root is produced and replanted the next season (see chapter 6, Carrot, "Root-to-Seed Method" on page 87). The seed-to-seed method is common in areas where the crop is easily overwintered and usually uses relatively small plants, which doesn't allow for bulb selection. The bulb-to-seed method is used in climates where overwintering the crop isn't possible or is used when the grower wants to perform selection on mature bulbs before replanting them.

Seed-to-Seed Method

In this method the seed is direct-sown or transplanted in mid- to late summer, depending on the onion type, its maturity class, and the climatic conditions. The planting date can vary considerably depending on the length of the growing season, as the onion plants do not need to form bulbs for this method to work. The goal is to grow large enough plants, plants that have matured past their juvenile stage, so they will physiologically recognize the cold period that triggers the vernalization response that will ultimately cause the plant to flower during the second season of the biennial life cycle. However, if the plant is not past the juvenile stage going into cold winter weather, then it won't vernalize fully and may not bolt during the second season of growth. If any number of plants in an onion population do not bolt in the second season, then in fact you will be selecting for easy-bolting genotypes or individuals. After a few years of doing this you can severely shift an onion (or any biennial) population to be more prone to bolting, and they will require fewer vernalization hours to bolt (see chapter 3, Understanding Biennial Seed Crops, "Vernalization" on page 17).

To ensure complete bolting in the second season the onions must be past the juvenile stage by the end of the first season so they will

begin to clock vernalization hours as soon as fall temperatures regularly bring temperatures below 50°F (10°C). The juvenile stage in many onion varieties is often when the plant is no bigger than a person's thumb. But the variation in the size of the plant that is necessary for it to be able to receive vernalization hours will vary from variety to variety. Also, the temperature range for optimum vernalization and the duration of the vernalization period needed for bolting can vary widely across all types of bulbing onions. All of the onions in a population that is planted for a seed crop must bolt, flower, and produce seed to maintain the inherent vernalization requirement of the population and prevent it from shifting.

It is helpful to know the history of the variety or type of onion that you are growing and to make sure your seasons and latitude are similar to its place of geographic adaptation. Cold hardiness can vary greatly, and it is definitely important to know the history of a variety before subjecting it to the low temperatures of a new region when overwintering using the seed-to-seed method. In most cases only short- and intermediate-day onions are overwintered. Most geographic regions suitable for producing a seed crop of long-day onion varieties are too cold in the winter to produce these varieties using the seed-to-seed method.

Sowing Seed in Season 1 of Seed-to-Seed: Seed for intermediate-day onions is often sown in early to mid-July at latitudes roughly between 34 and 40 degrees depending on the climate. Short-day onions are often planted in late summer through to early fall in some of the hotter valleys in lower latitudes. Seed is sown in rows 30 to 36 in (76 to 91 cm) apart. Because onion seed is small, it is planted at a shallow depth, and the soil must be kept evenly moist to encourage good emergence.

Thorough weed management is critical to get the plant to size up during this first season, especially at the seedling stage, as very young plants are the diameter of dental floss! Fertility must be adequate and weather conditions favorable to grow strong plants by the time they slow their growth in late fall. The plants are then left in situ for the duration of the winter. In some cases it is possible to mound soil around the onions to keep them protected from the lowest winter temperatures, but plants should not be completely buried, as then you stand the risk of losing some to rot.

Growth and Maturation in Season 2 of Seed-to-Seed: In spring, as active growth returns, it is imperative to ensure good growing conditions with adequate irrigation and regular cultivation. The use of the seed-to-seed method doesn't allow for selection of root characteristics except for color, vigor, or the presence of disease. However, it is always important to select against obvious defects, off-types, or sickly plants. As the plants bolt and start to open their flowers, it is also important to select against uneven bolting plants or plants whose flowers are not formed properly or lack robustness.

Seed grown in intermediate- or short-day onion seed production areas usually matures and is harvested earlier in the season than long-day onion seed, which often comes from transplanted bulbs using the root-to-seed method. Short-day seed crops can mature by early summer under optimum conditions, while intermediate types usually mature seed from mid- to late summer in most locations in the middle latitudes.

Bulb-to-Seed Method

The bulb-to-seed method is used primarily for two reasons: (1) it allows for selection of important bulb traits (see "Genetic Maintenance" on page 44); and (2) it allows you to grow an onion seed crop in areas where the winter is too cold to successfully overwinter onion plants in the field. When using this method the onion bulb crop is grown much like it is grown as a vegetable, though many growers prefer producing bulbs that are somewhat smaller than the marketable size. Generally the recommended size is from 2.25 to 3 in (5.7 to 7.6 cm) in diameter, which is considered a medium-sized bulb. There are several reasons for use of this sized bulb for seed production. First, it is possible to produce more bulbs per acre or hectare during the first season. Next, it is possible to store more bulbs for replanting in any given storage area. Also, medium-sized bulbs generally store better than fully grown bulbs. These bulbs that are grown for seed production are called mother bulbs.

Sowing Seed in Season 1 of Bulb-to-Seed: The seed that is sown for producing the mother bulbs is usually planted at the same time you'd plant if you were producing market bulbs, no matter which daylength type you're producing. The major difference is that the crop is usually sown at a higher density than would be used for the vegetable crop in order to produce medium-sized bulbs. Long-day types are sown in the higher latitudes in spring for late-summer-to-early-fall harvest. When seeding the intermediate and short-day types they are also planted at much the same time that these respective types normally would be planted for the vegetable crop. However, these latter types can sometimes be planted a few weeks later than usual, depending on the region and climate as well as their ability to produce nicely formed, medium-sized bulbs in a timely fashion. A number of selection steps can be practiced during the vegetative growth stages in this first year of the biennial cycle (see "Genetic Maintenance" on page 44).

Both leek and onion capsules open to reveal black seeds as the seed reaches full maturity. ▶

When the bulbs have matured and at least 50 to 60% of the tops have fallen over, then it is time to gingerly knock the rest of the tops over and allow the initial curing to take place in situ. If the weather is favorable, after 3 to 4 days it is advisable to pull the onions and allow them to cure in windrows for several more days before picking them up and topping them in preparation for storage. Whenever onions are left in the sun, revealing sides of the bulb that haven't previously been exposed to direct light, it is important to turn them frequently to avoid sunburn. Once the tops have cured to the point at which they can easily be twisted off, then twist them off by hand. Mechanical topping can bruise the bulbs; cutting the tops with shears is often used, but many farmers avoid it as well, as it can expose the bulb to rot organisms that may shorten the onion's storage life. Topping the crop is a perfect time to select the roots for the classic root traits like shape, color, width of the neck, and thickness of the skin.

Seed Harvest

In higher latitudes onions flower and form seed during the longest days of summer and are harvested during late summer. High-quality seed of these long-day onions is grown in climates where the chance of rain in late August into September in the Northern Hemisphere (late February into March in the Southern Hemisphere) has to be very low. In many intermediate and lower latitudes the intermediate- and short-day onion types will mature their seed during midsummer. Seed is often hand-harvested just when the umbels start to reveal their ripe black seed as the capsules begin opening. The ripening process can proceed quickly once it starts, and it is important to catch it as close to dehiscence as possible, but before any significant amount of seed has shattered from the heads. In open-pollinated onion varieties there will be a range of maturation over a period of a few weeks. As with all biennials, it is not advisable to gather seed only from the earliest-flowering plants; there is some evidence that the earliest plants to bolt are to some degree closer to being easy bolters, while the later-bolting plants are probably more bolt-resistant.

The harvest starts by cutting the scapes from 4 to 6 in (10 to 15 cm) below the umbels. These are harvested into buckets or muslin sacks; be careful to get the cut umbels into the container without shattering too much seed. The umbels are then carefully placed on breathable fabric or on specially made drying racks outdoors if dry weather prevails for at least 10 days. If dried and matured in the open then the umbels should not be piled any deeper than 6 in (15 cm) when they are spread out in the sun. They must also be turned regularly so that they dry evenly and don't get sunburned. When the seed has fully cured it will shed easily from the umbels.

If you are harvesting onions when wet or inclement weather threatens then it is best to cure the seedheads in sheds with good air circulation. The use of forced heated air to dry this valuable crop in a timely fashion is advisable when wet, cool weather sets in at the time of harvest. The seedheads will need to be turned even more frequently in this situation, especially during the first few days, as the umbels may have some dampness going into the dryer.

Seed Cleaning

When the umbels are dry—to the point where the seed separates easily upon rubbing a head between your palms—it is time to thresh the seed. The umbels can then be run through a stationary thresher or combine, but be very careful to set the rollers so that the rather delicate onion seed doesn't get damaged. The thresher's cylinder must be set at a distance that doesn't allow chipping or cracking of the seed coat. Seed-company threshing crews will often check the tiny seed for damage repeatedly with a hand lens or dissecting scope during the process to make sure they are not damaging any seed.

After threshing the onion seed, the screening and winnowed will usually eliminate most of the capsules and other debris. For particularly hard to clean onion seed lots, the seed can be put into a water vessel: the seed will sink and the trash will float and can be decanted off. The seed then needs to be spread out onto screens and dried quickly to avoid seed imbibation. This means good air flow and heat if necessary.

GENETIC MAINTENANCE

The genetic variation that is common to all cross-pollinated crop species is certainly found in onions and requires skill and patience from the seed grower to select and maintain a good open-pollinated onion population. Onions and the other related sexually propagated alliums are susceptible to inbreeding depression, so it is important to maintain genetic diversity by growing and harvesting seed from at least 120 to 200 healthy, well-selected onion plants each time you produce a seed crop. This will ensure that you maintain the genetic elasticity and diversity of any variety that you produce if you are not producing a large commercial crop.

As with all biennial crops, there are a number of times in the two seasons of producing seed when the grower can select for desirable traits to develop and maintain a good selection of the crop. During the seedling stage you can always select for good seedling vigor and general plant health. Early bulb formation is an important trait for selection in many varieties. Diseases such as downy mildew and pink root are endemic in certain climates, and growers always need to know the symptoms of any diseases that commonly recur in their specific area. When bulbs form and start to mature it is possible to select for uniformity of "fall down" of the tops or the ability of the tops or foliage to develop soft necks and collapse within a few days of one another and begin the curing process. Bulbs with thick necks that don't allow this neck closure at harvest, when at least 65 to 80 percent of the crop has the tops collapsing, should be eliminated. Plants that have any sign of neck rot, which is often caused by *Botrytis aclada*, or other foliar or bulb diseases should also be selected against.

After the bulbs are pulled and laid in windrows at the time of harvest, it is possible to select for bulb color, intensity of color, and bulb shape. As the bulbs cure and are put into storage, they can be selected again for their shape and color, as well as for the uniformity of skin formation. During storage bulbs should be checked periodically and selected for sprouting resistance, resistance to rot, retention of skins, and firmness of the bulbs. This selection for storability is very important for any varieties used in the bulb-to-seed method.

ISOLATION DISTANCES

Isolation to avoid crossing in onions and other related flowering *A. cepa* crops follows the general rule for cross-pollinated crops that are primarily insect-pollinated. Two crops within a given species of the same crop type, the same form, the same color, and the same daylength class need only be isolated by 1 mi (1.6 km) if grown in open terrain or 0.5 mi (0.8 km) if there are substantial physical barriers between the crops (see chapter 13,

Isolation Distances for Maintaining Varietal Integrity). An example of two seed crops of the same crop type that could use these isolation recommendations would be two red, globe-shaped long-day storage onions. The minimum isolation distance would definitely need to be increased to 2 mi (3.2 km) in open terrain if the second crop is an entirely different crop type (for example, a globe-shaped storage onion versus a short, flattened cipollini type) or if the two onion varieties have different colors (for instance, a yellow versus red or white onion). It would also need to be used between onion seed crops of different daylength types. This increased isolation requirement between two crops of the same species but different types can be cut to 1 mi (1.6 km) if there are substantial physical barriers between the two crops.

– 5 –

AMARANTHACEAE

The Amaranthaceae or goosefoot family contains several economically important species of crop plants and a number of the most prominent and pernicious weeds worldwide. The important vegetable crops in this family are beets, spinach, and Swiss chard. Minor vegetables include orach and Good King Henry. These crops are all from the Old World and have probably been used in agriculture in some form since ancient times. Several New World domesticates from this family have played an important role in the agriculture of the Americas, with quinoa historically being the most important grain crop in the Andean highlands, where it thrives at altitudes above 8,000 ft (2,440 m).

All of the vegetable members of the goosefoot family have leaves that have traditionally been harvested for their use as cooked greens or potherbs. The garden beet is also harvested for its fleshy, sweet roots and is an important root vegetable in temperate climates across the globe. With the increase in demand for cut leaves for bagged salads from local farmers and large produce growers alike, the demand for seed of all of these crops for their succulent and often colorful greens is rapidly increasing.

The plants of the Amaranthaceae are dicotyledons. The crop plants are similar in many of their characteristics, usually having alternate and simple leaves that form a large rosette in the vegetative phase of their life cycle. All of the cultivated chenopods are wind-pollinated, producing copious amounts of pollen that is capable of traveling several miles and remaining viable if environmental conditions are just right. An oft-repeated fact, that beet pollen was collected by an airplane at an altitude of greater than 3 mi (5 km) above sugar beet fields by beet researchers demonstrates the potential range of the pollen.

The Amaranthaceae are all dry-seeded crops, and their seed is most successfully grown with little or no precipitation during the flowering and seed maturation stages of their life cycle (see chapter 15, "Cool-Season Dry-Seeded Crops" on page 345). Most of the commercial seed production of these crops is done in cool temperate regions with dry weather in summer and early fall. Major seed production areas for beets, chard, and spinach include coastal regions of the Pacific Northwest in the United States, central Denmark, Southern Europe, and the South Island of New Zealand. However, smaller-scale production of regionally important varieties is common in a number of regions around the world where these crops are culturally significant. Beet seed production for many varieties popular in Eastern Europe and the Mediterranean Basin is undoubtedly extensive, even in areas with less-than-ideal conditions for growing *Beta* spp. seed.

FAMILY CHARACTERISTICS

Reproductive Biology

The cultivated Amaranthaceae include both annuals and biennials and consist of dry-seeded crops that most successfully mature their seed over an extended dry period in the late summer and early fall. The inflorescence of these crops has flowers that are small, inconspicuous, and borne in the axils of branches and leaves. Beets and Swiss chard have perfect flowers that almost always outcross due to a genetic trait for self-infertility. Spinach is naturally dioecious, with most open-pollinated populations having a balanced sex ratio of the two unisexual types. The dioecious habit in spinach is indeed a foolproof biological way to ensure cross-pollination.

The flowering habit of the Amaranthaceae is indeterminate, and once initiated, flowering will continue until harvest or frost. Hence seed will be maturing continuously throughout the rest of the season, with a certain percentage of it that is immature at harvest. Therefore you must monitor the relative maturity of your beet, chard, and spinach seed, understanding that at least 20% of the seed that has set won't be fully formed when you harvest the crop.

Life Cycle and Climatic Adaptation

The garden beet and Swiss chard are typical biennial plants producing a relatively large rosette of leaves and fleshy storage root during the first season of growth and one or more branched seed stalks during the second season. Both beets and chard thrive in cooler climates, with their most vigorous growth occurring in spring and fall. The goal in seed production is to produce a moderate-sized storage root in the late summer and fall of the first year that can be either overwintered in the ground or stored in a root cellar or coldroom for replanting in the spring. Swiss chard, though not grown as a root vegetable, does produce a substantial storage root. When grown for at least 90 days chard plants will produce roots that can be stored like beets for up to 5 or 6 months in a root cellar or cooler. Like all biennials, beets and chard require a period of vernalization (cold treatment) of at least 8 weeks for flower initiation during the second year of their life cycle. And they must have a substantial root with an intact apical growing point for both the shoot and the root. The three methods for overwintering beets and chard are the same as outlined for Apiaceae crops (see chapter 6), as indeed beets and carrots are treated much alike in most regards concerning seed production (see the Garden Beet section for specific differences).

Spinach is an annual that is almost always planted in spring as a seed crop. It must be planted early enough in the spring to produce a large enough plant to support a hearty seed yield (or produce a hearty amount of pollen if it's a male plant). Spinach flowering is initiated primarily by daylength. Days with at least 14.5 hours of light will initiate the development of the male and female flowers and inflorescences in many standard spinach varieties developed before the 1990s. Most newer spinach varieties are considerably more bolt-hardy. This means that most older spinach types will often be noticeably bolting by late May to early June at latitudes above 40 degrees, with the biochemical initiation occurring quite a bit earlier, especially at higher latitudes.

Seed Harvest

The chenopods share the trait of indeterminate flowering that translates to sequential ripening of the seed at the end of the season. A percentage of the seed that sets later in the season, easily 20% or more, will never mature. Therefore the timing of the seed harvest is a compromise of patience to get the highest amount of mature seed while not waiting too long and risking the loss of quality to inclement weather or loss of the most mature seed from shattering. As with most crops with indeterminate flowering grown in temperate climates, the earliest fraction of seed to mature is almost always superior—larger, with a higher germination percentage and greater seedling vigor.

Determining the time of harvest for beets, chard, and spinach is usually done via a visual assessment. In general the color of the seed is used as an important indicator of maturity. The seed of these three crops will turn tan to tannish brown in color if matured under dry, favorable conditions, and turn a darker brown when matured with unwanted late-season precipitation. Many growers will wait till between 60 and 80% of the seed on a solid majority of the plants in the field has turned brown to begin harvest. While this method can successfully result in a high-quality seed crop, there are several ways to check the endosperm of chenopod seed to get a more accurate picture of the relative maturity of the seed (see specific instructions under "Seed Harvest" for each crop).

At harvest mature chenopod seed crops are cut and laid in windrows on cleared, cultivated ground or on tarps or landscape fabric. Curing the seed in windrows can take anywhere from 5 to 10 days depending on weather. If wet weather threatens it may be necessary to move the crop to a drying shed, greenhouse, or seed dryer. Threshing is often done by manually feeding cured plants into self-propelled combines or stationary threshers. Cutting and swathing these crops is best done early in the day to take advantage of dew to reduce losses from shattering. This also reduces the amount of small pieces of stem the size of the seed that ends up in the seed fraction after the scalping step is completed.

Seed Cleaning

Chenopod seed is usually scalped as a preliminary step in the cleaning process. Scalping is the first step in the seed cleaning process where the bulk of the leaves, stems, and any debris is eliminated when the first cut of the freshly harvested material is run across a large screen with very large holes or slots. This should be done soon after threshing to eliminate any moisture that is held in the leaf and stem tissue prevalent in freshly threshed beet, chard, and spinach seed. Seed of these three crops frequently needs some degree of drying soon after harvest with the onset of cooler wet weather at the end of the season in the cool climates that are suitable for seed production. Chenopod seed is usually then rubbed in large, cannon-like cylinders that have a rotating shaft with large studs, which actually rub the seed against one another while under pressure in the shaft. These crops have seed that is frequently borne in clusters that can fuse together and are often also fused to pieces of the stem; rubbing is very important to break these clusters. In addition beet and chard seed has a corky outer surface from the dried flower bracts that fuse and form their characteristic multigerm seed balls. These are frequently rubbed until the multigerm seed balls are smoothed and rounded, a process called **decortication.** Decortication can also have the odd side benefit of damaging one or two of the multigerm embryos of the larger seed. This is viewed as a benefit in the seed business, as it may reduce the number of sprouts in the larger seed balls from three or four sprouts to one or two, thus making it less likely for a vegetable grower to get too dense a stand at planting.

ISOLATION DISTANCES

The crops of the Amaranthaceae have pollen that is known for its ability to travel great distances. These crops are predominantly wind-pollinated, and their pollen is relatively small, light, and easily carried by the wind. These crops also produce copious amounts of pollen. For these reasons, wind-pollinated crop species need a greater isolation distance between seed crops of the same species than insect-pollinated species. Certainly the chenopods require at least 2 mi (3.2 km) of isolation between crops of the same type when grown in the open and 3 or 4 mi (4.8 to 6.4 km) when growing different types of the same crop species or when isolating stockseed production from other seed production crops of the same species (see chapter 13, Isolation Distances for Maintaining Varietal Integrity). However, it should always be remembered that these isolation distances are not absolute, and there may still be a small amount of crossing even at these higher isolation distances, especially with these wind-pollinated crops.

GARDEN BEET

The humble vegetable known simply as the beet in North America is also known as the garden beet, red beet, table beet, or beetroot to distinguish it from both sugar beets and fodder beets. The sugar beet is grown for its high sugar content and the fodder beet, also called mangold, mangel, or mangel-wurzel, for its nutritional tops and roots, which are used as animal fodder. These three root crops, along with the leafy vegetable Swiss chard, are all members of the same species, *Beta vulgaris*.

Despite being distinguished by various sub-species designations by taxonomists, these four crops are all fully sexually compatible and will readily cross when flowering in proximity to one another.

The garden beet is thought to have originated in the western reaches of the Mediterranean region in southwestern Asia and was introduced into Europe during the Middle Ages.

The earliest cultivated beets were probably used as a leafy green vegetable. There is some

SEED PRODUCTION PARAMETERS: GARDEN BEET

Common names: beet, garden beet, table beet, beetroot
Crop species: *Beta vulgaris* L.
Life cycle: biennial
Mating system: cross-pollinated
Mode of pollination: wind
Favorable temperature range for pollination/seed formation: 65–75°F (18–24°C)
Seasonal reproductive cycle: summer (Year 1) through the following fall (Year 2) (~14 months)
Within-row spacing: seed-to-seed 16–24 in (41–61 cm); root-to-seed 18–30 in (46–76 cm)
Between-row spacing: seed-to-seed 30–36 in (76–91 cm); root-to-seed 30–48 in (76–122 cm)
Species that will readily cross with crop: Swiss chard, sugar beet, mangel, and fodder beet are all *Beta vulgaris* L. and readily cross with garden beet.
Isolation distance between seed crops: 1–2 mi (1.6–3.2 km) for beet crops of the same color; 1.5–3 mi (2.4–4.8 km) for beets of different colors; 3–5 mi (4.8–8 km) for different *B. vulgaris* crops. All stated distances are dependent on crop type and barriers present on the landscape. Distance between any *B. vulgaris* crops and transgenic or GM versions of this species should be a minimum of 10 mi (16 km).

The 'Yellow Intermediate Mangel' is a modern example of the mangel-wurzel, an old large-rooted form of field beet used primarily as animal fodder.

'Pronto' is a high-quality fresh-market beet with robust green tops. Its green tops need regular genetic selection in order to maintain this elusive trait. Photo courtesy of Micaela Colley

indication that the roots of this ancestral form of the crop were used for medicinal purposes. It has been speculated that the original form of the crop was an annual, probably a winter annual, and only through selection for a food plant that could be stored through the winter as the crop spread northward did the beet develop an enlarged storage root and become a biennial.

The first descriptions of garden beets with fleshy, edible roots are from Germany in the 16th century. It wasn't until the 18th century that written accounts of garden beet cultivation across Europe became common. By the 19th century beets were an important crop throughout Europe, North Africa, the Middle East, parts of western Asia, and North America. In 1885 Vilmorin-Andrieux devoted 15 pages to garden and fodder beets in his classic text *The Vegetable Garden,* which describes the common vegetable varieties used throughout Northern Europe at the time. By then both red- and yellow-colored beets were being described, and root shapes for garden beets were described as either "turnip-rooted" or long. The turnip-rooted types were globe-shaped or flattened globes; the long types were usually more pointy or cone-shaped than the modern cylindrical types. Vilmorin-Andrieux also lists several named "intermediate" types, "being intermediate between the garden and field types," which are described as being large and often forked like mangels, but "of good quality for table use."

With the onset of a thriving commercial seed industry in North America and Europe in the early 20th century, red, globe-shaped beets became the dominant form in these markets. Most garden beet varieties developed in the last 100 years have been either fresh market types with taller, robust tops or processing varieties with shorter tops and an emphasis on dark root pigmentation. Traditionally beet roots have been used as a cooked vegetable or pickled. Mature beet tops or "greens" have been cooked or occasionally used in salads. With the advent of prepared or bagged salads the use of beet greens grown for the baby leaf market has increased dramatically. This has greatly increased the amount of beet seed grown commercially, with a large increase in the organic beet seed produced.

CROP CHARACTERISTICS

Reproductive Biology

The garden beet is a biennial, normally requiring two growing seasons to produce seed. During the first year of growth it produces a rosette of leaves and a fleshy taproot. During the second year it develops a seed stem that is 3 to 5 ft (0.9 to 1.5 m) tall with extensive branching and can cover a diameter of more than 3 ft (0.9 m) on the ground.

Flowering is initiated by a cold treatment of the roots, or vernalization, after the first season's growth (see chapter 3, Understanding Biennial Seed Crops, "Vernalization" on page 17). The vernalization requirement for beets is approximately 8 full weeks of temperatures at or below 50°F (10°C). This, coupled with lengthening daylight in spring, will promote flower initiation and subsequent flowering. The garden beet is predominantly wind-pollinated and has extremely light pollen that is easily carried long distances by the wind. Perfect flowers are borne in most of the leaf axils of the large branching plant. They usually occur in clusters of two to five, with one bearing an extended bract that encircles the cluster. As this single bract dries it forms a corky, irregularly shaped multiseeded fruit. This is what is commonly called the seed ball or

multigerm seed in the seed trade. As this seed ball is capable of producing two, three, four, and even five seedlings, it is hard to achieve a well-spaced stand even with precision planting. Monogerm varieties have single flowers borne in bract axils, therefore no fusing of multigerm seed balls. Many vegetable growers have avoided monogerm cultivars, complaining of poor seedling vigor. If true, this is a serious hurdle for monogerm seed to cross in gaining acceptance by organic farmers.

The perfect flowers are protandrous, with the anthers of a particular flower maturing and shedding pollen for 2 to 3 days before the gynoecium of that flower matures and the stigma becomes fully receptive for up to 2 weeks. The small, greenish flowers are quite compact as the anthers have very short styles. They are not showy, as is normal for wind-pollinated species. Nectar is produced and insect pollination is common, although the wind accounts for most pollination events. Beets are self-incompatible, and therefore any individual beet plant will not receive its own pollen to fertilize its ovules and produce seed. Each individual plant must have pollen from a genetically different individual in the population (or another population) to produce viable seed. This encourages almost complete cross-pollination between individuals in a population, although rare individuals capable of self-pollination have been found and used in hybrid breeding programs.

Climatic and Geographic Suitability

The garden beet is a cool-season crop that thrives when grown at moderate temperatures between 60 and 72°F (16 to 22°C), which encourages steady, vigorous growth. When grown as a spring-planted vegetable, beets can tolerate hotter temperatures as they mature a root crop. The beet seed crop, however, is

much more sensitive to heat during the second seed-bearing season of the biennial cycle. In order to produce high-quality yields of vigorous seed with high germination rates beet seed crops should be grown in regions where summer temperatures don't usually exceed 75°F (24°C). The best beet seed production locales have cool, wet spring weather to establish a sturdy frame of vegetative growth, followed by cool, relatively dry summer weather to support optimum pollination, fertilization, and seed development. An extended dry period in late summer and into early fall is ideal for beet seed crops to minimize seedborne diseases and allow the crop to fully mature. Beet seed that avoids most precipitation as it matures also retains a healthy mahogany brown color at harvest. Important production areas include the Skagit Valley and the Salish Sea basin of maritime Washington and British Columbia. Beet seed is also grown commercially in the Rhône River Valley of France and on the South Island of New Zealand. As the garden beet is a very important crop across Eastern Europe and in many temperate regions of eastern Asia, there are undoubtedly a number of places where seed of regionally significant varieties is being produced for commerce and subsistence use.

SEED PRODUCTION PRACTICES

Soil and Fertility Requirements

Beets grown as a vegetable crop are best suited to a fairly deep, well-drained sandy loam. Growing the first-year roots on a friable soil of this quality will enable the seed grower to

select for the shape of the roots before using them as stecklings the following season if using the root-to-seed method (see chapter 6, Carrot, "Root-to-Seed Method" on page 87). Roots can be replanted in heavier soils in the second year for seed production. In fact, many growers claim that a heavier clay soil will produce a higher seed yield. The soil pH for beets is a narrow range, between 5.8 and 6.2, for optimum health. Beets are salt-tolerant; cropping is possible on soils that are too saline for many other crops. They are also very sensitive to boron deficiency in the soil, which can cause crown rot, blackening of leaf margins, or heart rot within the roots. Boron deficiency is most prevalent in alkaline soils. In seed crops, the amount of available nitrogen should not be too high in order to avoid excessive vegetative growth before bolting, as this promotes lodging of the plants during seed set. Well-maintained agricultural soils high in humus and microbial populations will supply adequate nutrients and water over the long seed production season.

Growing the Seed Crop

Beet seed is produced using the two basic methods, seed-to-seed and root-to-seed, that are used for carrots, parsnips, and other biennial root vegetables (see chapter 6, Carrot, "Seed-to-Seed Method" on page 86 and "Root-to-Seed Method" on page 87). The seed-to-seed method for beets has not been as successful as it has been for carrots and some of the other biennial seed crops, however, especially in the traditional beet seed growing regions. The seed-to seed method was used exclusively in the Skagit Valley of Washington (the largest garden beet seed production area in the United States) until the 1940s. The production practice of having the first-year beets overwinter in the field to produce seed in the same field, where they could easily infect any new beet crops being planted nearby with any parasitic organism (insect or plant pathogen) that they might harbor, proved to be a disaster. The Skagit Valley beet seed growers at that time were experiencing an increasing problem from beet mosaic virus (BtMV), a viral disease that causes mosaic patterns and zonate rings of leaves, and that will eventually cause foliar necrosis, which in turn leads to stunting and poor seed quality and yield. It was determined that this disease was spreading from the second-year seed-bearing beet fields to the new, first-year root crop in nearby fields via both the green peach aphid (*Myzus persicae*) and the black bean aphid (*Aphis fabae*). The BtMV and the aphid vectors were being harbored from one season to the next on the live plants overwintering in the fields. These plants were acting as a "green bridge" for both BtMV and the aphids, allowing them to pick up right where they left off in the previous season as soon as the growing season started. The solution was for all of the beet growers to agree to discontinue the use of the seed-to-seed method and to produce their roots outside of the valley. This would mean no beet plants to serve as a "green bridge" through the winter and no first-season beets anywhere in the valley to pick up any parasites in the first season to be spread in the second season.

Seed-to-Seed Method: The basic protocol for the seed-to-seed method for all biennial root crops is described in chapter 6, Carrot, "Seed-to-Seed Method" on page 86. The seed-to-seed method for beets can be problematic in areas where large-scale beet seed production is taking place, due to the "green bridge" effect described in the previous section. However, there are instances where farmers who are isolated from other beet (or other *Beta vulgaris* crops) production areas can use the seed-to-seed for production across a number of years, so long as they monitor

their crops for disease symptoms and utilize occasional seasonal breaks where they don't have any beets in the field. Another important aspect of potentially monitoring seed-to-seed cropping in beets is to remember that the aphids and other vectors move the viruses like BtMV, beet yellows (BYV), beet mild yellows (BMY), and beet curly top from field to field depending on weather patterns and air currents. Therefore, using as much isolation as possible in your own root and seed fields, even if you are isolated from other growers, is very important in controlling these pathogens.

As with carrots, the seed-to-seed method requires less labor and is simpler than the root-to-seed method. The drawbacks to using this method for garden beets are essentially the same drawbacks as exist for all of the biennial seed crops. The first is the fact that the roots must overwinter in the field. As table beets are generally less cold-hardy than some of the other biennial crops such as carrot, rutabaga, or cabbage, you must be knowledgeable about the ability of the crop (and the specific variety) to survive throughout the winter in your region. Most table beet varieties will survive temperatures down to a range of 24 to 27°F (−4 to −3°C) without permanent damage, depending on the duration of the cold and the severity of the freeze–thaw cycles.

The second drawback to the seed-to-seed method is the inability of the grower or the seed company to perform selection on the crop's roots to maintain varietal integrity. As is the case with all biennial root crops that are grown for seed, selection of the roots is very important in maintaining a true-to-type variety unless you're starting with well-maintained stockseed (see chapter 17, Stockseed Basics on page 364).

Root-to-Seed Method: The basic protocol for the root-to-seed method for all biennial root crops is described in chapter 6, Carrot, "Growing the Seed Crop" on page 86. As with all biennial root crops, this method allows you to evaluate the roots and select the best ones for planting stock. The roots must be acceptable phenotypically and must be free of any disease, pests, or physical damage in order to store properly under any of the storage methods outlined for beets. Phenotypic selection characters are discussed in the "Genetic Maintenance" section on page 60.

Beet roots are better suited to the root-to-seed method than many other biennial crops because the roots store so well and for relatively long periods of time if stored properly (see chapter 3, "Preparing Roots for Storage" on page 22). Beet roots are also ideally suited to having a piece of the side of the root cut off to select for internal characteristics like the intensity of color and the presence of zoning (see "Genetic Maintenance-Root Color" on page 61).

Seed Harvest

Seed Maturation: Beet seed formation usually starts anywhere from 6 to 10 weeks after flower stalk initiation. As the flowering habit is indeterminate, flowering and subsequent seed maturation will continue until harvest or frost. Because beet seed matures sequentially, the percentage reaching full maturity at the time of harvest will usually not exceed 75% of the total seed crop. The earliest seed to set will often mature several weeks before the bulk of the seed on any given plant has matured. This first seed set is usually of a high quality and has a high germination rate, but it may readily shatter as the bulk of the crop is maturing. You must determine when the maximum overall maturation has occurred, without losing a significant amount of the earliest-maturing seed. Cool, wet weather can often occur during the late season, and the seed maturation

Trimming beet roots for storage: (left) untrimmed beet root, (right) beet root trimmed properly for storage (compare to carrot stecklings on page 23). Photo courtesy of Jared Zystro

period for beets in the Pacific Northwest makes it even more important for growers there to closely monitor the seed maturity and not harvest too early or too late. Early harvest may result in a percentage of seed that is not fully mature. Harvesting seed past the optimum time period may result in reduced yield and seed quality due to seed shattering and an increased incidence of seedborne diseases.

A standard method used to judge maturity of the beet seed crop is a visual assessment of the color of the seed ball (a multiple-seeded fruit resulting from the fused dry corky bracts of two or more flowers that occur at the same node). Harvest should occur when between 60 and 80% of the seed balls on at least 90% of the plants in the field have turned a tannish brown shade, typical of mature beet seed. Unfortunately this method may sometimes be inaccurate due to the potential effects of the environment or the genetic variation of the particular beet variety being produced. In a number of environments the beet seed balls will turn a darker shade of brown, sometimes before they reach full maturity. This often occurs with higher-than-usual levels of precipitation during the final weeks of seed maturation. This darkening may be due to saprophytic bacterial or fungal growth on the corky bract tissue of the seed balls. Depending on the pathogen this darkening may not be harmful to the seed, but it seriously impedes visual assessment of beet seed maturity. There is also considerable variation among varieties for the degree of browning of the seed that occurs during the maturation process. Some beet varieties may retain greener hues than others, even when fully mature.

In order to make a more accurate assessment of the maturity of the seed it is best to check the relative maturity of the endosperm of the seed. The endosperm, which grows concurrently with the embryo, must be fully developed to produce viable, fully mature seed that will grow vigorously and maintain a high germination percentage through its expected storage life. The starchy endosperm can be monitored through the maturation process by cracking open any one of the several seeds that occur within each beet seed ball. The best way to determine the maturity is to squeeze a small amount of the endosperm out of the cracked seed and visually inspect it. If the endosperm is viscous and appears translucent or milky, it is not close to maturity and will require at least 3 to 4 weeks to mature. Sometime after this the endosperm will start to appear grayish and waxy, what is often called flinty, but it is still not close to maturity. Only when the endosperm becomes starchy, with a true solid white color and a firm texture, is the seed very close to maturity. A majority of the seed—at least 70 to 75% of the seed on a given plant—must be at this advanced starchy stage before you consider harvesting.

Seed Harvest: Beet plants in full flower can have a stature of 3 to 6 ft (0.9 to 1.8 m) in height and can often cover a diameter of more than 3 ft (0.9 m) on the ground. The commercial beet seed crop is rarely staked, so

▼ The beet seed of commerce is usually a multigerm dried fruit that is an aggregate "seed ball" composed of two to five seeds.

there is a tendency for the crop to lean in all directions, thereby making mechanical cutting or swathing nearly impossible without causing excessive amounts of seed shattering. Because of this, many growers elect to perform the initial harvesting of the crop by hand. Crews walk through the field pulling plants by hand and laying them into several windrows across the field. In some cases the harvesters use machetes to cut the remaining beet root and root mass from the plant, to minimize the chance of getting soil in the seed during the subsequent threshing. Other growers choose to leave the root mass attached, as the root may still supply some energy in the final ripening that takes place in the windrow. This initial harvesting process is ideally timed to coincide with a subsequent period of fair weather for the after-ripening of the seed in the windrow.

The seed plants should be placed in the windrows in such a way that they receive sufficient airflow to allow even drying, even with heavy dew or light rains that may occur during this after-ripening. This is important, as beet windrows are not commonly turned; seed can easily shatter during turning. Depending on the importance or value of a particular beet crop, it may be worthwhile to place the crop onto landscape fabric or some other comparable porous material in order to catch any seed that may shatter during this period. The use of porous material is important to ensure that any precipitation or moisture accumulation will quickly be drawn away from the seed.

▼ A mature beet seed crop that is swathed and drying in windrows in the Willamette Valley of Oregon.

59

The plants should remain in the windrows for 7 to 10 days. After this drying period threshing can be done using a belt thresher, a stationary rotating thresher, or a self-propelled combine that is manually fed. Threshing should be done early in the day, midmorning in most climates, as light dew on the plants will prevent much of the potential shattering that can occur when the plants are picked up for threshing. The moisture can also lessen the amount of stems that break and lessen the size of the broken pieces that

Manually feeding beet plants through a belt thresher in the field.

occur during threshing, thereby saving much time in subsequent seed-cleaning steps to remove them. Threshed seed should then go through an initial screen cleaning or scalping to remove large leaf tissue, stems, and soil clods that can hold moisture. Further drying of the seed crop should occur at this point in a well-ventilated warm space with supplemental heat as necessary.

GENETIC MAINTENANCE

Beets are a cross-pollinated species with lots of inherent genetic variation. In order to maintain adequate genetic diversity and elasticity in any open-pollinated beet variety, it is important to harvest seed from a minimum of 120 to 200 plants each time that you reproduce a variety. This will help ensure that you do not severely shift the genetic frequency of any important traits that may not be obvious in

your environment of selection and also ensure that the variety will maintain vigor and avoid inbreeding depression. The initial population size should reflect the intensity of selection activities. For instance, if a variety has been well maintained and requires very little selection to keep it true to type, then beginning with a population of 250 plants before selection will easily yield 200 selected plants. On the other hand, if a variety has a large percentage of off-type plants that need to be eliminated, it might require an initial population of 400 or more to derive a final population of 200 selected plants at a selection intensity of 50%.

Selection Criteria

The practice of trait selection of any crop is related to the needs of the farmer in a particular area, the environmental pressures of the production region, the cultural practices utilized, and market demands. While vegetable breeders normally concentrate on traits for farmers who grow the crop as a vegetable, it is also possible to select for traits important in the reproductive stage of the plant's life cycle. Therefore, when someone is developing and maintaining varieties for organic systems, the selection criteria should include traits that are necessary for both the organic vegetable farmer and the organic seed producer. Selection should be done at several points in the life cycle whenever possible, including at the seedling stage, at the market or eating stage, and during flowering, to maximize reproductive health and seed yield.

To select for root qualities, when approximately 80% of the beet roots in the first-year root nursery are of marketable size, the roots should be pulled and placed neatly on the beds for evaluation; keep them out of direct sun until they are replanted. Root selection on a cloudy, cool day is ideal. Selecting roots is best done with good knowledge of what "the norm" is for the phenotype (appearance) of each variety. Selection to a standard varietal type is the major objective in commercial production of established varieties. If a variety is adequately uniform at the outset of your seed production endeavor, then you can expect to discard approximately 10 to 20% of the roots, simply to maintain varietal integrity. If the variety is not uniform due to lack of prior selection (or poor selection), then expect to select out as much as 50% of the roots as off-types. Selection for improvements of market characteristics is always best done at the market stage.

Seedling Vigor: Seedling vigor and early robust growth are important to organic farmers, as these traits affect the plant's ability to compete against weeds, resist seedling diseases, and contribute to overall plant health. Vigor can be improved over several cycles of selection. Selecting for the quickest, earliest seedlings to emerge is the first step, but seedling selection should also include recognition of shape, size, color, and ability of the seedlings to grow under less-than-optimum conditions. This selection should occur soon after emergence and may be coupled with the initial weeding and thinning of the plot.

Leaf Size, Shape, and Color: While many seed growers do not pay attention to the leaf characteristics of beets, these traits are very important to vegetable growers who market bunching beets. Beets are generally classified as either short-top types, with tops that are often 6 to 8 in (15 to 20 cm) tall, or tall-top types, with tops 10 to 14 in (25 to 36 cm) tall. Selection of the tops for height within each of these categories is best done before the roots are pulled. Tops also vary in shape, from the narrower strap-leaved types to the wider heart-shaped-leaf types, and require regular selection, which is best done while the crop is standing. Leaf color can range from deep green to medium green to a largely red hue (for instance, 'Bull's Blood'). For bunching types the darker green color is desirable, as the lighter shades and reddish coloration are often associated with a product that isn't as fresh as the darker green type. Also, cooler fall weather will often promote a blotchy reddening of the leaves that many people associate with leaf diseases or decline of the plant. This trait is more prevalent in some beet cultivars than others, and selection for darker green leaves that resist reddening into the fall has been successful when practiced.

Root Shape: Table beets come in an array of shapes, from long cylinders to rounded spheres, flattened globes, and roots that are shaped like a toy top. The size and width of the taproot at the bottom of the beet will also affect the shape of the beet; for instance, a top-shaped beet often has a pronounced, thick taproot. If the shape is not regularly selected, then the unique shape of a variety can be lost within a couple of generations. When selecting for shape don't be overly concerned trying to get the perfect shape in each root; no two roots are the same and each shape is an approximation of the ideal shape for the variety.

Root Color: Table beets include red, pink, yellow, and white varieties. There can be differences in the shades of these colors and the intensity of the pigmentation. The color is somewhat apparent from viewing the exterior of the root, and you can make an initial selection in this way. However, if significant improvement in root color is desired, then you may select more precisely for the intensity of root color by cutting them open to see the interior. In order to do this and still be able to replant the beet, it is imperative that you do not cut into (or near) the apical growing points of the crown or the root. However, by cutting a slice of the root off the side of the beet (a "cheek"), you will be able to observe the interior color with no damage to the root. The piece of the root that is cut off can also be used for tasting and selection for flavor.

You can now leave the beet roots out in a cool, dry, shady spot for several hours to let the cut surface suberize or heal before it is planted. **Suberin** is a waxy substance that forms fairly quickly in damaged plant cells to prevent water from penetrating the tissue. The roots can then be replanted or stored for a short time before replanting. This evaluation for color of the cut roots should always be done at the end of the storage cycle soon before replanting the roots, as long-term storage of any damaged tissue is riskier than storing sound tissue.

Root Crown and Smoothness: The size and rough appearance of the crown of a beet can contribute greatly to its overall appearance and marketability. A large crown that does not have good definition and has rough outer margins that extend across much of the top of the root is not desirable in a fresh market beet. This characteristic also reduces the amount of harvestable tissue in the beet root for fresh market sale or processing. The outer tissue of the entire root may also have an unappealing roughness for the fresh market. This may be due in part to one or more of the common diseases of beets but also appears to be a heritable trait of some beet varieties. Selecting for a smoother root surface has been effective in some beet varieties.

ISOLATION DISTANCES

Beet pollen is capable of traveling several miles in the wind. The garden beet is predominantly wind-pollinated; its pollen is small and light and well adapted to being spread in this way. While insects are sometimes attracted to the sweet nectar and pollen that beets produce, they account for only a very small amount of the cross-pollination that can occur with the wind. Beets and the other chenopods have large numbers of flowers per plant and produce large amounts of pollen that can be seen blowing across seed production fields as a yellow dust during peak flowering. All of these factors make it necessary to increase the isolation

▼ Root color can be evaluated by cutting off a "cheek," or side, of the beet. This can be done shortly before replanting the roots in any of the versions of the root-to-seed method. In this case, it is possible to judge the color intensity of the "rings" of a 'Chioggia' type (root on right). Photo courtesy of Jared Zystro

distance over the recommendations for insect-pollinated crops. In the 1930s C. F. Poole of the USDA studied the distances that beet pollen can travel and concluded that under ideal conditions it can travel over 12 mi (20 km)! During the same decade other agricultural scientists collected sugar beet pollen at an altitude of 3 mi (5 km) above sugar beet seed fields.

In commercial beet seed production in the Salish Sea area of Washington the isolation between two different red beet varieties is 1 mi (1.6 km) if they are both open-pollinated (OP) varieties; while it's 2 mi (3.2 km) if they are both hybrid varieties or if one is a hybrid and the other is an OP. This disparity assumes that hybrid seed is intrinsically more valuable and deserves to be kept at a higher level of purity than seed of OP varieties. This is an outmoded idea that is common in vegetable seed production areas, where the practices have been dictated by the business-driven notion that hybrids are always superior. Well-maintained OP varieties that have improved adaptation and market traits that fit the bill for the organic farmer are capable of going toe-to-toe with hybrid varieties in many instances. Therefore, I am not recommending shorter isolation distances for OP varieties than for hybrids.

Another important consideration in avoiding cross-pollination in a beet seed crop is to be aware that during the first year of the biennial cycle, all *Beta vulgaris* crops may flower prematurely if exposed to excessive cold temperatures. This produces pollen from plots that are not normally considered in calculating isolation distances. Many vegetable growers know that a percentage of Swiss chard plants grown as a vegetable will bolt in the first season if planted early in the spring ('Rhubarb' chard is notorious for this). While this doesn't happen as often with sugar beets or garden beets, these crops will have an occasional beet plant that is bolting prematurely for the same reasons.

Isolation Distances for Red Beets: The minimum isolation distance between any two standard beet varieties with the same root color should be 2 mi (3.2 km) if grown in open terrain with few or no physical barriers. If physical barriers do exist, then it is possible to drop this isolation to 1 mi (1.6 km), but—given the nature of windborne pollen—with the wind blowing in the right direction during the peak pollination period, it is still quite possible to get a small but significant level of cross-pollination at this distance. That is why these recommendations are only for garden beets with the same root color. Separating red beet crops from other garden beet crops with different-colored roots will require greater isolation distances. Likewise, separating table beets from the other very closely related *Beta vulgaris* crops—Swiss chard, sugar beet, and mangels—will require even greater isolation distances than what is needed for separation from the different-colored table beets.

Isolation Distances Between Red Beets and Other Color Types: There are now a number of garden beet varieties available with roots that are orange, yellow, and white. There is also the 'Chioggia' type, which has concentric rings alternating between red and white in the interior of the root and is popular as a specialty market variety. (The rings of color are actually alternating xylem and phloem tissue.) The minimum isolation of any of these colored variants should be at least 3 mi (4.8 km) from any other different-colored garden beet variety in open terrain. This is the standard isolation distance used by the Willamette Valley Specialty Seed Association between different-colored beets. If sufficient physical barriers do exist on the landscape between seed production fields, then it is possible to grow two different garden beet varieties with a 2 mi (3.2 km) separation between fields.

Isolation for Different *Beta vulgaris* Crops and Stockseed: In considering the isolation distance necessary to minimize crossing between seed crops of any two of the four different *Beta vulgaris* crops (garden beet, sugar beet, fodder beet or mangel, and Swiss chard), most seed companies have agreed on 5 mi (8 km). Through experience, seed growers have discovered that this much isolation is indeed necessary to avoid almost all of the unwanted crosses between these crops. Any progeny that results from a cross-pollination between any two of these crops are easy to recognize, as they always have considerable hybrid vigor, are often more stemmy or leafy, and can look like a Swiss chard plant or an excessively large mangel or sugar beet. This is why crosses are so assiduously avoided. If sufficient physical barriers do exist on the landscape between seed production fields, then it is possible to grow two different garden beet varieties with a 3 mi (4.8 km) separation between fields.

Stockseed always requires a higher degree of caution in its production than commercial seed. Beet seed production companies that have maintained high standards of varietal purity in their beet varieties over long periods of time use a minimum 5 mi (8 km) isolation distance for all stockseed increases, without question. This distance is used as an absolute whether in open terrain or with physical barriers present, as genetic mixing in stockseed can be so detrimental to the integrity of the variety (see chapter 17, Stockseed Basics on page 364).

Isolation from Genetically Modified Versions of *Beta vulgaris*: The troubling issue of genetic contamination from genetically modified (GM) crops in certified organically grown seed is discussed at length in chapter 13, Isolation Distances for Maintaining Varietal Integrity. Sugar beets are the first of the *B. vulgaris* crops to be genetically engineered. A large share of the North American seed production for sugar beet is in Oregon's Willamette Valley and is owned by three or four large corporate agricultural companies. The independent seed growers producing organic garden beet and Swiss chard seed in the Willamette Valley are faced with GM sugar beets on all sides; even with 5 mi (8 km) isolation, some crossing is inevitable. Isolation distances between GM beet fields and organic seed production fields should probably be at least twice this far, 10 mi (16 km), if there is any hope of preserving the genetic integrity of certified organically grown crops. In all instances it must be remembered that crosses can occur even at these isolation distances, as is true with all crops under any field conditions (see chapter 13, "The Myth of Pure Seed" on page 334).

SPINACH

Spinach (*Spinacia oleracea*) is one of the most widely grown vegetables in temperate climates around the world. It is derived from a leafy, winter annual that evolved in and near the Fertile Crescent of the Middle East. Winter annuals are plant species that germinate in the cool of the fall and grow vegetatively until the short days and cold of winter slow their growth. In spring, winter annuals grow steadily until a combination of environmental factors prompts the reproductive phase of the life cycle. Spinach bolting is initiated primarily by daylength, and the ancestral forms of this crop bolt very early in spring with less than 14 hours of light per day. This allowed the predecessor of modern spinach to mature seed before the onset of the intense heat of summer in the Middle East. The seed then lay dormant, having evolved to only germinate with the onset of cool, wet weather in fall, and thus the cycle was repeated.

Modern forms of this crop have been selected to produce a lush and robust leafy vegetable that is widely adapted across environments and seasons. Various types of spinach still produce well as a fall-sown vegetable that can be harvested in fall, winter, or spring.

SEED PRODUCTION PARAMETERS: SPINACH

Common name: spinach

Crop species: *Spinacia oleracea* L.

Life cycle: annual

Mating system: cross-pollinated

Mode of pollination: wind

Favorable temperature range for pollination/seed formation: 65–75°F (18–24°C)

Seasonal reproductive cycle: early to mid-spring through late summer or early fall

Within-row spacing: seed-to-seed 16–24 in (41–61 cm); root-to-seed 18–30 in (46–76 cm)

Between-row spacing: seed-to-seed 30–36 in (76–91 cm); root-to-seed 30–48 in (76–122 cm)

Species that will readily cross with crop: All spinach seed growers should be conscious of nearby farm or garden plots of spinach grown for fresh market production as these crops can flower rapidly in late spring and summer before they are turned under.

Isolation distance between seed crops: 0.5–3 mi (0.8–4.8 km), depending on crop type and barriers present on the landscape

It is easily cold-hardy to 15 to 18°F (–9 to –8°C) but can survive much lower winter temperatures if insulated by snow. Many contemporary varieties have been developed to be spring-planted and produce a bountiful crop before the summer daylength causes bolting. Much recent breeding work has been devoted to develop fast-growing, upright plants that thrive at high population densities and are harvested at a juvenile or baby-leaf stage for bagged salads. Spinach leaf surfaces are smooth, semi-savoyed, or fully savoyed, depending on the amount of leaf curl. The savoyed curl is due to varied rates of growth of ground tissue parenchyma between leaf veins. Leaf shape ranges from the putative older form of triangular blade that is referred to as

◄ 'Viroflay' is a classic smooth-leaved spinach variety that is a progenitor of the modern North American fresh-market type.

having an arrowhead or Christmas tree shape, characteristic of Asian spinach varieties, to the rounder, less lobed leaf that has become the ideal in European and North American markets. Selection for color in variable spinach populations has created darker leaf pigmentation. Research investigating the nutritional quality of spinach has found that the dark green types have higher levels of a variety of phytochemicals with antioxidants such as the carotenoid lutein, which is important for the health of the macula in the human retina.

CROP CHARACTERISTICS

Reproductive Biology

Spinach is largely dioecious in its flowering habit (see chapter 2, Reproductive Biology of Crop Plants). The number of male plants to female plants is roughly equal in large populations. Occasionally, there may be monoecious plants in a spinach population that express both male and female flowers on the same plant, but these are less common. Male spinach plants exhibit two basic plant forms. The first males to flower are quite short in stature (often only 4 to 6 in/10 to 15 cm tall at full development), with suppressed leaf development on all upper nodes, but prodigious in staminate flowers at all nodes. These extreme males, as they are often called, flower only for a short period of time and die after flowering but are capable of producing large amounts of pollen, ensuring that plenty of pollen is available for early-flowering females. The second type of

Spinach is dioecious, with separate male and female plants. In this photo a female plant with immature seed already forming is on the left and a robust vegetative male at peak flowering is on the right. ▶

male plant is known as a vegetative male and is a more robust, larger plant with both staminate flowers and leaves at all nodes. Vegetative males initiate flowering a week to 10 days after the first appearance of extreme males. They are longer-lived than extreme males and flower over a longer duration, ensuring ample pollen for the female plants throughout the period of pollen receptivity. Female plants are similar in stature to vegetative males and bear pistillate flowers and leaves at all nodes. They are long-lived and usually begin flowering within 3 to 7 days of the vegetative males.

Flowering is initiated primarily due to daylength. Heat can play a role by speeding the metabolic rate of the spinach plant, accelerating the flowering process once flowering has been initiated. Spinach is predominantly wind-pollinated and has extremely light pollen that is easily carried long distances by the wind. Both male and female flowers are very small and lack petals. Both types of flowers are borne in groups in the leaf axils of their respective plants. The calyx of the male or staminate flower has a sepal below each of the four stamens. In turn, two anthers are borne on each stamen. In the female, or pistillate flower, the calyx can have two or four sepals that persist after fertilization and combine with the pericarp to create a hard protective casing for the single-seeded fruit. The calyces of several fruits that are borne together can fuse and form tight clusters of seed that will require special seed-cleaning techniques when milling the seed.

The spinach seed coat may be either prickly or smooth. Historically, the prickly-seeded types were generally the large-leaved winter varieties (presumably, the more ancestral form), while the smooth-seeded types were the spring and summer varieties. The prickly-seeded trait is increasingly rare today, and there does not seem to be a genetic association between the prickly seed coat and either winter hardiness or large leaves.

Climatic and Geographic Suitability

Specific environmental conditions are required to produce high-quality yields of high-germinating, large-seeded spinach crops that are free of seedborne pathogens. Spinach is a cool-season crop that is very sensitive to temperatures above 75°F (24°C), both as a vegetable and as a seed crop. For this reason,

A closer view of a female spinach plant (left) and two extreme male plants (right). Photo courtesy of Organic Seed Alliance

there are few locations where spinach seed can be grown for commercial use. The two most important spinach seed production areas worldwide are the Skagit Valley of Washington and a large swath of central Denmark. Both areas have cool, wet springs followed by dry, cool summers (temperatures usually not exceeding 75°F/24°C) and relatively dry fall weather for harvest. The cool, wet spring weather of these ideal climates enables the spinach plant to establish a robust, large vegetative rosette of leaves before flower initiation under the longer days of late spring. Summer weather that exceeds 85°F (29°C), especially during pollination and early seed development, can dramatically lower germination rates, seed size, and seed yields. As is common to all dry-seeded vegetables, an extended dry period in late summer is favored for producing a seed crop that is relatively free of disease and disease-causing inoculum.

SEED PRODUCTION PRACTICES

Soil and Fertility Requirements

Spinach grown for seed can be planted on a variety of soils, but the soils must be well drained to avoid root rot problems. Soil pH should be maintained above 6.0, as spinach is sensitive to acidic soils. In seed crops, the amount of available nitrogen should not be too high in order to avoid excessive vegetative growth before bolting, as this promotes lodging of the plants during seed set. Well-maintained agricultural soils that are high in humus and microbial populations will supply adequate nutrients and water over the long seed production season. Spinach is somewhat tolerant of soil salinity and very tolerant of alkaline soils, although foliar fertilizer applications may be needed on alkaline soils to counteract the reduction in availability of micronutrients such as manganese under high soil pH.

Growing the Seed Crop

The optimum temperature range for germination of spinach seed is 45 to 75°F (7 to 24°C). Spring-sown spinach can be planted as early as the ground can be worked, though waiting another week or two till true spring weather prevails to plant will ensure vigorous seedling growth. Planting should not be delayed too long, however, as the key is to get as large a vegetative rosette as possible before flower initiation. This rosette of leaves (or frame, in the parlance of seed growers) will not grow appreciably after the flower stalk emerges and will be the photosynthetic factory responsible for producing the seed crop.

Sometimes spinach seed crops are planted in late summer or fall and overwintered for seed production the following summer. This eliminates ground prep and planting during inclement spring weather. It also enables the spinach to develop a larger vegetative frame to optimize the seed yield potential of the crop. Unfortunately, the overwintered crop may act as a green bridge, providing a live, vegetative host allowing a number of diseases to survive through the winter (see chapter 16, Seedborne Diseases).

The planting density for spinach seed production requires much wider spacing than for producing spinach as a vegetable. For seed production, spinach plants are generally spaced 8 to 12 in (20 to 30 cm) apart within rows. Standard row centers are normally 22 to 26 in (56 to 66 cm) apart, but in raised beds spacing between rows can be dropped to 12 to 14 in (30 to 36 cm) apart. Wider plant spacing increases airflow through the crop, reducing disease pressure. Spinach seed crops are still often grown with overhead irrigation in the Pacific Northwest, although its use is limited to the early part of the season before bolting, flowering, and seed maturity. Avoid it during these reproductive phases of the life cycle, as the incidence of foliar and possible seedborne diseases will certainly increase as a result. Drip or furrow irrigation has the advantage of not wetting the foliage and can be used throughout the season.

Early-season weed control is critical for optimum establishment. If plants become leggy due to early competition they will be difficult to evaluate in the selection process and be more prone to lodging when full-sized. Spinach seedlings grow slowly and do not compete well with weeds. Starting with a stale seedbed is advisable, as well as avoiding fields with high weed pressure. Pre-emergence flame weeding can be effective. Thinning and blocking by hand are usually required in organic production, as is hand hoeing until plants are large enough to not be smothered by mechanical cultivation.

The fact that spinach is dioecious is an advantage in hybrid spinach seed production. The female plants can be used to develop all female flowering lines. This is possible because female spinach plants, if isolated from spinach pollen during flowering, will eventually "sex reverse," subsequently producing male flowers in addition to their female flowers in order to achieve pollination. The subsequent generation will inherit the female genes. This sex reversal enables breeders to increase seed of a female line. During the development of female lines for hybrids, the breeder must perform meticulous selection on the female plants, ensuring that the lines uniformly produce only female flowers for at least 5 to 6 weeks before sex reversing. Then, when used as female parents in hybrid seed production, these lines will have adequate time to receive all of their pollen from the male parental line of the hybrid in the field. Female plants can also be selected to

have a monoecious flowering habit to serve as the male parent. This eliminates the variability of having both extreme males and vegetative males as your male pollen source and produces much better phenotypic uniformity in the resultant vegetable crop, especially as it just begins to bolt (when spinach is often still harvested). All of this can be done without resorting to chemical growth regulators and is therefore possible in organic production systems.

Seed Harvest

Spinach seed, as with all members of the Amaranthaceae, is formed and matures in an indeterminate growth pattern, beginning on older, lower branches and continuing up the flower stalk through the season. Due to this sequential maturation, only a portion of the seed continuously being set will reach maturity by the end of the season. In most cases only about 75% of the seed on any given plant will reach maturity by harvest. One of the most difficult steps in successfully growing a spinach seed crop is judging when this optimum amount of seed is mature on a majority of plants in the field. A standard method of gauging the maturity of a spinach seed crop is to make a visual assessment of the percentage of seed on most of the plants that has turned a tan-brown color, typical of mature spinach seed, and then harvesting the crop when 60 to 80% of the seed has become this color. This is an unreliable method of judging maturity, however, due to the considerable effects of genetic and environmental variation. Environmentally, the seed may prematurely lighten in color in the presence of the Stemphylium/Cladosporium leaf spot disease complex (see the table in chapter 16, Seedborne Diseases on page 353). Genetically, some spinach cultivars have much greener seed, even at full seed maturity.

Spinach seed maturing on the plant. ▶

The most important trait to monitor in gauging the maturation of spinach seed is the relative maturity of the seed endosperm. The endosperm and embryo, which mature concurrently, must be fully developed to produce vigorous, vital seed. The endosperm, which is primarily composed of starch, is easily monitored by cracking open the seed and squeezing it to visually inspect the stage of endosperm development. Appearance of the endosperm will transition from translucent or milky early in development to what is described as flinty, with a grayish, waxy appearance at the midpoint of development. At these stages, the endosperm can be squeezed out of the developing seed for inspection. When seed in the middle of the stalk is at the flinty stage, irrigation may be stopped to speed the maturation and drying process. This can be important if there is risk of maturation extending into the wet, disease-promoting conditions of fall.

At a mature stage of development, the endosperm turns a starchy white color, which can be observed by cracking open the seed. Once the endosperm turns to this solid, true white color and is firm, the seed is mature and ready to harvest. When the majority of the plants in the field have about 75% of the seed at this advanced, starchy white stage, it is time to cut the plants near the base of the stems to stack into windrows. This should be done preferably during a warm, dry period. The windrowed stalks will be ready to thresh in 4 to 10 days, depending on the weather. Rotating stalks in the windrows facilitates uniform drying of the seed. Cutting or windrowing the crop in wet weather should be avoided. If a crop is ready to cut and an extended wet weather pattern is forecast, cut the crop and bring it into an airy, dry shelter to cure. Once dried, seed stalks are usually threshed and then cleaned by screening and fanning.

Threshing is best accomplished with a belt thresher to break up the clusters of seed that form when the calyces of several spinach fruits fuse while maturing. Many seed companies then clean away any remnant of this fused material, as well as prominent burrs in prickly-seeded types, with a rotary deburrer in a process that approximates decortication in beets and chard.

GENETIC MAINTENANCE

When planting spinach with the intent of performing selection on a population, the initial planting density that I use is anywhere from six to eight plants per foot (30 cm) within the row.

Planting six to eight spinach plants per foot (30 cm) for the initial stand before practicing selection establishes a population that is large enough to perform multiple selection events from the time of emergence to flower initiation and ensures that you will still have an adequate population to maximize your seed yield at the end.

As with all crops, seedling vigor and early robust growth are always important traits to select for in organic production systems. Spinach prospers from increased vigor as the spring crop is often planted into cold, wet soils and must make rapid growth before longer daylength and hot weather induce bolting. Selecting for vigor may be accomplished by eliminating any slower, poorly formed seedlings after all (or most) of the seedlings have emerged. This can be repeated in another 7 to 10 days, eliminating any slowpokes or malformed plants.

There are various stages at which leaf size can be selected depending on the leaf shape desired by the vegetable market you serve. As an example, much of the modern baby-leaf and teenager-leaf spinach is selected for rounded leaves. Alternately, some prefer the triangular or arrowhead-shaped (also called Christmas tree) spinach leaf for shape variation in prepared salad mixes. Leaf textures are characterized as savoyed (extreme leaf curl), semi-savoyed, or flat, and are very important in determining the market class of spinach varieties. Open-pollinated spinach varieties can have considerable plant-to-plant variation in their degree of savoy curl, and it is important to select for uniformity of this trait. Anyone selecting for this trait should also realize that the degree of curl varies with the stage of growth and seasonal temperatures. Likewise, variation in leaf color is common in open-pollinated spinach populations, and gain from selection is possible. Darker leaf color is usually preferred and is correlated to higher levels of nutritionally significant carotenoids. Selection for color can be done across growth stages.

Plant stature is another important trait for spinach growers, especially for harvest of leaves cut for salad mix. The ability of the plant to hold its leaves in an upright position can reduce the amount of soil that may get trapped in the underside of leaves. Upright foliage also lessens the amount of fungal and bacterial pathogens that can be splashed onto leaves due to rain or irrigation.

Incidence of disease should be monitored and identified accurately for proper field management and for selection purposes. Routinely selecting for plants free of disease or with relatively low incidence of disease symptoms can help develop partial or horizontal resistance to a particular pathogen. This requires familiarity with the symptoms of spinach diseases endemic to your region.

ISOLATION DISTANCES

Spinach, like all of the chenopod crops, has pollen that is known for its ability to travel great distances. It is almost exclusively wind-pollinated, and its pollen is relatively small,

Matthew Dillon, the author, and Micaela Colley evaluate spinach varieties in a field trial.

light, and easily carried by the wind. The male plants of this dioecious crop will usually produce so much pollen that it is possible to see yellow puffs of it moving across a spinach field with a gust of wind at the peak of pollination. Depending on the direction and force of prevailing winds and the relative humidity, spinach pollen can pollinate other flowering spinach crops within 2 mi (3.2 km) if grown in open terrain. While this may represent a rather small amount of outcrossing to another spinach crop at this distance, it is especially important when the two adjacent crops are distinctly different types of spinach or are foundation seed or stockseed, where purity is very important.

If a substantial physical barrier is present (see chapter 13, Isolation Distances for Maintaining Varietal Integrity), then the minimum isolation distance needed between two different spinach crops can be lowered to 1 mi (1.6 km). However, it should always be remembered that these isolation distances are not absolute, and there may still be a small amount of crossing even at these higher isolation distances, especially with wind-pollinated chenopods.

SWISS CHARD

Swiss chard is one of the four crops shared by the species *Beta vulgaris* and is probably closest to the ancestral form of this species that grew wild in the Mediterranean Basin. A number of researchers agree that the earliest cultivated beet-like vegetable was probably grown for its leaves. Continued selection for larger, more succulent leaves resulted in the crop that is usually called either Swiss chard or silver beet. As this progenitor crop also had a substantial root (something typical of many biennials), there were farmers that selected it for a more refined, fleshy, edible root, which became both the table beet and mangel.

Chard's common name is a corruption of the Old French *cardon*, for "cardoon," a similar-looking but unrelated vegetable (*Cynara cardunculus*, a close relative of globe artichoke) with origins in the Mediterranean region. Chard, like cardoon, has long been prized for its succulent, celery-like thick stalks or petioles. The *Swiss* part of its moniker is

SEED PRODUCTION PARAMETERS: SWISS CHARD

Common name: Swiss chard, silver beet, leaf beet

Crop species: *Beta vulgaris* L.

Life cycle: biennial

Mating system: cross-pollinated

Mode of pollination: wind

Favorable temperature range for pollination/seed formation: 65–75°F (18–24°C)

Seasonal reproductive cycle: summer (year 1) through the following fall (year 2) (~14 months)

Within-row spacing: seed-to-seed 16–24 in (41–61 cm); root-to-seed 18–30 in (46–76 cm)

Between-row spacing: seed-to-seed 30–36 in (76–91 cm); root-to-seed 30–48 in (76–122 cm)

Species that will readily cross with crop: Swiss chard, sugar beet, mangel or fodder beet; all of which are *Beta vulgaris* L. and readily cross with garden beet

Isolation distance between seed crops: 1–2 mi (1.6–3.2 km) for beet crops of the same color; 1.5–3 mi (2.4–4.8 km) for beets of different colors; 3–5 mi (4.8–8 km) for different *Beta vulgaris* crops. All stated distances are dependent on crop type and barriers present on the landscape. Distance between any *Beta vulgaris* crops and transgenic or GMO versions of this species should be a minimum of 10 mi (16 km).

also inaccurate, as the crop's origins are definitely farther south and probably from coastal areas of the Mediterranean. Chard's modern association with Switzerland is mired in mystery, although there does appear to be evidence that there was a preference for it in Switzerland over some other leafy greens more popular in other parts of Europe.

Chard's popularity is growing thanks to the expanding market for fresh, local produce. It is also now widely used as a component of salad mixes of all kinds. Chard is now frequently planted thickly and cut as a baby leaf for bagged salad mixes. This has raised its profile as a vegetable crop, and there has been a flurry of plant breeding to create the various-colored chards that have entered the market in recent years. A number of new colors in the stems of Swiss chard have been added to the more traditional colors of red and white. While yellow pigmentation has long existed as a color in *B. vulgaris* crops, the number of newer varieties with shades of yellow and gold has skyrocketed. Yellow and gold are complemented with other shades of chard, including pink, magenta, orange, and intermediate hues of these as well. A number of seed companies have also designed multi-colored mixes with many or all of these colors, which have become popular in many markets across North America.

CROP CHARACTERISTICS

Reproductive Biology

The reproductive biology of Swiss chard is virtually identical to that of the garden beet, as they are the same species. The only noticeable morphological difference is that chard's flowers are borne on internodes that are usually spaced farther apart than the flowers borne on most beet varieties.

Climatic and Geographic Suitability

Swiss chard seed is grown in many of the same regions that are ideally suited to beet seed production (see Garden Beet, page 51). However, chard seed is sometimes grown in districts with summer high temperatures that can be as much as 6 to 10°F (3 to 6°C) warmer than what is normally advised for beet seed production. Chard seems to thrive at these higher temperatures, and some growers believe that the extra heat units gained by growing chard in these warmer districts may contribute to increased seed yields at season's end.

SEED PRODUCTION PRACTICES

Soil and Fertility Requirements

Swiss chard prefers the same soil and fertility regimen as beets.

Growing the Seed Crop

Growing the chard seed crop is virtually identical to growing the beet seed (see Garden Beet, "Growing the Seed Crop" on page 55). There are only a couple of small variations in the manner of how you treat the chard seed crop, which are discussed below.

Many chard varieties can achieve a much larger overall size and stature (height and width) when flowering and producing seed than is normally achieved for a beet plant during the reproductive phase. Some chard varieties can easily achieve a height of 6 to 7 ft (1.8 to 2.1 m) or more. Chard roots can be described as unrefined beet roots. They probably look much like the roots of the ancestral biennial plant from which the modern table beet is derived. Chard roots may have multiple growing points on the taproot and many adventitious roots coming from the root area adjoining the taproot. The root crown is often larger and more irregularly shaped than the crown of the more refined beet. Both of these features make chard roots somewhat more susceptible to rotting when stored in coolers or root cellars, especially in the high-humidity conditions that root storage requires. Fortunately, if stored properly (see chapter 3, Understanding Biennial Seed Crops, "Overwintering the Crop" on page 19), chard roots can have just as good a survival rate as beets.

When chard roots are overwintered in the field for seed production, they can be somewhat more cold-hardy than beets in mild temperate climates. Like beets, the growing point of a chard plant will become damaged during winter in many temperate climates. Some garden beet varieties will suffer permanent damage to their growing point and not resume growth in spring when exposed to temperatures of approximately 25°F (−4°C) and below. In contrast, I have witnessed a number of chard varieties where a percentage of the plants will survive temperatures down to 14°F (−10°C) if they are not subjected to long periods of time at these temperatures, or where cold weather of this intensity doesn't recur multiple times throughout the winter. The recurring freeze–thaw syndrome at these temperatures can be very hard on plants that might otherwise survive a couple of skirmishes at these lows. Some of the methods used to cover a standing root crop with a straw or soil mulch may also prove beneficial in this situation (see chapter 3, "Preparing Roots for Storage" on page 22).

Seed Harvest and Seed Cleaning

Seed harvest and seed cleaning for chard is practically identical to the process for beet seed. The only possible difference would be the fact that the second-year chard plant is usually significantly larger than the second-year beet plant, and therefore, when piling the plants into windrows, you must be careful to pile the plants in short enough piles that they will still dry quickly in the field.

GENETIC MAINTENANCE

Genetic differences within a chard variety are apparent at multiple stages in the crop's life cycle. Upon sowing the crop, you have an opportunity to select for seedling vigor as the crop emerges from the soil. Within the first few days after the emergence of the seedlings, it is usually quite easy to select against the seedlings that are later emerging and obviously less vigorous. At this stage it is also quite easy to identify seedlings that are not true to type for petiole or stem color. The common white or pale green petioles are recessive in their expression to all of the other colors: red, magenta, pink, orange, yellow, and gold. Seedlings resulting from crosses between the deeper colors and the white and pale green types can be expressed as intermediate pale colors. It is easy to thin out the white, light green, and pale intermediate segregants in the seedling stage. Expression of the petiole color in the seedling stage is relative and has a strong correlation to the degree of pigmentation in the adult plant. Hence, any selection for intensity of petiole color at any stage in chard's vegetative growth will increase the color for use of the crop as either a baby leaf or bunching variety.

Selection for the intensity of the green leaf color is also important, as chard's **ground** or underlying leaf color can range from very pale green to a deep forest green. Most farmers marketing the crop agree that a darker green, especially in contrast with a rich petiole color, is very attractive in the marketplace. Likewise, maintaining a solid green color into the cooler weather of fall, when the leaves of many chard varieties start to turn a rusty brown or chocolaty color, is considered important by many growers. While colder temperatures trigger this change in color, there is almost always genetic variation within different chard varieties, allowing room for selection toward greener plants.

Considerable variation also exists in many chard varieties for the width of the petiole. Many European white-stemmed types that are exported around the world have very

wide petioles. This is also true of varieties from Australia, New Zealand, and South Africa. Little or no selection for petiole width has been practiced on most North American types, and genetic variation for petiole width is almost always present. Because growers that are bunching their crop usually prefer wider petioles, it is an obvious trait to select for as when producing a seed crop.

Foliar diseases are usually not severe but are often present in most temperate climates. Cercospera leaf spot (*Cercospera beticola*) and phoma (*Phoma beta*) can be found in many districts where chard is grown, and there is often plant-to-plant variation for symptoms for both of these leaf-spotting diseases. Breeders have been successful incorporating partial resistance to both of these maladies by carefully selecting over generations for plants with less severe symptoms. Downy mildew (*Peronospera farinosa*) also affects chard in much the same way it affects beets (see the Garden Beet section). Much like beets, chard roots coming out of storage can have a grayish growth, typical of latent downy mildew, on the crown and growing point, which can later affect the flowering structures and seed yield. Selection has been successful in developing good levels of partial resistance in both beets and chard.

ISOLATION DISTANCES

Isolation distances are the same as with beets, with similar increases in distance between colors. Increased isolation distances must also be used between chard and the other crops of *Beta vulgaris*: beets, mangels, and sugar beets (see Garden Beet, "Isolation Distances" on page 62).

◄ Any of the intensely pigmented Swiss chard varieties like this golden type require selection during seed production to maintain good color.

– 6 –

APIACEAE

The Apiaceae family includes several vegetables, most notably carrots and celery, that are widely consumed by many cultures around the world. It also includes several of the most prominent herbs—dill, cilantro, fennel, and parsley—used to flavor dishes in many cuisines. The plants of this family, formerly known as the Umbelliferae, all share an inflorescence in the form of a compound umbel, so named for its umbrella-like appearance. The small perfect flowers are borne on small umbellettes, with pedicels that arise from a single point, which is in turn attached to larger pedicel rays that are joined at the apex of the flower stalk (hence its compound nature). All members of this family have leaves, flowers, seed, and roots that are aromatic due to volatile oils that impart very distinctive flavors to the cultivated members.

The Apiaceae are all dry-seeded crops, and their seed is most successfully grown when there is little or no precipitation during the flowering and seed maturation process. Almost all of the commercial seed production for the most widely grown Apiaceae crops—carrot, celery, and parsley—takes place in temperate regions with dry summers, the so-called Mediterranean climate. However, smaller-scale seed production of regionally important varieties of these crops, as well as production of unusual vegetables in this family like celeriac and parsley root, is widespread in places like Eastern Europe and Asia Minor

◄ The primary or king umbel of a carrot plant in full flower.

(and many carrot varieties throughout Asia). And while there is considerable seed produced commercially of the minor Apiaceae herb crops such as cilantro, dill, cumin, and fennel, much of the seed for these crops is grown regionally for sale and subsistence use across Southern and Eastern Europe, northern Africa, and all of Asia.

FAMILY CHARACTERISTICS

Reproductive Biology

The cultivated Apiaceae include both annuals and biennials and consist of dry-seeded crops that most successfully mature their seed over an extended dry period in late summer and early fall. The cultivated Apiaceae have perfect flowers with five stamens, five petals, and five sepals. They are largely cross-pollinated due to insect activity but can also self-pollinate. However, a number of important crops in this family, including carrot, celery, cilantro, and parsnip, have an elegant device to discourage inbreeding. On any particular flower the fertile pollen will shed before the female stigma is sexually receptive, thus decreasing the chances of a self-pollination on that flower, thereby increasing the chances that pollen from that plant will only pollinate the receptive stigma of a slightly older flower on another plant. This is important as some species in this family, most notably carrot and celery, frequently suffer from inbreeding depression.

The flowering pattern of most of the crop species in this family is distinct. It begins with one central inflorescence borne at the terminus of the main floral stem, which is known as the primary or king umbel in carrot. This is followed by secondary umbels that are located at the terminus of the branches arising from the main stem. In most temperate climates the seed borne on the flowers of the primary and secondary flowers is the seed that most reliably matures. Depending on the length of the growing season, subsequent third- and fourth-order umbels will form, but the opportunity for the seed to fully mature on these later-setting flowers is greatly reduced in most temperate locations. The ovary of most of the cultivated crops in this family has two locules, each producing a separate seed. This pair of seeds is joined at maturity. In crops such as carrot, celery, and parsley these paired seeds easily separate when the seed is threshed and cleaned. But in a crop like cilantro they remain joined as a round seed ball, which is preferred by many for ease of planting but often results in two tightly spaced seedlings when the germination rate is high.

Seedling emergence and early seedling growth is notoriously slow in most crops of the Apiaceae. Also the days to germination and initial growth rate of the seedlings can vary quite a bit from seed to seed within a given variety and even within a particular seed lot. Research in carrots suggests that seed from the primary and secondary umbels that is uniformly mature and of a high germination rate will produce seed that germinates more uniformly than seed produced by later-appearing umbels, in less time, and ultimately producing a more uniform stand of carrots. This is probably true with other crops in the Apiaceae, all of which have the same flowering pattern. Seed growers working with these crops must be conscious to harvest the most mature seed possible without exposing the crop to excessive wind and rain.

Life Cycle

The vegetable crops of the Apiaceae are usually biennials, while herbs such as dill and cilantro have an annual life cycle. When grown as a seed crop the biennial vegetables are routinely direct-seeded sometime during midsummer of the first year to produce the first year's storage root. The goal is to produce a moderate-sized root the first year that can then be stored, either in the ground or in a root cellar or cooler for replanting the following spring. Whether the Apiaceae crop is a leafy crop like celery or parsley, or a traditional root crop such as carrot, celeriac, or parsnip, it is very important to have a substantial root with an intact growing point to survive the long storage period and vigorously grow in spring.

There are three methods used for overwintering biennial Apiaceae crops:

1. Plant the crop in the field of intended seed production. Overwinter the crop in the field with flowering and seed production in situ in Year 2.
2. Grow the crop in a nursery plot in the first year; lift the roots or plants in the fall, transplanting them to overwinter in next year's seed production field.
3. Grow the crop in a nursery, lifting it in the fall for storage in a root cellar or cooler until replanting it to the field in the spring of Year 2 for seed production.

The advantage of the first method is that it doesn't involve the time and labor of transplanting; however, it does require more land area in Year 1, as the crop is planted and thinned to the wider spacing that is required for seed production right from the beginning. For root crops such as carrots, celeriac, parsley root, and parsnip this method is known as the seed-to-seed method. The great disadvantage of using the seed-to-seed method with root crops is that you don't get an opportunity to select for all of the root characters important to each particular crop (see each individual crop profile for a description).

The second and third methods for overwintering biennials use a nursery plot to raise the first-year plants before storing them or transplanting them directly for seed production in the second season. In general the nursery plot will only consume between a tenth and a fifth of the space of the seed production field, since the plants must be replanted at a wider spacing to accommodate the much larger flowering plants. Both of these methods when used with root crops are considered the root-to-seed method as the roots are lifted and selected before replanting for flowering and seed production. This gives the grower the ability to genetically maintain and improve the important root characteristics of these root crops.

The last major element to consider when choosing an overwintering method for seed production is to match the cold hardiness of the crop with the severity of your winter. Several of the prominent Apiaceae crops, including carrot, celeriac, parsley, and parsley root, are generally cold-hardy, if gradually hardened off, to temperatures of 15 to 20°F (–9 to –7°C). This also is largely dependent on the particular variety of each crop and the degree of freezing and thawing that occurs over the course of a particular winter season. Snow cover can ease this with its insulating effect; in essence it acts like a 32°F/0°C thermal blanket if fully covering the aboveground parts of the plant. Parsnips on the other hand can take temperatures down to about 5°F (–15°C) and possibly colder when overwintered, making them an even more forgiving candidate for the seed-to-seed method in many temperate climates.

Seed Harvest

The maturation of Apiaceae crop seed begins with the primary umbel and advances through the secondary and tertiary umbels over time. The timing of harvest involves a compromise on the grower's part between harvesting the

▼ Mature carrot seedheads.

highest amount of mature seed and not waiting too long to harvest and thus risking the loss of quality to fall rains or loss of seed from shattering. While there are varied opinions as to precisely when to harvest the seed of these crops, growers have a variety of signs that they look for to determine maturity. In general the color of the seed is used as an important indicator of maturity. Most Apiaceae seed will turn an earthen shade of brown when mature. Some growers will harvest when the seed of the primary umbels is fully brown and the secondary umbels are midway in the process of turning brown. Others wait until both primary and secondary umbels are brown and the tertiary umbels are turning brown. Some growers judge maturity based on when seed on the primary umbels starts to drop. All of these maturity guides depend on the crop species and the climate in which they are being produced. Most importantly, each grower must get to know the relative maturity of the crop species as well as the specific variety for their climate. Then it is possible to know when to harvest the crop before losses from fall precipitation or shattering of the most mature seed from the primary umbel occurs.

At harvest maturity Apiaceae seed crops are usually swathed and laid in windrows on cleared, cultivated ground or on tarps to allow for further drying and finishing (final maturation of the seed). Windrows should not be piled higher than 2 ft (0.6 m) deep to allow for good airflow. Windrows should be allowed to cure for 5 to 10 days during warm, dry weather. If poor weather threatens and the crop is not too large, then placing the crop in an airy barn or greenhouse may preserve its quality. In large-scale production, mobile combines are often used for the threshing and initial cleaning of the seed. For smaller-scale commercial production, seed is often threshed by manually feeding several plants at a time through a stationary plot thresher or into the header of a grain combine with a cylinder that has rubber-covered bars. Swathing and threshing are best done early in the day to take advantage of morning dew in reducing losses from shattering. Making sure that the plant material to be threshed is not too dry and brittle is also important in minimizing the excessive breakage of the plant stems, especially the small stem-like petioles of the umbellettes. These are easily broken into small pieces about the same size as the seed and can be very hard to clean out of the seed crop later in the cleaning process.

Seed Cleaning

The seed of most Apiaceae crop plants have ribs that run the length of the seed. Carrot and dill have spines or beards that emanate from these ribs. If not debearded these spines can cause clumping of the seed in planters, especially precision planters. They are generally removed after the initial cleaning (scalping and screening particulates) by milling the seed with a decortication cylinder that gently rubs the beard free without damaging the seed. The removal of the fine, broken stems of the umbellettes is also required in order to produce a seed lot that will flow satisfactorily through most planters. A percentage of these stems can be removed by running the seed through a series of seed screens varying in either size or shape of the holes, or by winnowing. However, the smallest of these stems are best removed using a piece of equipment known as an indent cylinder. This device has a rotating drum with a series of indented depressions, which can catch the seed, while the larger stem pieces are eliminated. Final screening and sizing of the seed is then done. Sizing of carrot seed is extremely important to most discerning growers, as it ultimately influences the size and uniformity of the carrot roots.

CARROT

Carrot (*Daucus carota* L.) is one of the most recently domesticated of the common vegetable crops. The earliest conclusive documentation of the use of carrot as a vegetable is from the 10th century in present-day Afghanistan. From its center of origin in the Middle East the carrot spread north to Europe and east to China and Japan, presumably via the Silk Road, over the next several centuries. From the earliest documentation in the Middle East up until the 17th century in both Europe and Asia cultivated carrots were less refined than our modern crop and were either purple or yellow in color.

Paintings of kitchen scenes by European realist artists in the 17th century offer the earliest conclusive record of orange carrots. The arrival of an orange-rooted type, possibly as a variant of yellow, stimulated a rash of carrot breeding in Northern Europe, concentrating on improvements in the crop's eating quality and on the intensity of orange color. The popularity of orange carrots grew quickly from this effort. White carrots also appeared in Europe at around this time and have remained popular in Belgium and France as a unique ingredient in soups and stews, as well as their use to feed farm animals. Orange carrots soon spread around the world; over time this has become the predominant color of carrots in most growing regions.

SEED PRODUCTION PARAMETERS: CARROT

Common name: carrot
Crop species: *Daucus carota* L.
Life cycle: biennial
Mating system: largely cross-pollinated, with some self-pollination
Mode of pollination: insect
Favorable temperature range for seed germination: 65–85°F (18–29°C)
Favorable temperature range for pollination/seed formation: 65–80°F (18–27°C)
Seasonal reproductive cycle: summer through the following late summer or early fall of the next year (12–14 months)
Within-row spacing: seed-to-seed 4–8 in (10–20 cm); root-to-seed 8–18 in (20–46 cm)
Between-row spacing: 22–36 in (56–91 cm)
Species that will readily cross with crop: Wild carrot or Queen Anne's lace (*D. carota* var. *carota*).
Isolation distance between carrot seed crops: 0.5–2 mi (0.8–3.2 km), depending on crop type and barriers present on the landscape

The orange carrot variety in the foreground is a uniform selection of 'Nash's Nantes' that was harvested before full maturity (from Nash Huber's farm stand at the Port Townsend Farmer's Market). Photo courtesy of Micaela Colley

Carrots can be placed into two basic types. The western type, characterized as a cool season crop, which includes orange, white, purple, and yellow forms, originated in the Mediterranean basin and Europe. The second type, more often subtropical carrots, were grown and selected for centuries under the hot growing conditions in China, Japan, and Southeast Asia. The subtropical type may be yellow, purple to reddish purple, or red in color and usually has leaves with a fine pubescence and a grayish green color with leaflets that are less divided and more compound than the more common western type. They also have a propensity for early bolting or flowering before producing an edible root in both North American and European production areas. This has created a problem for seed companies wanting to improve some of these subtropical types with unusual colors for use in the temperate north. Selection to improve bolting tolerance will usually take several breeding cycles, especially if the breeders don't want to narrow the genetic base of the population they are improving for this exceedingly important trait.

Ever since the onset of breeding work on orange types began, the other colored types have gotten little breeding attention and have often been relegated to a lower culinary use status or for use as fodder crops until very recently. For this reason many of the unusually colored carrots—purple, red, yellow, and white—have a more "rustic" raw carrot flavor, which is often more complex and less sweet but can be rife with off-flavors from the terpenoids (flavor compounds), which haven't been tamed by plant breeding. Some breeding of these types that started in the 1990s has created some colored varieties that are more akin to our modern orange types, but breeding for flavor in carrots is not easy due to the complexity of this trait. For now, though, most of these colored types offer us an opportunity to rub shoulders with our ancestors and experience the rustic flavor of carrots as they tasted in the past!

SEED PRODUCTION PRACTICES

Climatic and Geographic Requirements

Carrots as a vegetable crop are produced over a wide range of environments, from the subtropics to cool climates in Northern Europe and southern Argentina. The production of good-quality roots requires fertile soil with a good deep tilth and an even supply of water throughout the season. The geographic and climatic requirements for carrot seed production are more exacting than for root production.

Geographically the carrot seed grower must be sure that there is no Queen Anne's lace (*Daucus carota* var. *carota*) within a 1- to 2-mi (1.6- to 3.2-km) area of the potential production field (see "Isolation Distances," page 91). This common weed, also called wild carrot, has spread to temperate climates around the world as an established weed. It is especially prolific in the eastern half of the North American continent (east of the 100th parallel) and in much of Europe, and it readily crosses with cultivated carrot. Indeed, P. W. Simon of the USDA has found that it frequently exchanges genes with cultivated carrot and is truly part of the carrot gene pool, for better or worse. Unfortunately, it universally carries the single dominant gene for white root color; it is very undesirable to allow it cross with your seed crop, as large "gnarly" white roots with many adventitious or branching roots will appear in the next generation.

In most regions where Queen Anne's lace occurs it is so ubiquitous and evenly spread across the countryside that it is impossible to destroy all of the plants within a 1- to 2-mi (1.6- to 3.2-km) radius. The only effective way to produce pure carrot seed in the vicinity of Queen Anne's lace is with well-maintained pollination cages equipped with a mesh material specifically designed for insect-pollinated crops. These cages are too expensive for commercial seed production and are generally only used by breeders or seed banks for research or preservation needs, respectively. To produce high-quality carrot seed commercially it is imperative to grow the crop in a geographic area free of this weedy variant of cultivated carrot.

The best climate for carrot seed production includes a fairly mild spring and a dry summer with temperatures that routinely reach 85°F (29°C) and above, but generally don't reach temperatures above 92 to 95°F (33 to 35°C) until late in the summer, after the seed is set and maturing. As for all dry-seeded vegetable crops, a dry, Mediterranean climate with low precipitation amounts from late June until mid-September is best, because it significantly lowers the chances of any diseases forming on the seedheads as they mature. However, this means that there needs to be reliable water applied with either a drip or furrow irrigation system; overhead irrigation may promote disease in the seedheads.

Growing the Seed Crop

Carrot seed is produced using two different methods, the "seed-to-seed" and "root-to-seed" method, depending on your goals. The seed-to-seed method is the most efficient if you have well-selected stockseed and feel confident that you can produce a genetically uniform crop without root selection. The root-to-seed method, on the other hand, affords you the chance to select the crop based on the root characteristics, replanting only the roots that meet your selection criteria, thus improving the variety. Root-to-seed also allows you to plant the root crop in a smaller nursery plot during the first season of growth and then transplant the roots to a larger field for seed production in the second season of the biennial cycle.

Seed-to-Seed Method: The vast majority of large-scale carrot seed production is done using the seed-to-seed method, which requires that the crop be sown in mid- to late summer in the field in which the seed crop will be produced the following year. The crop is left in the field throughout the winter and is able to grow and flower in situ during the next growing season until the seed is harvested the following August or September. This exposure to cold also serves to vernalize

the crop, which will induce flowering the following spring. Vernalization is a cold treatment for a given number of hours below a temperature of approximately 50°F/10°C (see chapter 3, Understanding Biennial Seed Crops, "Vernalization" on page 17). In some warmer temperate areas you must make sure that there are enough vernalization hours during the course of the winter or carrots won't fully bolt in the spring. Unless this is taken into consideration, harvesting seed from the carrots that do bolt becomes the equivalent of selecting for easier-bolting carrots, a trait that is extremely undesirable.

There are at least two main challenges to the use of the seed-to-seed method. First of all, the carrots must be overwintered without excessive damage from freezing or rodents. One of the main reasons for sowing the crop as late as mid- to late summer is so the carrots won't be too large to go through the winter successfully. Experience has shown that carrot roots with a diameter of approximately 0.75 in (2 cm) or less at the shoulder survive winterkill due to cold temperatures better than full-sized carrots. It is generally agreed that carrots at this size can withstand temperatures at or slightly below 14°F (−10°C) if there isn't excessive freezing and thawing throughout the winter. Though it should be noted that there are varietal differences in the ability to withstand winter conditions among carrots, with subtropical types being especially sensitive to these low temperatures.

Some growers will mechanically throw a layer of soil ("soil mulch") onto the row in fall for winter protection, although this may hinder regrowth in the spring and promote rotting in the crown. On a smaller scale, some growers have mulched the crop with straw or other organic materials. However, in some cases this has contributed greatly to the second challenge of overwintering carrots (or other biennial roots). The mulch can create a welcoming environment for rodents to build nests amid an ample winter food supply, and you won't know your losses until you uncover the roots in spring. Also, the mulch needs to be removed promptly in spring so as not to promote rot.

In terms of varietal integrity the main drawback with the seed-to-seed method is not being able to evaluate and select the roots that will make seed. Many of the open-pollinated carrot varieties that are available are of poor varietal uniformity due to the lack of selection through numerous generations of seed-to-seed production. If you're going to use the seed-to-seed method, it is imperative that you use well-selected stockseed for each crop you plant. Otherwise, the "garbage in–garbage out" rule applies.

The result of using less-than-ideal planting stock will often be an amplification of any genetic flaws inherent in the population. This is because flawed roots are often less refined (hairy roots, large crowns, or outcrosses with hybrid vigor) and often make bigger flowering plants that produce proportionately more pollen and seed than the average carrot in the population. This is why open-pollinated varieties among the cross-pollinated crops are so easily run into the ground when selection isn't practiced.

Root-to-Seed Method: The root-to-seed method requires that the carrot roots be lifted, selected, and either replanted soon after the selection process or stored in a cooler or root cellar and replanted the following spring. The advantage of this method is that it affords the grower a chance to evaluate each root and decide if it is worthy of contributing to the

▼ Carrot flowers in the field.

next generation. You must make sure that all the roots you retain: (1) receive adequate vernalization hours in storage in order to bolt the following season; (2) conform to the standards set by the selection criteria; and (3) are free of growth cracks, splits, disease, or any insect tunnels or damage. It is important to remember the old adage "one rotten apple spoils the whole barrel," as rot in stored root crops can work much the same way.

The root-to-seed method has several modifications depending upon the climate and goals of the seed grower. If your situation necessitates replanting the roots immediately after lifting and selection (thus overwintering them in the field), then it is best to grow roots of a comparable size to the seed-to-seed method. This is especially true in areas where the winter lows approach a temperature of 15°F (–9°C), where even smaller-diameter carrots are damaged by the cold. But if you are going to store the roots throughout the winter months or if you're replanting in a mild-winter location (temperatures not dropping below about 29°F/–1.7°C), then you may want to grow the carrots to the size in which they are normally harvested for eating. The great advantage of this is that you can then select the carrot traits at precisely the point in their life cycle that corresponds to their true use. Many carrot varieties used on the organic market are larger-rooted types like Nantes, Chantenay, Flakkee, and Kuroda that don't fully attain their distinctive shape until they are mature. Maturity is measured by the extent to which their tip—the point from which the taproot emanates—has filled and their tips appear blunt. This blunting, as carrot breeders refer to it, along with the overall size and taper of the root, determines the characteristic shape of each carrot type. Hence, using the root-to-seed method and selecting at full size and tip fill for each variety is ultimately the only way to maintain good

shape in a carrot variety. Other traits that are best selected for at full maturity include color, bolt tolerance, foliar disease tolerance, and very importantly flavor. (See "Maintenance and Selection of Genetic Stocks," page 92.)

Cold Storage of Roots: Proper storage of carrot roots for the root-to-seed method is very important, as some growers may need to store the roots for upward of 5 months before replanting in the spring. Roots of biennial crops such as carrots that are stored for seed production are known as stecklings. Environmental conditions for carrot stecklings stored for seed production are essentially the same as for carrots stored for consumption. The storage temperature should be between 35 and 38°F (1.7 to 3°C) to slow the roots' physiological processes, but avoiding freezing with errant temperature swings when the cooler is set closer to 32°C (0°C). This exposure to cold temperatures must be for at least 8 weeks in order to fully vernalize the roots and promote flowering and subsequent seed production in most carrot varieties (see chapter 3, Understanding Biennial Seed Crops, "Vernalization" on page 17).

Carrot roots must also be stored at a high relative humidity of about 90 to 95% to remain in good condition for subsequent growth when replanted. These two environmental requirements—temperature and relative humidity (RH)—are most easily satisfied with a cooler with humidity control. If you're using this type of cooler, you can store clean, sound stecklings in wooden totes with small openings between the slotted planks on the sides to allow the free flow of humidity to reach the roots. At the Alf Christianson Seed Company I found that covering these totes with a 2- to 3-in (5- to 7.6-cm) layer of *clean* wood shavings (not sawdust!) will keep the roots free of standing water that can accumulate on the surface of the uppermost roots under high-humidity conditions. This helps eliminate a source of potential rot. Cedar wood shavings may be superior for this purpose, as cedar is reported to have a higher level of antimicrobial factors than most other woods.

Carrot stecklings can also be stored in traditional root cellars, which benefit from cold temperatures and high humidity of the earth or soil at varying depths in many temperate zones. In root cellars carrot roots or stecklings are traditionally stored in moist, clean sand or clean, undecayed deciduous leaves (in New England people swear by oak leaves, which have relatively lower tannins), where the roots are laid carefully between layers of this material so as not to touch. This can slow the spread of rot through the lot of roots. It is crucial that you wait until the temperature of the cellar is about 40°F (4°C) or below for proper storage, which will also ensure a higher humidity.

Alternatively, stecklings can be placed into plastic bags that will create a high-humidity environment for the roots. Make sure that there is a series of small penny-sized holes (three to four rows of six to eight holes for an 18×36 in/0.5×1 m bag used for 25 lbs/11 kg of carrots). Into each bag put three large handfuls of wood shavings, trying to spread them evenly among the carrots. These shavings will soak up much of the condensation that forms, and the holes will allow for excess moisture to slowly escape from the bags. Care should always be taken not to place warm roots from the field into any of these various cold storage situations directly, as this will cause excessive condensation that may promote rot.

Preparation of Stecklings for Storage: Special care should be given to stecklings of all biennial root crops prepared for storage. Carrots must be cleaned gently of all soil clinging to the roots *without* the use of water. This is best achieved by harvesting the roots during clear, dry weather, preferably when the soil has had a chance to dry to the point where it readily crumbles and falls away from any root surface. Don't use a hard brush and avoid rubbing any soil vigorously against the surface of the root; stiff bristles, or any sand or small rocks in the soil may abrade and damage the surface of the carrot, thus allowing an avenue for fungal or bacterial infection during the long storage period. Never wash the roots and avoid getting them wet in any way before storage, as this will only encourage the growth of rots. If roots do get wet due to unanticipated rain, let them air dry thoroughly. This can take several hours, so if the stecklings are small they will need to be checked frequently so that they do not become too dry. Healthy soil has many beneficial microbes that will discourage the growth of root rot under most conditions; therefore we do not need to be concerned with getting all the soil off before storing the stecklings.

Proper removal of the tops from carrot stecklings is very important for long-term storage. The leaves and petioles (stems holding the leaves) are collectively called the tops and are often the most susceptible part of a carrot plant to rot. Removing as much of the tops without destroying the growing point within the crown is crucial to your success in storing stecklings of any root crop (see chapter 3, Understanding Biennial Seed Crops, "Preparing Roots for Storage" on page 22). For the 8- to 10-week storage period before transplanting (which is often used primarily to vernalize the crop) the tops are literally twisted at about 1.5 to 2 in (4 to 5 cm) above the crown, which tears the petioles but doesn't usually promote significant rot in this shorter storage time. However, if you are going to store the stecklings throughout the winter, which can be for upward of 5 to 6 months, then you should trim much more of this easily decaying tissue. Trimming will require a sharp knife, a steady hand, and a little bit of knowledge of where the new top growth will come from when the steckling is transplanted. All new shoot growth upon replanting emanates from the crown or the stem apex. This is the point where the petioles connect to the carrot and the shoot apical meristem creates all new foliar growth. It is possible to trim most of the petioles off, about 0.25 to 0.33 in (0.6 to 0.8 cm) above the crown, and not damage the growing point (hence not hindering subsequent leaf and flower stalk development). You should then be able to see very small fern-like leaves below the point where you have cut and not see the round outline of an orange-white stem. If you see the outline of a stem, either you have cut too close or that particular root has already started flowering and you have cut off the emerging flower stalk. It is a good idea to practice this technique on roots that you don't intend to use for seed production until you get the hang of it.

Planting and Transplanting

Planting: Initial sowing/planting for the seed-to-seed or root-to-seed method should only be done into well-prepared soil with good moisture-holding capacity, just as you would for the vegetable crop. As the crop will be planted in summer in most locations you will need to ensure even, adequate moisture to achieve a good stand. In desert production of carrot roots in California, where seeding is frequently done in late summer at temperatures around 100°F (38°C), growers use daily overhead irrigation for evaporative cooling in the top horizon of the soil until germination and emergence are achieved.

For the seed-to-seed method, seed is sown sparingly so that only a cursory thinning is necessary to achieve the desired in-row spacing. Thinning can be achieved by blocking with a

hoe or by cross-cultivating with a spring-tooth harrow or tine weeder across the field at a perpendicular angle to the rows of carrots when they are young but well-established seedlings. The goal is to achieve an in-row spacing between roots of 4 to 8 in (10 to 20 cm). The spacing between rows can be from 22 to 36 in (56 to 91 cm). Many growers have gone to the closer row spacing of 22 in (around 55 cm) for seed-to-seed production in recent years.

With the root-to-seed method carrots are sown much as a grower would for producing the vegetable. As the crop is planted in midsummer, you should be particularly careful to plant at a spacing that will ensure that roots achieve their optimum characteristic size and shape for selection upon lifting in the fall. Upon replanting stecklings in the spring the in-row spacing should be 8 to 12 in (20 to 30 cm) between plants, as fully grown stecklings often produce larger plants. The spacing between rows is routinely between 24 and 36 in (61 to 91 cm).

The spacing between plants within the row is partly dependent on the type of carrot you're producing. If you are producing seed of Amsterdams, Paris Market types, or true Nantes types (or inbred lines for hybrid production) it is possible to plant them at a closer spacing, as they all have flowering plants that are of a smaller statue than most other carrots. These types are frequently planted 4 in (10 cm) apart within the row for the seed-to-seed method and 8 in (20 cm) apart for the root-to-seed method. Other types of carrots, like Imperators, Chantenays, Flakkees, or Kurodas, usually have larger frames and are planted at 8- to 12-in (20- to 30-cm) spacing, respectively, for the two methods.

The density of the planting also influences which class of umbel yields the highest proportion of seed. The carrot seed produced by the primary and secondary umbels is universally regarded as superior to later-forming seed of the third- and fourth-order umbels, due to its size and degree of maturity. In many temperate regions third- and fourth-order umbels commonly have a large percentage of their seed that is small, hasn't fully matured, and has a lower germination rate. Under higher-density plantings the development of the later-forming third- and fourth-order umbels is restricted, thereby benefiting the development and quality of seed from the primary and secondary umbels.

Transplanting: Carrot stecklings are usually prepared for replanting in two steps. First the tops should be removed at harvest (see "Preparation of Stecklings for Storage," page 89). This cuts the transpiration flow from the plant and preserves moisture in the root. Upon transplanting the steckling, the carrot will regrow new foliage soon after reestablishing new feeder or adventitious roots. The next preparatory step is to cut about one-third of the root off at the bottom or taproot end of the carrot. This gives you a chance to see both the intensity of the color of the carrot and the size and color of the core for selection purposes (see "Maintenance and Selection of Genetic Stocks," page 92). The cut should be made cleanly, with a sharp knife, and at a slight angle laterally (15 to 30 degrees from a straight diagonal cut). This allows for quicker healing of the wound, and the root will meet with less resistance when it is pressed into the soil upon transplanting. These cut roots are then placed, one deep, in a cool, airy area for several hours to allow the wound to heal or suberize, forming a scar-like protective layer on the cut surface. Stecklings must be at least 3 to 4 in (8 to 10 cm) long to successfully regrow and are often 6 to 8 in (15 to 20 cm) long. This length easily accommodates replanting into furrows as compared with full-sized carrots.

ISOLATION DISTANCES

Carrots are insect-pollinated and, like most crops in the Apiaceae, are very attractive to a wide diversity of pollinators. This is relevant as carrots grown for seed in most temperate-climate settings will attract insects from a broad perimeter around the field in which they are planted. However, this also means that you will need to be careful in determining isolation distances from other carrot crops and any wild carrot (Queen Anne's lace) that may be present in your region.

The standard isolation distance of 1 mi (1.6 km) between carrot crops of the same crop type, without physical barriers on the landscape, should be observed (see chapter 13, Isolation Distances for Maintaining Varietal Integrity, for a full discussion of physical barriers). In cases where significant barriers dot the landscape it is possible to diminish this distance to 0.5 mi (0.8 km). Crop types in carrots would include Nantes, Imperators, and Chantenays; each of the unusually colored carrot types (purples, reds, yellows, and whites) would constitute a major group. As it is always possible to have some crossing at these distances, two carrot varieties within a particular group will suffer less varietal damage if crossed than if a Chantenay were crossed with an Imperator in the production of commercial seed.

If you are producing carrot seed of a different type than the nearest neighboring carrot seed field or a known patch of wild carrot (Queen Anne's lace), you will have to double the distance to ensure a high level of genetic purity. Between different carrot types you will need 1 mi of isolation with barriers and 2 mi (3.2 km) without barriers. This is important to observe across the different color classes and also serves as the minimum distances to

Harvesting high-quality carrot stockseed by hand.

be used for any carrot stockseed production (see chapter 17, Stockseed Basics).

Maintenance and Selection of Genetic Stocks

Maintaining the genetic integrity of open-pollinated carrot varieties that are good performers requires attention to detail, knowledge of carrot traits, and a higher level of commitment than for many other vegetables. As with many cross-pollinated species, even well-selected carrot varieties with a high degree of uniformity are genetically heterogeneous and may have an appreciable amount of variation in the traits you're selecting. Remember that when selecting a crosser like carrots your goal is to establish the amount of variation that is acceptable for each trait of interest and not select so narrowly that you jeopardize the genetic integrity of the variety. Carrots are more prone to inbreeding depression than most cross-pollinated crops. Many a good carrot variety has been ruined by an overzealous person selecting a relatively small number of "perfect" roots for a seed increase and thereby narrowing a genetically healthy carrot population (variety) into a shadow of its former self. The health and genetic resiliency of a good carrot variety relies on keeping a broad genetic base. The art of the selection process is having an appreciation of the breadth of variation for the traits of interest that can be retained while still producing a variety that is uniform enough to satisfy the farmers who will produce the crop and the markets they serve.

Selection for Foliar Characteristics: Selection in carrots begins before the roots are lifted. As with all root crops, carrot tops are an important constituent of the crop, especially for the organic grower. Selection for vigorous seedling growth and the early establishment of a robust full set of leaves can be instrumental in developing a carrot for the organic market that is able to compete with weeds early in the season. Selection for this foliar canopy that will better compete with weeds can be done early in the season, long before the roots are fully formed. Also important is the stature of the carrot's foliage. Selection for carrot tops that stand erect with minimal contact with the soil is important in slowing the potential spread of both fungal and bacterial foliar diseases that are soilborne. Foliar diseases can decrease the yield and eating quality of the carrot root crop by diminishing the amount of energy (and sugars) that is fixed through photosynthesis. They can also severely affect the yield and quality of the seed crop (see chapter 16, Seedborne Diseases).

If foliar diseases are present when producing the carrot roots for seed production it is often possible to select for horizontal resistance to these pathogens. When present, there is usually a continuum of disease symptoms that can be observed from plant to plant within the crop. In this situation it is advisable to rogue out (eliminate from the population) the most susceptible individuals. This can be done as early as symptoms arise to slow the spread of the pathogen and can be repeated throughout the season, including during the flowering cycle in Year 2.

Selection for Root Characteristics: With the root-to-seed method of seed production you will have a golden opportunity to select for many root characteristics. There are as many and varied a number of root characters to select for as there are colorful characters who grow carrot seed in this world. I will list the most important traits that are universally selected for, realizing that for any particular carrot type or carrot variety there may be traits of import for that specific type or the market it serves.

The shape of a carrot variety is probably the most recognizable feature of that carrot. As stated earlier, the shape, relative size, degree of taper, and extent of blunting of the tip determine the characteristic shape of each carrot variety. While the environment exerts a sizable amount of influence on these traits, the seed grower must make the overall shape of the roots a top priority in the selection process, as the shape of a carrot variety will become

Nash Huber and Micaela Colley evaluating and selecting for root characteristics in a carrot trial in Sequim, Washington.

much too variable in very few generations if ignored. Another trait that is usually quite variable is the degree of smoothness of the surface of the root. The predominant features on the exterior of a carrot root are the lateral root scars that appear as pale, off-white lines in small indented regions running perpendicular to the length of the root. Selection for small lateral root scars with a minimum of indenting will make the carrot appear smoother. It is also important to select against the presence of small adventitious roots that occasionally emanate from these root scars.

Carrot roots are very prone to producing additional lateral roots, a tendency growers call **forking.** Sometimes this forking will result in as many as three, four, and even five additional lateral roots. There is a lively debate among researchers as to whether these extra roots are due to early damage of the developing taproot from either a root rot organism or a nematode feeding, from hardpan or hitting a rock. Regardless of the cause, the consensus among carrot breeders is that the genetic background of a carrot can influence the degree of forking in a particular variety; hence it is possible to select against forking. Also common in carrots is the propensity of certain carrots to crack while growing and then fully heal. These growth cracks result in ugly scars in the mature crop that make the carrot unmarketable. Alternatively shatter cracks may occur at the time of harvest from any impact sustained by the root during the harvesting process. They appear as long, vertical cracks along the length of the root and also make the affected roots unmarketable. While both types of cracking are thought to occur due to excessive uptake of water, it is agreed that genetic selection against both forms of cracking over time can lessen the extent of the problem.

Breeding for color in carrots is a time-honored tradition, as described in the development of orange carrots. When the intensity of the color in a variety is relatively low, as it is with older orange varieties and in most of the open-pollinated unusually colored carrots, then it is possible to select based on a visual inspection of the exterior of the root. For improved accuracy in selecting for root color, however, it is important to make a smooth lateral cut and visually evaluate the internal color (as described in "Preparation of Stecklings for Storage," page 89) to more accurately compare the intensity of color between roots. This is a highly heritable trait that will improve rather quickly through several cycles of selection. The added bonus is that all of the carrot pigments are nutritionally significant, and these increases represent a health benefit to anyone eating the variety!

Selection for flavor is very important; we all know how many bad-tasting carrots there are in this world. However, selecting for carrot flavor isn't easy, for a couple of reasons. First, it requires tasting each root in the carrot population you're reproducing. This is usually done with smaller breeding populations of 300 to 400 roots (see chapter 13, "Cross-Pollinated Crops" on page 338) that are grown using the root-to-seed method, where it's possible to taste the piece that's cut from each root in preparation of the stecklings. (You quickly learn to spit out each bite that you taste to avoid the stomachache that can come from over 300 bites of carrot.) Second, carrot flavor is biochemically quite complex, and there is no consensus among carrot lovers on what constitutes the ideal. Many people want the sweetest, mildest-flavored carrots, while others prefer a strong "carroty" flavor that isn't too sweet. And the genetics for human ability to discern flavors is also quite variable. So all those selecting for carrot flavor must be confident in their own ability to assess the relative merits of a particular carrot.

CELERY & CELERIAC

The domestication story of *Apium graveolens* L., which includes celery, celeriac, and the ancestral type, smallage, is incomplete, but it is undoubtedly the result of selection by several agricultural societies living between the eastern Mediterranean and the Himalayan foothills. Originally marsh species, wild forms of these crops are still found in wetlands throughout the warmer temperate zones of Europe and western Asia. The earliest use of this species was for the medicinal properties of its seed in ancient Egypt and in the Roman Empire. The seed and seed extracts were used as a diuretic, for intestinal ailments, and as an aphrodisiac.

As a vegetable the earliest domesticated version of celery and its allies was probably a leafy plant with smaller, non-succulent stems, not unlike smallage, or leaf celery, which is still cultivated widely across Asia and is used primarily as a salad green, potherb, and aromatic herb in cooked dishes.

The development of celery as the crop that we know in the modern era probably started in Italy and France beginning in the 16th century. The most important traits that were selected by farmer-breeders in transforming the crop at this time were increased fleshiness of the petioles (leaf stems) and reduced levels

SEED PRODUCTION PARAMETERS: CELERY & CELERIAC

Common names: celery, celeriac, cutting celery
Crop species: *Apium graveolens* L.
Life cycle: biennial
Mating system: largely cross-pollinated, with some self-pollination
Mode of pollination: insect
Favorable temperature range for seed germination: 65–85°F (18–29°C)
Favorable temperature range for pollination/seed formation: 65–80°F (18–27°C)
Seasonal cycle: summer through the following late summer or early fall of the next year (12–14 months)
Within-row spacing: seed-to-seed 4–8 in (10–20 cm); root-to-seed 8–18 in (20–46 cm)
Between-row spacing: 22–36 in (56–91 cm)
Species that will readily cross with crop: All versions of celery and celeriac are fully sexually compatible and will easily cross.
Isolation distance between carrot seed crops: 0.5–2 mi (0.8–3.2 km), depending on crop type and barriers present on the landscape

'Redventure' is an open-pollinated red celery from Wild Garden Seeds in Philomath, Oregon. For seed production it is important to select against the open habit exhibited by this plant.

the primary part of the plant used in this European version of the crop, celery began to spread across Europe and beyond, becoming a prominent vegetable in many temperate areas of the world. Celeriac or celery root appeared in European agriculture sometime in the 17th century as an apparent variant of stalk celery, eventually becoming important for its role as a winter storage crop in many of the colder reaches of Europe.

Celery is not as cold-hardy as many of the other Apiaceae seed crops and is therefore grown in temperate regions with milder winters than many crops in this family. Much of the commercial celery seed crop is grown in the coastal valleys of California near San Luis Obispo and the warmer reaches of the Rhône River Valley of France. Celeriac seed is also produced in these areas, and there is a significant amount of localized production still done in Eastern Europe, where the crop is a staple vegetable that is used daily in many areas as a soup or stew ingredient. Seed of smallage and the many other variants of leaf celery are still grown on-farm in almost all of the agricultural areas from the Mediterranean to eastern Asia, where these crops are essential to the regional cuisine as vegetables and as aromatic condiments and where they are more extensively cultivated than celery. Because leaf celeries are

of strong or aromatic compounds; these developments would redefine the use of this vegetable. As these petioles or stalks became more heat-tolerant than other crops in this species, their seed can be produced throughout a much wider climatic area than stalk celery or celeriac.

CROP CHARACTERISTICS

Reproductive Biology

Celery is a biennial that sequesters most of the stored energy from its first season of growth in its petioles or stems. Therefore, to successfully produce an acceptable yield in the seed crop, you must produce a plant in the first season of adequate size and stored food to produce a robust flowering plant in the second season. If allowed to grow to its full size where it is sown it will produce a sizable taproot as well; however, if the first-year plant is grown in a nursery and transplanted using the equivalent of the root-to-seed method (see chapter 6, Carrot, "Root-to-Seed Method" on page 87), then the long taproot will be severed in transplanting. Celeriac produces a more traditional biennial taproot and is better suited to the root-to-seed method.

After vernalization and flower initiation the celery plant grows to a height of 3 to 4 ft (0.9 to 1.2 m) during the second season of growth. Celery and its allies produce a highly branched plant that contains a series of compound umbels. These umbels are smaller and considerably less compact than the umbels of carrots and most other commercial Apiaceae crops.

A flowering celery plant.

97

◀ The compound umbels of the celery inflorescence.

The flowers are white and usually shed pollen in the morning soon after the petals open for the first time. After 3 to 5 days the style becomes fully extended and the stigma becomes receptive, therefore encouraging cross-pollination. Celery and its allies, however, do not have any barriers to self-pollination and will produce a certain amount of selfed seed, based on insect movement from one umbel to another. Celery produces abundant nectar, and many species of insects, including wasps, syrphid flies, and bees, take part in the pollination. The seed that is produced, like that of the carrot, is borne in pairs and results when the schizocarp fruit splits at maturity.

Climatic and Geographic Suitability

Commercially produced celery seed is grown in milder winter regions than many other Apiaceae seed crops, when it is overwintered using the seed-to-seed method. Celery is more tender than most of the biennials in this family. Many stalk celery varieties will survive frosts from 28 to 32°F (–2 to 0°C) if they are hardened off gradually but will suffer severe damage at lower temperatures, which restricts the temperate regions where seed can be grown. Celeriac and leaf celery can usually withstand temperatures that are slightly colder, but they are also more easily damaged than many other species in this family. You can use floating row covers to protect all of these crops from light frost, but experiment with them with before relying on them to protect the plants below 30°F (–1°C).

The celery seed crop must be grown in temperate zones that have regular intervals of cold temperatures between 40 and 50°F (4 to 10°C) during the winter months between the seasons in order to be fully vernalized. If celery plants are stored in a coldroom between seasons, they will need temperatures in this range (or colder) for several weeks before they can be transplanted and expected to bolt. As with most biennials the duration and temperature range needed to promote flowering vary greatly among varieties, and you must have good observational skills to determine if the variety is receiving adequate vernalization for the entire population to bolt and produce seed.

SEED PRODUCTION PRACTICES

Soil and Fertility Requirements

Celery and its allies were domesticated from wild marsh plants that grew in the peat soils in and around the Mediterranean. The modern crop still thrives in organic soils and loams with a high percentage of organic matter, though a wider range of fertile soils can be used for the seed crop. Growing a successful crop has always demanded relatively high levels of macronutrients and a steady source of moisture throughout the season. Nitrogen levels for the seed crop should be carefully monitored, as excessive amounts of N can be problematic in producing a flowering celery crop. Celery also requires a relatively high pH, between 6.5 and 7.5.

Growing the Seed Crop

Seed-to-Seed Method: Commercial celery seed production is usually done using the seed-to-seed method in a similar fashion to its use for root crops like carrots or beets (see chapter 6, Carrot, "Seed-to-Seed Method" on page 86). The crop is direct-seeded or transplanted as seedlings in mid- to late summer of the first growing season into the field where the plants will stay through the end of seed production.

Celery has been traditionally planted on 3 ft (0.9 m) centers for seed production, though some growers now plant as close as 2 ft (0.6 m) between rows. The final in-row spacing between plants can be from 1.5 to 2 ft (0.5 to 0.6 m). When using the seed-to-seed method the in-row spacing between plants can be much closer than this through the early stages of growth, as this will afford you several stages to perform phenotypic selection before it is necessary to settle on the final population at the appropriate spacing as the crop begins to flower in the second season.

Root-to-Seed Method: When using the equivalent of the root-to-seed method (the plant-to-seed method in the case of stalk or leaf celery), you can either direct-seed or transplant seedlings into a high-density nursery or produce the plants as you would the vegetable crop. The crop is planted in mid-summer (early summer if you want to grow them to full vegetable "maturity" for selection purposes). If the climate is mild enough then the plants can either remain in a nursery until spring before transplanting or they can be transplanted in the fall into the seed production field. In either of these cases they can be transplanted into the final spacing given for the seed-to-seed method.

Alternatively, if your climate experiences winter lows below the 28 to 32°F (–2 to 0°C) range, you will need to lift and select the plants in the fall and place them in cold storage until you transplant them the following spring in the seed production field. Storing stalk and leaf celery is somewhat different from storing the Apiaceae root crops. Storing celeriac is done in a similar fashion to storing carrots or parsnips. However, for celery and leaf celery you should leave much of the soil that clings to the roots upon lifting the plants, as the plants' roots need to stay buried in moist soil

for the duration of the storage period. The temperature in the coldroom should be maintained as close to 34°F (1°C) as possible. The relative humidity of the room should be close to 75% if possible, as humidity is necessary to keep the petioles and foliage from desiccating. If the humidity goes above this level, though, it can cause freestanding water to form on the plants' surfaces, which can promote rot. As with all coldrooms there should be good ventilation; air movement in the vicinity of the plants is very important, as plants will probably be packed together tightly and need the flow of air to avoid decay.

The duration of the storage period needed to initiate vernalization does vary greatly for the different varieties of celery and its allies, but most growers working with these crops agree that a period of at least 8 to 10 weeks of cold treatment is necessary to fully vernalize the entire population. Celeriac may require at least 2 to 4 more weeks of cold than the other celeries to become fully vernalized, as it is a colder-season version of these crops; it has been selected for longer-term storage to fit the more northern growing regions and has become adapted to these colder growing seasons.

At the end of the storage period all of the propagules of the celery allies must be checked for decay and viable apical growth when you sort them before transplanting. Outer celery stalks that have decay can be eliminated, as should any vegetation that has desiccated or gotten too leggy. Any pruning that is done at this time should take into account that the apical growing tip must be left fully intact to expand into the flowering plant.

Seed Harvest

Seed harvest for celery and its allies is similar to other Apiaceae crops. Celery plants are pulled or cut right at ground level when the variety has matured a majority of the seed on most of the plants. A majority of the seedheads will have a brownish cast, and the plants will turn yellow as they senesce. As with all Apiaceae crops there is a delicate balance between waiting for a majority of the seed to mature and losing some of the earliest-maturing seed present on any given plant during this wait due to shattering. Celery seems to be more prone to shattering than carrots; therefore keeping close vigil when observing celery seed maturation is important. As with all dry-seeded crops, the timing of harvest and dry-weather cycles is part of what distinguishes good seed growers from the rest.

Because celery seed may shatter more readily than the seed of other related crops, there is good reason to lay mature celery plants onto landscape fabric (or some other type of permeable cloth that will allow water to drain through it in case of precipitation) when making windrows. Windrowed plants should be allowed to dry for at least 4 to 7 days, being turned once during the drying process. The crop needs to be monitored closely and threshed through a stationary thresher or fed through a combine before significant shattering takes place. These processes are always best done in midmorning after the dew has lifted but while the stems are still pliable.

GENETIC MAINTENANCE

Because the celery seed crop is often planted at a later time in the season than the vegetable crop would be planted, it is important to do initial selection on the crop when you grow it to full maturity as a vegetable crop. By doing this it is easier to select for all of the traits associated with the fully grown petioles (stalks) and determine which plants have the desired traits for the cultural and market conditions under which they will be grown. Traits

include height, color, petiole size and ribbiness, stem base, flavor, texture, and resistance to diseases or insect pests. The height, stature, and leafiness of the plants are important, and the overall plant should appear symmetrical and cylindrical. The color of the stalks should be uniform. Selection is for either the light green or yellowish self-blanching types or the greener types that are favored in North America. Selection for size, shape, and tightness of the stalks around the heart is important. The stalks should also be smooth with an absence of pithiness, stringiness, and large ribs. Petiole cracking and the presence of disease should also be monitored and selected against.

The flavor variation in celery can be great as with all Apiaceae crops, though many modern plant breeders ignore flavor in their selection, assuming that all celery is bland and basically tastes alike. On the contrary, most chefs consider celery an important flavorful aromatic, and indeed, once you start to taste the different varieties and familiarize yourself with the differences that exist in the distinctive celery flavor, it will be easy to select for flavor within an open-pollinated variety or segregating population.

Celeriac is often grown to its full size as a vegetable and stored over winter in a root cellar to be transplanted for the seed crop in many parts of the world. This is possible because celeriac stores so well. The roots are stored and transplanted in a similar fashion to carrots (see Carrots, "Cold Storage of Roots" on page 88 and "Planting and Transplanting" on page 90), however the roots themselves do not require cutting before transplanting them. Growing the crop to full edible root maturity allows you to select for root characteristics such as size, shape, root smoothness, crown size, and the amount and extent of secondary roots. Foliar characteristics like height, width of petiole, attitude, and color are often also selected for in the first season before lifting the crop for storage or transplanting. As with all biennials it is important to eliminate any plants that show any sign of bolting in the first season, as well as any plants that bolt so late in the second season that they are unable to mature seed in a timely fashion. As always, unhealthy or malformed flowering plants should be eliminated at any stage throughout the second season.

ISOLATION DISTANCES

The standard isolation distance of 1 mi (1.6 km) should be used between celery crops of the same type when there are no physical landscape barriers between two seed crops. In cases where there are significant barriers it is possible to reduce this distance to a minimum of 0.5 mi (0.8 km). As it is always possible to have some crossing at these distances, two celery varieties of the same type will suffer less varietal damage if crossed in the production of a commercial seed crop. Celery types can be categorized into classes by height, time of maturity, and whether the variety has stalks that are self-blanching or green (or pink/red). If you are producing seed of a different celery type than the nearest neighboring celery crop, then you will need to double the isolation distance to a minimum of 2 mi (3.2 km) to ensure a high level of genetic purity if there are no barriers on the landscape.

This increased minimum distance of 2 mi (3.2 km) between seed crops is also very important to use if you are producing either a celeriac or a leaf celery seed crop in the same locale as a celery seed crop. When there are significant barriers on the landscape the isolation distance can be decreased to a minimum of 1 mi (1.6 km) for all of the diverse types of this species.

CILANTRO

Cilantro (*Coriandrum sativum*) is one of the oldest known culinary herbs and is native to the Mediterranean basin. Its origins are obscured by the fact that it apparently spread through agricultural societies of the Mideast before recorded history. Seed dated to 6000 BCE was recovered in a cave dwelling in modern-day Israel. Sanskrit writings refer to its use as far back as 5000 BCE. It made the trip to China via the Silk Road and traveled north to Europe with Roman conquest. The herb then traveled to the Americas with the Spanish conquistadores. Its use is woven into the fabric of the cultures of North Africa, the Middle East, and Asia Minor. The herb is known as cilantro when used in the Spanish-speaking countries of North, South, and Central America; in other regions it is referred to as coriander leaf or as Chinese parsley. It is also used extensively in the cuisines of Southeast Asia, India, and China.

It is hard to determine when the leaves of cilantro were first used as an herb for flavoring, but cilantro has undoubtedly had a long culinary history and continues to serve as two distinct, very important crops worldwide. It is grown in many cultures for its very aromatic foliage, which is used as both a cooked and raw condiment. Cilantro leaves give salsa its distinctive flavor. Its seed is used as a spice, coriander,

SEED PRODUCTION PARAMETERS: CILANTRO

Common names: cilantro, coriander
Crop species: *Coriandrum sativum* L.
Life cycle: annual
Mating system: largely cross-pollinated, with some self-pollination
Mode of pollination: insect
Favorable temperature range for seed germination: 65–85°F (18–29°C)
Favorable temperature range for pollination/seed formation: 65–80°F (18–27°C)
Seasonal reproductive cycle: early to mid-spring to mid- to late summer
Within-row spacing: seed-to-seed 4–8 in (10–20 cm); root-to-seed 8–18 in (20–46 cm)
Between-row spacing: 22–36 in (56–91 cm)
Species that will readily cross with crop: All cilantro or coriander seed growers should be conscious of nearby farm or garden plots of cilantro grown for fresh market production, as these crops often flower before they are turned under.
Isolation distance between carrot seed crops: 0.5–2 mi (0.8–3.2 km), depending on crop type and barriers present on the landscape

Cilantro (*Coriandrum sativum*). ▶

which is an important ingredient in curries, chutneys, and breads across much of the Old World. The roots are used to add an intense flavor to soups and curries in several Asian cultures. The use of cilantro has grown in recent years as many of these culinary traditions have expanded into North America and Europe.

The great majority of the cilantro seed used across the world for planting is grown locally by farmers. There are undoubtedly thousands of varieties adapted to the unique climates and culinary needs of the people who cultivate these varied strains around the world. Large-scale commercial seed production is done in Mediterranean climates with seasonal dry weather occurring during the seed maturation stage. In North America the cool coastal valleys of Washington State produce good yields of high-quality seed.

CROP CHARACTERISTICS

Reproductive Biology

Cilantro is a cross-pollinated annual that stands upright with a height of 1 to 1.5 ft (0.3 to 0.5 m) in the vegetative phase of its life and extends sometimes to a height of 3 ft (0.9 m) during flowering and seed production. Leaves of the vegetative plant transition from a trifoliate ovate shape with deep cuts on the margins to a more frilly, highly pinnate form emerging from the flower stalk that's almost feathery in appearance. The inflorescence is a compound umbel that is typical of the Apiaceae, yet smaller than the umbels of carrots or dill. The flowers are borne on umbellettes and are asymmetrical, with longer petals on the outer side of the umbel. Cilantro is self-fertile, though as with all Apiaceae cross-pollination is the norm, with insects of many species working the flowers.

Climatic and Geographic Suitability

The climatic requirement for cilantro seed production is essentially the same as for the other important vegetable Apiaceae. The main difference is that cilantro can mature a crop somewhat earlier and under cooler conditions than is possible with carrots or parsley. This is borne out in the seed production areas around the Salish Sea of Washington in the United States and British Columbia, Canada; parsley and carrot will not fully mature a satisfactory commercial seed crop in the cooler maritime valleys of the area, requiring warmer interior valleys to the east and south of the Salish Sea basin. In contrast, cilantro plantings in these coastal valleys do produce perfectly acceptable commercial seed crops with a high percentage of seed maturing. In fact, this cool coastal climate allows cilantro the chance to produce a larger frame (basal rosette of leaves) in spring that supports a large complement of flowers and, therefore, a better yield of seed. Another unique quality of cilantro, one that isn't shared by many of the other cool-season crops, is its ability to produce good-quality seed in hotter climates. While it may not produce optimum yields in hotter climates, where it doesn't put on as much spring growth before flower stalk initiation, it is still capable of producing high-quality seed. This is not surprising when you consider that much of the cilantro that has been cultivated for at least 2,000 years has grown in hot climates.

SEED PRODUCTION PRACTICES

Soil and Fertility Requirements

Cilantro seed crops will tolerate a wide range of soil types as long as adequate fertility and moisture levels are maintained. However, soils must be well drained to avoid losses from root rot organisms. Lighter soils are especially important for establishing early plantings. Cilantro can produce a good crop from moderate levels of balanced fertility. Fertility from compost, legume plow-down, or other organic sources should produce a satisfactory seed crop. Excessive foliar growth early in the season due to an overabundance of nitrogen can cause a leggy seed crop, in which plants may lodge (fall over) during seed production.

Growing the Seed Crop

It is important to plant cilantro early enough in the season to establish a good foliar frame during the cool weather of spring. This will support an optimum seed set once the plants begin to bolt. Cilantro is frost-tolerant and can be planted as soon as the ground can be worked, but to ensure good vigorous seedling growth the crop should be sown after the weather has settled. Temperatures between 50 and 85°F (10 to 29°C) provide optimum growing conditions. The main guiding principle is to produce a large full frame of foliage during the cool weather of spring before hot weather induces bolting. This is why the best-yielding cilantro crops are grown in cool-season seed-growing areas, even though the crop will produce successfully in hotter climates.

Cilantro is frequently planted in rows with 18 to 24 in (46 to 61 cm) centers. As plants rarely exceed a height of 18 to 24 in (46 to 61 cm), under optimum soil fertility regimes for seed crops it is possible to space the crop as closely as 14 in between rows. (Large commercial production is done at 22 to 24 in/56 to 60 cm row centers.) In-row spacing is usually done at 4 to 8 in (10 to 20 cm), so it is easy to eliminate any early bolters that may not be easily seen at a tighter spacing. When the seed is planted there are frequently "doubles" or two seedlings that will appear where each seed is planted. This is because the ribbed, globe-like seed of cilantro is really a two-seeded fruit. Some growers will split this schizocarp fruit by gently running the seed between rollers before planting, therefore increasing the number of single seeds planted. If this is not done the grower will often go through and thin any doubles to single plants to achieve a good plant density.

◄ Asymmetrical cilantro flowers with longer petals on the outside of the umbel.

Seed Harvest

When a majority of the cilantro seed is mature the plants should be cut and placed in windows to cure during dry, settled weather. The timing of this is particularly critical with cilantro as discoloration of the seed can easily occur with precipitation. This discoloration is not well understood and hasn't proven to be detrimental to the seed quality (it's probably due to saprophytic fungi or bacteria). However, growers and seed companies perceive this discolored seed as being of poorer quality. The best way to combat this is by growing the seed under the driest conditions possible at the time of seed maturation. As with all dry-seeded vegetables, overhead irrigation should never be used during the later stages of seed production while the crop is maturing. When the plants are cut for harvest during dry weather it is a good idea to swath them in the morning while there is still dew as it will minimize the shattering of the most mature seed from the plants.

GENETIC MAINTENANCE

Selecting against early bolting is one of the most important things a seed grower can do in a fast-growing leafy vegetable crop (spinach and arugula are two other good examples). Otherwise a crop like cilantro will invariably become faster bolting over several generations due to genetic drift. More than one cilantro variety released specifically as a "slow-bolting" variety in the last decade has degraded to a significantly faster-bolting type due to a lack of selection for this trait during repeated increases of the seed crop used as stockseed.

Other traits that may be considered for selection include seedling vigor, leaf type, and upright stature. All of these can be visually selected during the normal seed production cycle. Variation for flavor is common in almost all cilantro populations. Selection for uniformity of flavor requires a discerning palate and an infinite amount of patience but has been done. Selection for improved flavor is possible if a good range of flavor variation exists in the population in question.

ISOLATION DISTANCES

Cilantro requires the same standard isolation distance as other members of the Apiaceae. If grown in open terrain, isolation between seed crops should be 1 mi (1.6 km); in an area with a substantial number of barriers on the landscape the isolation can be decreased to 0.5 mi (0.8 km) (see chapter 13, Isolation Distances for Maintaining Varietal Integrity). An increase in isolation may be necessary when two cilantro crops with significantly different characteristics are grown in the same region. For instance, if one crop has a distinct type of foliage, flavor, or stature that is different from another cilantro seed crop that will be grown nearby, then the isolation distance should be increased to 2 mi (3.2 km) in open terrain or 1 mi (1.6 km) with substantial barriers. This increased isolation distance might be especially necessary if one field is a fast-bolting coriander type grown for its aromatic seed and the second field has a slower-bolting, leafy cilantro type.

A real threat from unwanted cross-pollination also exists if there are any diversified vegetable farms or home gardens with cilantro growing in the vicinity of any commercial seed crops. This is because a portion of almost any cilantro grown as a leaf crop will invariably bolt and flower before it is turned under.

PARSLEY

Parsley (*Petroselinum crispum*) originated in and around the Mediterranean basin and was grown by the Greeks and Romans more than 2,000 years ago. It was originally cultivated for its medicinal properties, being used to improve digestion and increase assimilation of food. It also served as an important remedy for ailments of the bladder and kidneys, and its efficacy for the health of these organs has been borne out by modern scientific investigation. It is unclear when parsley was first used as a vegetable and herb, but it appears that its culinary importance grew as its use spread geographically.

By the 15th century parsley was being grown extensively in Western Europe; from there it spread to many parts of the world over the next three centuries. During its domestication, selection centered upon profuse leaf development and increased volatile flavor compounds that have made it the popular herb it is today. Human selection also produced various forms of the crop based on the degree of curling of the leaves, from the flat Italian types (*P. crispum* var. *neapolitanum*) preferred by many chefs to the double, triple, and moss curled types that are of greater economic importance in Northern Europe and North America. A form of the crop (*P. crispum* var. *tuberosum*) has also been developed for its fleshy taproot and is known as Hamburg parsley or turnip-rooted parsley.

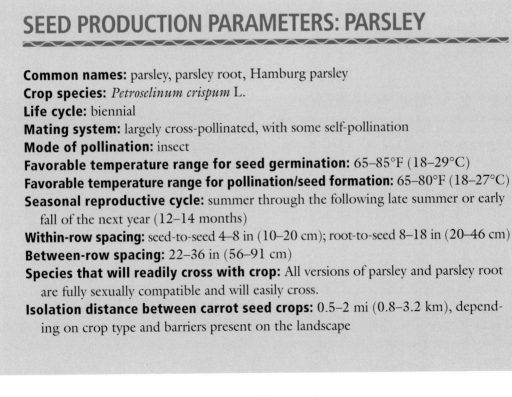

SEED PRODUCTION PARAMETERS: PARSLEY

Common names: parsley, parsley root, Hamburg parsley

Crop species: *Petroselinum crispum* L.

Life cycle: biennial

Mating system: largely cross-pollinated, with some self-pollination

Mode of pollination: insect

Favorable temperature range for seed germination: 65–85°F (18–29°C)

Favorable temperature range for pollination/seed formation: 65–80°F (18–27°C)

Seasonal reproductive cycle: summer through the following late summer or early fall of the next year (12–14 months)

Within-row spacing: seed-to-seed 4–8 in (10–20 cm); root-to-seed 8–18 in (20–46 cm)

Between-row spacing: 22–36 in (56–91 cm)

Species that will readily cross with crop: All versions of parsley and parsley root are fully sexually compatible and will easily cross.

Isolation distance between carrot seed crops: 0.5–2 mi (0.8–3.2 km), depending on crop type and barriers present on the landscape

Italian flat leaf parsley (*Petroselinum crispum* var. *neapolitanum*). ▶

It is used like any other root vegetable, and is a staple ingredient in soups and stews in parts of Northern and Eastern Europe.

Both as a foliage and root vegetable parsley is grown as an annual, planted anytime from early spring through midsummer depending on the climate or seasonal harvest needs. Once the crop enters its reproductive phase in the second year of the biennial cycle, the quality of its leaves or roots for food is greatly diminished.

Commercial parsley seed is grown in many of the same areas that have traditionally been important in the production of carrot and parsnip seed, including the Rhône River Valley of France, the Sacramento Valley of California, and the agricultural valleys of central Oregon and eastern Washington. Parsley seed is also grown regionally by smaller producers in most of the cultures of Eastern Europe and western Asia, where this crop is used extensively in their cuisine. Hamburg parsley seed is also grown as a commercial crop in all of the major production areas to a limited degree, but the majority of cropping is done on a more localized scale in or near the areas of intensive demand.

CROP CHARACTERISTICS

Reproductive Biology

Parsley is a biennial, producing a prolific rosette of leaves in the first season before vernalization promotes flower initiation in the second season. Seed stalks are usually 3 to 5 ft (0.9 to 1.5 m) tall and bear compound umbels that are less dense than most other Apiaceae crops. Each bisexual parsley floret has five pale yellowish green petals with five stamens, two styles, and a two-celled ovary. When normally fertilized the two-celled ovary produces two seeds. While parsley flowers are less showy than carrots or cilantro they appear to get just as much activity from wild insect pollinators, especially wasp and hoverfly species.

Seed set is predominantly from cross-pollination, though the crop is self-fertile. Crossing is encouraged, as stamens will usually ripen and shed viable pollen before the style of the same floret fully elongates and its stigma becomes receptive.

Climatic and Geographic Suitability

Parsley is a cool-season crop that thrives at temperatures between 50 and 61°F (10 to 16°C) during the vegetative stage of its life cycle, though it can tolerate temperatures considerably higher than this once it is established. When grown as a seed crop parsley is probably similar to carrot in requiring temperatures above 68°F (20°C) for good flowering and seed set. Average temperatures at or above 80°F (27°C) seem to be best for maturing high-quality seed late in the season. Parsley seed is slow to mature and requires a longer season and more heat units than many of the other Apiaceae seed crops. This is why it's often grown in the warmer reaches of the seed production climates of these crops. In North America this includes the Willamette Valley and Madras area of Oregon, both of which usually have sustained warmer temperatures at the end of the season when compared with the Apiaceae production areas of British Columbia and Washington.

SEED PRODUCTION PRACTICES

Soil and Fertility Requirements

All Apiaceae crops respond well to light soils with good tilth, similar to the soil favored by carrots or parsnips. Parsley can be grown successfully across a broader range of soil types, however, as long as good drainage, tilth, and fertility are assured. Parsley seed is slow to germinate and requires a soil that doesn't easily form a crust to ensure a good stand. Both regular and Hamburg parsley have a substantial taproot that thrives in well-drained, healthy soils. Balanced fertility without excessive nitrogen is important in growing a solid frame to support seed production without too much foliar growth.

Growing the Seed Crop

The parsley seed crop can be grown using either the seed-to-seed or the root-to-seed methods in a similar fashion to carrots (see chapter 6, Carrot, "Seed-to-Seed Method" on page 86 and "Root-to-Seed Method" on page 87). Both of these methods require planting the crop in midsummer to ensure the production of a large enough plant to properly evaluate in the vegetative state. For the seed-to-seed method the crop is usually sown in mid- to late summer, while the crop that is grown using the root-to-seed method is usually planted before the summer solstice as its growth is delayed when it is lifted, evaluated, and transplanted in the fall. The crop must be sown into a fine seedbed that is well drained but is easily kept moist during warm weather; parsley can take up to 2 weeks to germinate and emerge.

Seed-to-Seed Method: In the seed-to-seed method the crop is sown in rows that are spaced 22 to 36 in (56 to 90 cm) apart; plants within the row are ultimately thinned to a spacing that can be anywhere from 10 to 14 in (25 to 36 cm) apart. This method works exceptionally well with all types of leaf parsley that can be adequately selected for horticulturally important traits via selection of foliar characteristics. Selection can be done over time during the first season as the leaf characters become apparent while the plant matures. Selection can also be done after the overwintering period when the plants that have less winter damage and better spring growth can be favored while eliminating any poorly performing plants.

Root-to-Seed Method: The crop can be sown in nursery beds with rows spaced 14 to 18 in (36 to 46 cm) apart with plants thinned to 1.5 to 2 in (4 to 5 cm) apart in the rows in preparation for transplanting using the root-to-seed method. Nursery beds planted at this density will easily produce enough plants to transplant into an area 10 to 15 times larger than the nursery and still give you an opportunity to select for leaf characteristics before the plants are pulled to transplant in the fall, or the following spring depending on your needs. When the plants are pulled and sorted, a selection for vigorous, disease-free roots is also possible and should be practiced with all types of parsley. For Hamburg or root parsley, selection for solitary, nicely formed roots with little or no branching is advised. All selected plants should have most, if not all, of their foliage clipped before transplanting to minimize respiration in the newly transplanted plant, being careful not to damage the apical growing point in the process.

Seed Maturation: Plants can get as tall as 3 to 5 ft (0.9 to 1.5 m) while flowering and may require staking in windy locations. Maturing seed on parsley umbels may look deceptively green for some time as the plant approaches full maturity. Also, parsley seed shatters easily at maturity and must be harvested promptly when ripe or losses will occur. Therefore, for many growers it may be a good practice to check the degree of starchiness in the developing endosperm of the seed to become familiar with other outward cues that reveal the time to harvest the plants.

Seed Harvest

Cutting, windrowing, and threshing the crop is identical to the process used for carrot.

GENETIC MAINTENANCE

Parsley is a biennial, and its seed is traditionally produced from plants that have been overwintered in the field. Any plants that flower or show any signs of flower initiation (apical stem elongation) or premature bolting during the first year should be viewed as suspect and should be rogued out (eliminated from the population). Selection should also be done in the first season for leaf type (intensity of leaf curl), leaf color, plant stature, and overall vigor, as these traits are most easily spotted during the vegetative growth period. Plants that exhibit disease symptoms, excessive insect damage, and poor overall health can be eliminated at any point during the biennial cycle. If you're growing a root parsley type it is best to lift the roots either in the fall of the first year or in early spring of the second year in order to select for root characteristics. Root parsley has a rather unrefined, primitive root that commonly has some number of adventitious roots that wouldn't be acceptable in its more sophisticated relatives, carrot and parsnip. Selection in this crop should be against roots with an excessive number of these adventitious roots and for well-filled roots with little or no discoloration.

ISOLATION DISTANCES

The standard isolation distance of 1 mi (1.6 km) should be used between parsley seed crops of the same leaf type when there are no physical landscape barriers. In cases where there are significant barriers on the landscape between two crops of the same leaf type, it is possible to diminish this distance to 0.5 mi (0.8 km). As it is always possible to have some crossing at these distances, two parsley varieties of the same leaf type will suffer less varietal damage if crossed in the production of commercial seed.

If you are producing parsley seed of a different leaf type from the nearest neighboring parsley crop, then you will need to double the isolation distance to 2 mi (3.2 km) to ensure a high level of genetic purity if there are no barriers on the landscape. This increased distance is also necessary if you are producing a leaf parsley seed crop in the same district as a root parsley seed crop. When growing these divergent types the isolation distance can be decreased to 1 mi (1.6 km) when there are significant barriers on the landscape between crops.

PARSNIP

The parsnip (*Pastinaca sativa* L.) is a native of Eurasia, from the region between the western Mediterranean and the Caucasus Mountains, where it has been grown for at least 2,000 years. It was used in both the early Greek and Roman civilizations but did not spread as far and wide via trade as many of the other Mediterranean crops that were grown by these societies. Its popularity is strong in some parts of Northern Europe and the UK, and it has been an important vegetable in parts of New England and eastern Canada since they were settled in the 17th and 18th centuries. It is especially popular in the cooler temperate zones, where it grows to its full potential in climates where lengthy, cool autumn weather produces sweeter, heavier roots.

Parsnip is a robustly growing biennial with a strongly aromatic taproot that is often larger than most types of carrots. Parsnips are grown, harvested, and stored much like carrots, though unlike carrots they can be left in the ground through the winter in most climates where they are grown. They are often harvested fresh from the ground for market until the ground freezes. Parsnips have an

SEED PRODUCTION PARAMETERS: PARSNIP

Common name: parsnip
Crop species: *Pastinaca sativa* L.
Life cycle: biennial
Mating system: largely cross-pollinated, with some self-pollination
Mode of pollination: insect
Favorable temperature range for seed germination: 65–85°F (18–29°C)
Favorable temperature range for pollination/seed formation: 65–80°F (18–27°C)
Seasonal cycle: summer through the following late summer or early fall of the next year (12–14 months)
Within-row spacing: seed-to-seed 4–8 in (10–20 cm); root-to-seed 8–18 in (20–46 cm)
Between-row spacing: 22–36 in (56–91 cm)
Species that will readily cross with crop: There are areas where parsnips have escaped from cultivation and exist as a fully fertile feral or wild version of the crop that will readily cross with the parsnip seed crop. Fortunately, this feral version of parsnip is not nearly as prolific or widespread as wild carrot.
Isolation distance between carrot seed crops: 0.5–2 mi (0.8–3.2 km), depending on crop type and barriers present on the landscape

'Turga' is a stout-rooted variety of parsnip that originated in Hungary.

ability to convert stored starches into sugars with cold temperatures, and as the roots are relatively low in moisture they are able to freeze solid and be harvested for food or allowed to regrow as a seed crop when they thaw in spring.

Parsnip plants can achieve a height of 5 to 6 ft (1.5 to 1.8 m) and produce prolific numbers of compound umbels on primary, secondary, and tertiary branches on this indeterminate flowering species. Their small, greenish yellow flowers are less showy than the flowers of many other Apiaceae crops, but they are very adept at attracting a wide number of pollinating insect species. Their seed is fairly flat and slight and is famously short-lived, as its germination rate will usually start to drop within 2 years of harvest.

Parsnip seed is grown in climates similar to those preferred for carrots, though high-quality parsnip seed can be grown in climates that are somewhat cooler and shorter. Parsnip seed is grown in the cooler coastal areas of the Pacific Northwest of the United States and Canada, the Rhône Valley of France, New Zealand, and numerous small pockets across the parsnip-growing areas of Northern Europe, New England, and Canada, where growers maintain regionally important strains.

CROP CHARACTERISTICS

Reproductive Biology

Parsnips are biennials that bear a single primary umbel at the terminus of the main floral stem, which is followed by highly branched stems with many successive secondary and tertiary umbels that continue to develop throughout the second season of reproductive growth. Perfect flowers are greenish yellow and are borne on broad compound umbels and have five petals, five sepals, five stamens, and two styles with nectaries at their base. Like carrots and other Apiaceae crops, parsnips are protandrous, meaning that fertile pollen will shed before the female stigma of the same flower is sexually receptive, thus decreasing the odds of self-pollination. Parsnips are largely cross-pollinated due to insect activity, but they will self-pollinate.

In most temperate climates the seed borne on the flowers of the primary and secondary flowers is the parsnip seed that most reliably matures. Depending on the length of season subsequent third- and even fourth-order umbels will form, but the opportunity for the seed on these later-setting flowers to fully mature is greatly reduced in many temperate regions.

Climatic and Geographic Suitability

Parsnip roots are best produced in moderate climates where they can both germinate and produce much of their initial growth under moderate temperatures below 77°F (25°C). Soils must be kept moist for up to 3 weeks after sowing as parsnip seed can take from 2 to 3 weeks to germinate even at optimum germination temperatures of 65 to 68°F (18 to 20°C). As with other root crops, parsnips require an even supply of water throughout the season to produce good-quality roots where it is easy to distinguish the phenotypic characteristics of the variety. The climatic characteristic that sets parsnips apart from carrots and all of the other major biennial root crops is the fact that parsnip roots grown for seed can be left in the ground throughout the winter to freeze solid in most climates where they are grown.

The best climates for parsnip seed production across the two seasons of this biennial are similar to carrot, though successful seed crops can be matured during the second season in climates that are cooler than those required to get the best yields in carrots. In fact, climatic conditions during the second season are more akin to the conditions best suited to many of the brassica crops, where there are long mild springs and the summer temperatures don't routinely top 80°F (27°C) (see chapter 8, Brassicaceae, "Climatic Adaptation" on page 156). As with all dry-seeded vegetable crops, parsnips thrive in the dry Mediterranean climate with low precipitation amounts from late June until mid-September. This climate significantly lowers the chances of any diseases forming on the seedheads as they mature. However, this means that reliable water must be applied with either a drip or furrow irrigation system; overhead irrigation may promote disease in the seedheads.

◄ Parsnip seedheads, showing the differential maturation of primary and secondary umbels versus the smaller, later-forming tertiary umbels.

SEED PRODUCTION PRACTICES

Soil and Fertility Requirements

Parsnips are best produced in deep, rich, lighter loam soils with good water-holding capacity. Heavier or stony soils will have a tendency to produce crooked, branching, or malformed roots. Parsnips are similar to table beets or chard in their fertility requirements, responding to relatively high levels of fertility and a pH of 6.5 to 7.0 to produce the healthiest roots.

Growing the Seed Crop

The steps to produce parsnip seed are very similar to the steps used in carrot seed production (see Carrot, "Growing the Seed Crop" on page 86). Both the seed-to-seed and the root-to-seed methods are executed in the same way as is used for the carrot seed crop. The major difference is that it is much easier to use the seed-to-seed method for parsnips, as they are highly tolerant of freeze damage. This allows growers in much colder climates to use the seed-to-seed method with parsnips, and it means you don't need to be as concerned with having smaller roots of a certain size to successfully go through the winter. This frees you from being as concerned about the timing for planting the crop in the first year. That said, remember that the roots from all root crops are best overwintered when they have not grown too large, as most older, larger roots have often accumulated more environmental damage (cracks, insect bites, and so on) and often have a lower chance of making it all the way through to producing a seed crop.

The other major difference in planting parsnips is the spacing that is used for seed production. Parsnip plants are somewhat bigger than carrot plants, both in the vegetable stage and during the reproductive stage, and therefore need to be spaced somewhat farther apart than do carrots for both parts of their life cycle.

Seed-to-Seed Method: As with carrot, seed should be sown sparingly in order to make thinning easier. Thinning can be achieved by blocking with a hoe or by cross-cultivating with a spring-tooth harrow across the field at a perpendicular angle to the rows when the parsnips are young but well-established seedlings. The goal is to achieve an in-row spacing of anywhere from 6 to 18 in (15 to 46 cm) between plants, depending in part on how early you plant them (hence, how large they'll become in the first season) and in part on how tightly spaced you want them to be during the second reproductive phase of their life cycle. Traditionally, parsnips have been spaced 18 to 36 in (46 to 91 cm) apart in the row for the reproductive phase of their life cycle, but in recent years many growers have been spacing the roots somewhat closer, 12 to 18 in (30 to 46 cm) apart for flowering. So the grower can either plant and thin to this wide spacing during the first season or thin the crop to a closer 6 in (15 cm) spacing, allowing for a second thinning to every other or every third plant at the beginning of the second season. This allows the seed-to-seed grower to at least perform selection for the most vigorous, healthy parsnip plants coming out of winter. The spacing between rows can be anywhere from 24 to 48 in (61 to 122 cm) with the seed-to-seed method.

Root-to-Seed Method: The root-to-seed method is also performed for parsnips much as it is for carrots. With all root crops, the root-to-seed method has two basic versions:

1. The first version is to pull the roots in the fall and store them in a coldroom over winter for replanting them for seed production in the spring.
2. The second version is to overwinter the roots in situ, pulling them in the spring to replant into a seed production field.

Both of these versions of the root-to-seed method allow you to evaluate and select the best roots to replant for the seed crop. The great advantage of the cold hardiness of parsnip roots is that it is possible to use the second version of this method in a much wider range of climates than is possible with almost all other biennial root crops. Therefore, there are few growers who use the first version of the root-to-seed method, essentially choosing either the second version of this method or using the seed-to-seed method, both of which overwinter the roots in the soil as a form of storage. Because parsnip roots seem to overwinter successfully even in soils that freeze solid during the depths of winter, it is unclear what weather conditions will actually damage the apical growing point for the shoot, which is located in the crown of the root. In other root crops that can withstand a certain amount of freezing, it is sometimes the repeated freezing followed by thawing events that can damage this apical growing point; however, it has been hard to find any written reference or handed-down wisdom as to what specific set of challenges will significantly damage a parsnip root to the point where it won't regrow in the spring.

Upon replanting the roots in the spring for the root-to-seed method, the roots are often planted at anywhere between 18 and 30 in (46 to 76 cm) apart within the row, depending on the eventual full size of the variety and the between-row spacing. The between-row spacing in the root-to-seed that was once used between plants was as wide as 6 ft or 72 in (1.8m/183 cm). Today most growers use a spacing of 36 to 48 in (91 to 122 cm) between the parsnip rows.

Seed Harvest and Seed Cleaning

Most aspects of seed harvest and seed cleaning are virtually identical to carrot (see Apiaceae, "Seed Harvest" and "Seed Cleaning" on pages 81 and 82). The only significant difference is that parsnip seed will shatter from the umbels more readily than carrot seed when both are equally mature. Therefore, care must be taken to cut and harvest the seed in a very timely fashion. Growers will often windrow the plants onto landscape fabric that is porous to precipitation when curing the seed before threshing.

Caution: A word of caution when harvesting and handling parsnip plants, especially at the time of seed harvest. Parsnip stems and leaves contain a compound, furanocoumarin, within the sap that causes extreme rashes in many people. This rash is akin to what's experienced when people come into contact with poison ivy or poison oak. This chemical is photosensitive and may be intensified when the affected skin is also exposed to sunlight. There is evidence that furanocoumarin in the sap is more concentrated in the relevant tissue during the reproductive phase of the parsnip's life cycle. The resultant rash can often last for weeks, if not months. All workers handling parsnip plants at the time of seed harvest should wear long-sleeved shirts, long pants, and gloves and wash these items after use.

GENETIC MAINTENANCE

The characteristics that are important when maintaining parsnip varieties are much the same as many of the traits important for carrots (see Carrot, "Maintenance and Selection of Genetic Stocks" on page 92). Traits such as root shape, root smoothness, and root length, as well as foliar characteristics like top height, top size, erect growth habit, foliar color, and low incidence of disease, are important. The major root disease of parsnips, canker, if present in your area, can give you an opportunity to select roots that exhibit a lower incidence of this widespread malady. Another important trait to be sure to check for is root shape. Parsnips have a unique tapered shape that needs to maintained. Also important to monitor is the smoothness of the surface of the roots, as they can be rough due to large lateral root scars. Selection for smooth roots should occur every time roots are pulled and evaluated using the root-to-seed method to maintain a reasonable level of smoothness in any variety that you are producing.

ISOLATION DISTANCES

Parsnips are insect-pollinated, and like most crops in the Apiaceae they are very attractive to a wide diversity of pollinators. This is relevant, as all members of this family grown for seed in most temperate climate settings will attract insects from a broad perimeter around the field in which they are planted. This is especially true of organically grown seed crops on farms with crop species diversity. However, this also means that you will need to be careful in determining isolation distances if you plan to grow more than one parsnip variety for seed or if there is any wild parsnip in the area.

The standard minimum isolation distance of 1 mi (1.6 km) for insect-pollinated crops should be used between parsnip crops of the same crop type, when there are no physical barriers on the landscape. In cases where there are significant barriers on the landscape it is possible to diminish this isolation distance to 0.5 mi (0.8 km). The distinction of different crop types among parsnip varieties is minimal, as most commercial parsnip varieties at present are much alike in their phenotype. There are, however, some differences in the degree of taper and the length of the parsnip root between different varieties that may warrant increased isolation distance when two seed crops with different shapes are produced in the same area. The other major difference that can exist among parsnip varieties is in their disease-resistance profile. There has been some breeding work in recent years to incorporate canker resistance into a number of parsnip varieties. If seed of two parsnip varieties with differences in their levels of canker resistance or in their shape were going to be produced in the same district, then it would be wise to double the minimum isolation distance used between crops to 2 mi (3.2 km) without barriers and at least 1 mi (1.6 km) if there are significant barriers on the landscape to interrupt the movement of pollen.

– 7 –

ASTERACEAE

The Asteraceae is the largest family of flowering plants in the world. It includes over 1,500 genera and more than 23,000 species of woody and herbaceous plants. Considering the size of this family there are relatively few vegetable crops of major importance, specifically lettuce (*Lactuca sativa*), endive (*Cichorium endivia*), and the chicories (*C. intybus*). This family also includes sunflower, however, which is one of the five most important oilseed crops worldwide, and globe artichoke, which does not come true from seed and is usually propagated asexually. The Asteraceae also includes the minor vegetable crops burdock root, cardoon, Jerusalem artichoke, salsify, and scorzonera.

The Asteraceae are dry-seeded crops, and their seed is most successfully grown with little or no precipitation during the flowering and seed maturation process (see "Cool-Season Dry-Seeded Crops" on page 345). Almost all commercial seed of lettuce, endive, and the chicories is grown in temperate regions with dry summers and moderate temperatures, the so-called Mediterranean climate. However, small, localized seed production of regionally important varieties does occur with both the major and minor crops of this family. Decentralized seed production is quite common in types of chicory crops like radicchio and catalogna in parts of Southern Europe, where many farmers still maintain their own varieties with a healthy amount of variation and a high degree of adaptation to the local environment. In radicchio these farmer-bred varieties are prized and are often quite unique to a region.

FAMILY CHARACTERISTICS

Reproductive Biology

All species within this family have flowers borne in head-like flower clusters. Each head,

A lettuce capitulum appears to be a single flower, but is actually a cluster of florets.

121

◀ Lettuce flowers are borne in a corymb-like flower cluster, in which flower stalks along the stem of the plant all grow up to approximately the same height.

called a **capitulum,** appears to be a single flower but contains a group of single florets that are protected by a whorl of bracts, called an **involucre.** Lettuce, endive, and the various chicories are all members of the tribe Lactuceae and have florets with five fused petals and five fused anthers that form a column surrounding the pistil. A branching, corymb-type, racemose inflorescence will blossom indefinitely in all three of these species, usually until the end of the season in most temperate climates. Seed will mature sequentially, and—as in the case of lettuce—the seed is sometimes harvested multiple times during the season to optimize yield and quality.

Life Cycle

Lettuce and endive are annuals and can produce a seed crop in a single season. As with most annual seed crops, it is crucial to produce a vigorous frame of vegetative growth during spring to support a healthy, fully realized flowering plant in summer. These stages must happen in a timely fashion in an appropriate climate to produce an adequate yield of fully mature seed.

Some regions with appropriate climates in all other regards may simply have too short a growing season for some of the varieties of these annual species to mature sufficiently and make a good seed crop. Also, in both lettuce and endive there are varietal differences in the number of days it takes to mature a seed crop. Certainly, many lettuce and endive seed growers will stress how important it is to get these crops started and growing in the field as early as possible to realize a good yield. In most temperate climates appropriate for the production of seed of these crops, it

is important to first test all prospective seed crop varieties of these annual Asteraceae species in a small suitability test plot before committing a larger field to commercial seed production of a variety that may not mature in your region. While the indeterminate nature of the flowering in these crops means that you will never harvest the full complement of seed from any Asteraceae crop, you should expect to harvest at least 60 to 70% of the seed that matures on a fully flowering plant if the crop is to prove commercially viable in your particular environment.

The chicories (*Cichorium intybus*) are essentially biennials, requiring a period of vernalization, or cold treatment, to promote flowering (see chapter 3, Understanding Biennial Seed Crops).There are great differences in the vernalization requirement between different chicory crops and within the varieties of each crop type. When produced as vegetables there is always the danger of any of these crops bolting during the first season if planted early in the season, when there are plenty of cool nights with temperatures that routinely fall below 50°F (10°C). For seed production these crops should be planted or transplanted in midsummer when danger of premature bolting has passed. They will then produce a reserve of energy in their roots, petioles, and leaves, which is stored for winter and used in the next season after flower initiation for production of seed.

Climatic Adaptation

These three species of the Asteraceae are indigenous to the greater Mediterranean Basin and Middle East area. The ancestral form of lettuce most probably had a very short vegetative phase in early spring and flowered long before the onset of high temperatures in summer. The early forms of endive were probably winter annuals, producing a vegetative frame during the cool winter months and also completing their reproductive phase before the heat of summer. The biennial ancestors of the chicories took advantage of the mild winters in the area and put on most of their growth during the cool of winter, also flowering early enough to produce seed before the hottest part of the summer. The modern-day forms of these crops produce inferior seed in hot weather that exceeds 86°F (30°C), although some fluctuation above this can occur in later stages of development (after initial seed set and early rapid growth of the seed), if it is not excessive and sustained. These crops all excel in temperate climates with dry, warm summers and falls, mild winters, and cool, moist springs that encourage luxuriant foliar growth.

Seed production for these three species is generally best accomplished in warm temperate zones that are warmer than the cool coastal regions where the cool-season crops are grown, and they usually require a longer season to mature a full seed crop than many cool-season crops (see chapter 15, Seed Crop Climates).

Seed Harvest

Because the seed of lettuce, endive, and the chicories is borne on plants with indeterminate flowering, it is important to remember

A lettuce seed field.

that the seed crop for all of these species does not mature uniformly. Hence, they are all harvested when approximately 50 to 70% of their seedheads are mature, depending on the species, specific variety, and environment in which they are produced. Seed growers must be attentive to harvesting the seed when there is an optimum amount of mature seed before the first seed to set (which can often be the highest-quality seed of the lot) starts to shatter and is lost.

When the plants of these species have reached a maximum maturity they are usually cut and windrowed in the field, allowing the crop to dry and further mature in the sun. Threshing follows several days later, by either manual or mechanical means. Some seed growers avoid mechanical threshing, as immature seed (which isn't easily released from immature seedheads when the crop is threshed manually) will be mixed in with the fully mature seed using mechanical methods. Since much of this immature seed can be the same size and nearly the same weight as the mature seed, it is not easily separated from the mature seed in the cleaning process. This can mean the difference between a seed lot that germinates in the 80 to 90% range and one that germinates well into the 90s.

Mechanical harvesting of lettuce seed in one step is sometimes used in larger seed production operations. It is possible to combine the standing crop directly from the field if the crop is sufficiently dry. However, this method has two inherent flaws in the production of high-quality commercial seed. First, there can be considerable waste through shattering when the crop is cut by the moving combine. Second, this method bypasses the curing of the seed in a windrow that is usually done, as a significant fraction of "almost mature" seed at the initial time of harvest will mature during the curing process.

Seed Cleaning

Asteraceae seed should be spread out and dried quickly after harvest, as small pieces of stems, leaves, and flower parts can hold a significant amount of moisture in the threshed material. This should be done in a shaded area, since direct sun (and heat) can cause dormancy in these species.

The bulk of the cleaning is accomplished through some form of winnowing that can easily remove the lighter plant debris from the seed. Slotted screens are also useful for removing immature flower buds. A gravity table can be used to separate the light, less mature seed from the good seed.

GENETIC MAINTENANCE

Both lettuce and endive are self-pollinated species, so they are easily maintained as true breeding for most traits with simple roguing. The chicory crops are cross-pollinated and therefore require considerable attention to ensure that their inherent variability remains within the range that the producer deems allowable for that particular crop (see chapter 13, Isolation Distances for Maintaining Varietal Integrity).

The traits that are important for selection in all of these leafy crops include leaf shape, texture, color, color variation, and plant stature. Also universally important is selection for resistance to tipburn, premature bolting, lettuce mosaic virus (LMV), aster yellows, and sclerotinia (*Sclerotinia sclerotiorum*). The presence of sclerotinia in seed crops of these three species is quite prevalent and can be destructive. Several organic seed producers in the western United States have made considerable progress increasing the level of resistance in a number of lettuce varieties through selection for horizontal resistance (HR) to this fungal pathogen. Significant levels of resistance to tipburn and premature bolting, as well as HR for LMV, have also been achieved through careful selection during genetic maintenance of lettuce stocks. Consideration of leaf thickness, waxiness, and presence and degree of leaf pubescence are also traits to be monitored in the selection, especially in the chicories.

ISOLATION DISTANCES

The isolation distances for the self-pollinated crops in this family, lettuce and chicory, are frequently debated, as crossing can occur between different varieties in each of these two species. Recommended isolation for growing two lettuce varieties will usually specify a minimum of 6 to 12 ft (1.8 to 3.7 m) between varieties or whatever distance is necessary to avoid physical mixing of plant material at harvest. Endive's is often less specific, stating that it is "similar to lettuce." As with all cases of potential cross-pollination in self-pollinated species, the incidence of crossing will go up in certain environments and when the crop is grown organically, depending on the type of pollinators present (see chapter 13, Isolation Distances for Maintaining Varietal Integrity). While lettuce and endive growers in many climates report very little crossing at the standard isolation distances of 6 to 12 ft (1.8 to 3.7 m) between varieties, this is not recommended as standard practice, since some percentage of crossing (usually well below 1%) can occur in almost any climate and will probably be higher than that in some climates.

CHICORY

The name *chicory* distinguishes two types of crops belonging to the species *Cichorium intybus* L. The first type, the salad chicories, are all used as leafy vegetables and are classified into four major crop groups: (1) radicchio (Chioggia, Treviso, and Castelfranco types); (2) *pain de sucre* (sugarloaf chicory); (3) witloof chicory (Belgian endive); and (4) catalogna (Italian dandelion or asparagus chicory). The second type of chicory is known as root chicory, which produces a large fleshy root that is ground, roasted, and mixed with coffee. The salad chicories are grown to some extent in all temperate regions around the globe, producing their most favorable crops when their culture includes some seasonal cool conditions.

Both the salad chicories and chicory root are biennials, requiring a cold vernalization to produce flowers and subsequent seed.

The probable center of origin of this species stretches from the northern reaches of the Middle East to the southern Balkan Peninsula, from whence it eventually moved throughout the Mediterranean basin and northward across Europe. Over time and through selection in diverse agricultural settlements there developed a diversity of forms, giving us the rich number of forms within and across the four major crop types. With the spread of this species also came the spread of the wild form of *Cichorium intybus,* often just called wild chicory in English, which is a major weed of

SEED PRODUCTION PARAMETERS: CHICORY

Common names: radicchio, sugarloaf, witloof chicory, catalogna, chicory root
Crop species: *Cichorium intybus* L.
Life cycle: biennial
Mating system: cross-pollinated
Mode of pollination: insect
Favorable temperature range for pollination/seed formation: 60–74°F (16–23°C)
Seasonal reproductive cycle: summer (Year 1) through the following fall (Year 2) (~14 months)
Within-row spacing: 1–1.5 in (2.5–4 cm)
Between-row spacing: 22–30 in (56–76 cm)
Species that will readily cross with crop: Wild chicory (*C. intybus*) is a feral form of the species, and any wild populations will easily cross with the cultivated chicories.
Isolation distance between seed crops: 0.5–2 mi (0.8–3.2 km), depending on crop type and barriers present on the landscape

fencerows and roadsides in any number of temperate regions around the world.

Ethnobotanists speculate that *pain de sucre,* a rustic leafy crop that resembles romaine lettuce and is sometimes called sugarloaf chicory in English, may be the oldest cultivated vegetable in this species, though its cultivation is largely confined to the border regions of Southern Europe where France, Switzerland, and Italy meet. The first cultivation of radicchio, a group of chicories with red leaf pigmentation, may derive from the 'Rossa di Treviso' types grown in northern Italy in the 15th century. Different radicchio cultivar groups may have been derived from these types, including early- and late-maturing Treviso types, round red Chioggia types, and the largely green-leaved, Castelfranco types that are variegated with red spots.

Although it is still a mainstay of Italian cuisine, chicory's use has spread across Europe and beyond. There is speculation that the enlarged apical bud that is witloof chicory (commonly called Belgian endive in English) was first selected by Belgian farmers from overwintered chicory roots stored in pits for roasting in the 1870s. The tender white leaves from these blanched yellowish leaf buds that had enlarged during

▼ 'Catalogna a Foglie Frastagliata' is a Catalogna-type chicory, commonly known as "Italian dandelion" in North America for its deeply indented leaves and upright growth habit.

winter slowly became popular in Belgium as farmer breeders developed varieties that would consistently produce tight buds that suited the burgeoning demand for this off-season leafy crop. Catalogna is probably of Italian origin. In English it has historically been called asparagus chicory, but as its leaf shape is reminiscent of dandelion, it is often sold under the name *Italian dandelion* in North America.

These various leafy chicory crops are most widely grown in Europe, but with a few exceptions they have not gained wide acceptance in the rest of the world, occupying a specialty market status. The most notable exception to this has been the increase in use of radicchio and catalogna in cool-season premixed salads in industrial societies over the last decade.

Most of the world's chicory seed is produced in Europe, with France and Italy the most important production areas. There is also production of some of the hybrid witloof and radicchio seed in other parts of Europe. All of these crops require a vernalization period for flowering. The open-pollinated types are usually overwintered in the field in milder climatic zones to satisfy their vernalization requirement, whereas the parental stocks for hybrid production are often stored in controlled environments between the first and second seasons of their biennial life cycle and therefore have a wider geographic range for seed production.

CROP CHARACTERISTICS

Reproductive Biology

The morphology of all reproductive parts of chicory (*C. intybus*) is very similar to that described for endive (*C. endivia*) (see Endive & Escarole, "Reproductive Biology" on page 136), although there can be slight differences in length of the **peduncles** (flower-bearing axillary shoots) and in the number of flowers borne in each capitulum. The endives also seem to physically close their flower clusters earlier in the day than the chicories. Otherwise, the emergence of the stigma through the channel formed by the anthers is very similar to endive at anthesis. The bottom line is that the two species are sexually compatible with each other and therefore must be treated as if they were the same species, respecting isolation distances when producing seed of any member of the genus *Cichorium* (see "Isolation Distances," page 132).

The main reproductive difference between these two species is that *C. endivia* is highly self-fertile, and very little crossing between plants takes place, whereas *C. intybus* has a very strong sporophytic incompatibility, which limits self-fertilization to far less than 1% of opportunities to self under field conditions. In other words, endive seed is almost always the result of self-pollination, while chicory requires cross-pollination in order to produce seed, as an individual chicory plant will not receive its own pollen. Because the chicories must receive pollen from other individuals, they are much more apt to outcross than endive. For this reason the chicories must be isolated farther from endive than the reverse (see "Isolation Distances," page 132).

The second major difference between chicory and endive is their life cycle. The various endives are all annuals, while the chicories are biennial and may initiate flowering after some combination of vernalization and the lengthening hours of daylight in late spring. The vernalization requirement in chicory can vary widely, and certain crops and cultivars within *C. intybus* will often run to seed by midsummer if they are exposed to temperatures at or below 50°F (10°C) for extended periods of time during early stages of growth. As annuals, all endive crop types (*C. endivia*) will readily

Chicory flowers (*Cichorium intybus*) from a Castelfranco type (left) and from a radicchio (right).

flower during their first season of growth, regardless of environmental conditions.

Climatic and Geographic Suitability

The crops of the *Cichorium* genus are quite cold-hardy, easily overwintering in most parts of their center of origin (Mediterranean basin and southwestern Asia). There is ample genetic variability for tolerance to cold between crop types and specific varieties among the chicories, making them ideal for overwintering in many of the more moderate climatic regions of the temperate zone. Most chicories will come through winter temperatures between 20 and 25°F (–7 and –4°C) with little damage, especially if they are at a relatively early stage of growth. I have seen several different radicchio varieties with young plants bearing 8 to 10 sets of true leaves that had a greater than 50% survival rate when the winter low temperature reached 15°F (–9°C), without any thermal protection or snow cover.

Like endive, the chicories are best suited to temperatures between 58 and 68°F (14 to 20°C) for optimal vegetative growth. During the reproductive phases of flowering and seed growth they flourish at higher temperatures, but temperatures shouldn't regularly exceed 86°F (30°C) for high-quality seed. Cool springs that encourage good vegetative growth followed by seasonally dry summer weather are desirable in order to attain adequate yields and harvest clean, disease-free seed. The growing season is somewhat longer than that needed for many of the cool-season brassica and chenopod seed crops. Much of the commercial seed crop is produced in the Netherlands, Belgium, northern France, and northern Italy. The eastern part of the Po River Valley in Italy is an important seed production region, especially in the regions of Emilia-Romagna and the Marche. In fact, at the time of this writing, there are still many farmers in these regions that maintain open-pollinated family strains. In North America the Willamette Valley of Oregon and a number of interior valleys of Northern California have proven to be well suited to producing high yields of high-quality chicory seed.

SEED PRODUCTION PRACTICES

Soil and Fertility Requirements

Chicories can be grown on a wide range of soils for seed production; however, soils must be well drained to minimize the chances of a number of diseases becoming a problem. Chicory does well on lighter sandy or silt loams as a vegetable crop. The plants have rather shallow lateral feeder roots and require a steady supply of water and nutrients throughout the season. When growing chicory as a seed crop many growers prefer somewhat heavier soils, like a clay loam or silty clay loam that can more easily deliver nutrients and water to the crop over

the course of the much longer season that is required to produce seed. These soils can be especially important for their ability to hold water during hot weather in late summer as the chicory seed crop is maturing.

Adequate soil fertility that is sufficient to produce vigorous spring growth to establish a good frame or basal rosette of leaves to support a bountiful seed crop is mandatory. Soils rich in available phosphorus are desirable, since phosphorus is essential for early plant development. This is especially important in areas with cool springs, as the mineralization of phosphorus in soils that are managed organically is slow in cool weather. Nitrogen levels should be moderate, as excessive foliar growth can be detrimental to producing a stout and sturdy flowering plant that produces a satisfactory seed yield.

Growing the Seed Crop

Unlike their close relatives endive and escarole (*C. endivia*), the chicories (*C. intybus*) are all true biennials and must be overwintered at temperatures at or below 50°F (10°C). With the exception of the witloof chicory and hybrid radicchio produced in Northern Europe, the chicories are overwintered in the field. As most chicories will easily withstand temperatures between 20 and 23°F (−7 and −5°C) and often colder, with little damage at the peak of the winter cold, they have a somewhat greater geographic range for seed production than many other biennial crops.

In the first year of the 2-year cycle for seed production it is important to determine what the best size is for overwintering the particular chicory type and variety that you intend to grow in your climate. Many of the chicories that form heads, such as the radicchios and *pain de sucre* types, are more cold-hardy before they have formed a full head. It is important to get these and other chicories large enough in fall so that they will be able to form a substantial frame of a plant in spring, yet not too large going into winter to suffer undo freezing damage. This size depends on both the variety and the environment, and seed growers in colder winter areas will have to experiment in their own environment with the different varieties to find which chicory type or variety is able to successfully overwinter and produce seed.

Generally the chicories are sown into flats 3 to 5 weeks before the date when they will be transplanted into the field in mid- to late summer. Plants can start to be selected for color, leaf type, and vigor when they have four to six true leaves. Seedlings are planted 14 to 20 in (36 to 51 cm) apart within the row depending on the size of the flowering plant, which can vary considerably. Spacing is between 24 and 36 in (61 to 91 cm) between rows.

Seed Harvest

Seed harvest for the chicories is done successfully much like the single harvest method described for lettuce (see Lettuce, "Seed Harvest" on page 145). The chicories, however, require somewhat more heat and a longer season to mature their seed than is needed for many of the lettuce varieties. As with lettuce, the seed of the chicories can easily shatter if the capitula are ripe and excessively dry at the time of the pulling of the plants, the laying into windrows, or the final threshing. For this reason it is important to closely monitor the maturation of these seed capsules and pull the plants at exactly the right time.

GENETIC MAINTENANCE

Selection for the maintenance or improvement of several key characteristics can be done at different stages in the various chicory crops. This is based on selecting for: (1) horticultural characteristics, including form, shape, color, or flavor; (2) agronomic traits such as growth rate, maturity, or disease resistance; and (3) the plant's suitability to the seasons of production. As with all leafy green vegetable crops, it is possible to select at various stages of the growth cycle for color, leaf shape, leaf curl, and overall vigor of the plant. This is especially true among the open-pollinated radicchio types, *pain de sucre,* and catalogna types, which are often still quite genetically variable for both the horticultural and agronomic traits listed above. This makes sense in that these types come from a large and very diverse genetic pool in and near their ancestral center of diversity—the Mediterranean basin. On the other hand, the witloof or Belgian endive types were derived from a much narrower genetic base, as this species was transported to Northern Europe. Indeed, this latter type has been intensively selected and is most often sold commercially as hybrid varieties.

Selection for the color, degree of variegation of leaves, and sometimes flavor can be done early in the life cycle once you learn what constitutes the typical version of these traits for any particular variety. The most successful selection for form and shape of heading versions of chicory crops is done when the crops have formed heads. Flavor can be strongly influenced by temperature or by challenges of the environment.

This species also has a remarkable ability to put on vigorous growth during cool conditions when temperatures don't exceed 65°F (18°C), even putting on growth at or below 55°F (13°C). Because chicory is a biennial that is often overwintered in the open in the field (with the exception of witloof types), most seed crops will easily withstand temperatures between 20 and 23°F (−7 and −5°C) with little damage at the peak of the winter cold (see "Growing the Seed Crop," page 130). In some cases, based on the size and stage of development, there are some varieties among the different chicory crop types that can survive temperatures down to 14°F (–10°C) in the open field. Depending on the specific crop type and variety it is possible to select for increased cold tolerance at winter temperatures between 14 and 20°F (–10 to –7°C), through a combination of natural selection (plants with apical buds destroyed) and conscious selection (plants that you eliminate due to worse-than-average frost damage), thereby increasing the survivability of the crop in a cold climate over a number of cycles.

As with all biennial vegetable crops there is almost always new vegetative growth in the spring to support the coming reproductive growth of the second year. Many growers have noted the great variation in the vigor, rate of growth, and extent of growth that occurs from one chicory variety to another at this point. This makes it possible to select for the most vigorous or bountiful types at this point in the season; this has obvious advantages to a vegetable grower who is harvesting an early-spring crop, but this additional growth will also give the seed grower an improved frame for the seed-producing plant that can increase the potential seed yield.

ISOLATION DISTANCES

The first hurdle you must get over before growing seed of the chicories is to become comfortable distinguishing the chicories (*C. intybus*) from the closely related endive or escarole types (*C. endivia*). As these two species will only rarely cross, the important reason to be able to recognize their differences is so you will know when the recommended isolation distances are necessary and when they are not. Both have very similar blue-petaled, typical Asteraceae flowers that are hard to tell apart from one another unless you are quite familiar with the *Cichorium* genus. Even when you become quite familiar with the genus, the easiest way to distinguish between these two species is to look at the phenotypic differences between plant types of an endive/escarole and the various types of chicory, since there really are no chicories with the more lettuce-like leaves of endive or escarole. All seed growers who endeavor to grow either of these species need to be comfortable detecting these distinctions, as it can mean the difference between growing a pure chicory seed crop or one that could have considerable outcrosses in it. In other words, make sure that you definitively identify the species of any *Cichorium* crop growing anywhere in the vicinity of your seed crop. Because *C. intybus* and *C. endivia* are not usually sexually compatible, there is little concern with growing seed of these two species in close proximity. In experiments conducted in the Sacramento Valley of California in the 1950s, Charlie Rick found that crosses between these two species could occur at a very low rate but that it is difficult to produce a cross even when one is desired. Therefore, if there are in fact two crops, but they are from these different species, then using the basic isolation distances given for the self-pollinated endive/escarole crops should suffice to minimize the chance for any crossing between these two crop species (see Endive & Escarole, "Isolation Distances" on page 138).

If you are growing seed of one of the chicories (*C. intybus*) in a region where there are other chicory seed crops, then it is important to follow the general rule for isolating cross-pollinated crops that are primarily insect-pollinated. For commercial seed crops, if you have two chicories of the same crop type, the same form, the same maturity class, and the same color, then they need only be isolated by 1 mi (1.6 km) if grown in open terrain, or 0.5 mi

(0.8 km) if there are substantial physical barriers between the crops (see chapter 13, Isolation Distances for Maintaining Varietal Integrity). An example of this among the chicories would be two red, round-headed, Chioggia-style radicchio varieties. However, even if you were growing two red radicchio varieties, but one was a more pointed, taller Treviso type and the other a Chioggia type, then the isolation would need to be increased, just as it would if you were growing two entirely different chicory varieties, as described below.

The isolation distance definitely needs to be increased to 2 mi (3.2 km) in open terrain if the second chicory seed crop is an entirely different crop type—for example, a radicchio and a witloof chicory type—or even if it is two different red, round-headed Chioggia types that are in two different maturity classes. This increased isolation requirement between two crops of the same species, but different types, can be reduced to 1 mi (1.6 km) if there are substantial physical barriers between the two crops.

ENDIVE & ESCAROLE

Endive is the catchall name frequently used in English for the loose-headed leafy salad crops in the species *Cichorium endivia* L. This includes the various crop types listed as endive, escarole, or frisée in many seed catalogs. All of these crops have distinctly ruffled to curled, pale green to yellow leaves with serrated edges. In the vegetative stage the outer leaves of the basal rosette are tougher and usually bitter, while the inner leaves often fold into one another as the plant matures and have a milder flavor and a lighter green, yellow, or cream color due to the partial blanching of this "heart." Escarole varieties have broader, somewhat flatter leaves than the true endive type and are sometimes confused with green leaf lettuce by both the vegetable buying public and grocers. Endive has narrower, more curled leaves and includes the super-curled frisée varieties that are often used in salad mixes. Endive types are sometimes referred to as curly endive; escarole is also known as Batavian endive or simply Batavian in order to distinguish between these two types in the marketplace. It is important to understand that the degree of leaf curl is essentially all that distinguishes these two crop types from each other, and that they are fully sexually compatible.

SEED PRODUCTION PRACTICES: ENDIVE & ESCAROLE

Common names: endive, escarole, frisée
Crop species: *Cichorium endivia* L.
Life cycle: weak biennial
Mating system: highly self-pollinated
Mode of pollination: perfect flower; no need to stimulate pollen shed
Favorable temperature range for pollination/seed formation: 60–74°F (16–23°C)
Seasonal reproductive cycle: late spring through late summer or fall (4–5 months)
Within-row spacing: 1–1.5 in (2.5–4 cm)
Between-row spacing: 22–30 in (56–76 cm)
Species that will readily cross with crop: Endive or escarole can cross with both wild and cultivated forms of chicory (*C. intybus*), but only at a very low rate— endive is highly self-pollinated, and this would be an interspecific cross with a high rate of failure even when all other environmental conditions are favorable.
Isolation distance between seed crops: 10–50 ft (3–15 m), depending on the crop type and barriers that may be present on the landscape

'Eros' escarole (*Cichorium endivia*).

Endive's ancestral origin is possibly from an area spanning the eastern Mediterranean basin and extending into the Middle East and southwestern Asia. There are several wild *Cichorium* species that shared this area with both *C. endivia* and *C. intybus* during the period of their domestication, and there are ongoing molecular taxonomic studies to try to clarify the relationships between them and their wild relatives. Unlike chicory (*C. intybus*), which has a weedy form that has followed humans to many temperate climates across the globe, endive is almost always only found as a cultivated crop (though occasionally feral populations are found in Italy). There has long been a theory that endive may have originated from a cross between chicory, which is undoubtedly the older cultivated form, and one of the wild *Cichorium* species. Endive's origins as a crop may trace back to ancient Egypt. It was probably used as both a medicinal and a food plant in ancient Greece and in the Roman Empire. Endive spread to Northern Europe and was well established when it was described in the 16th century in England. Endive has long been used as both a cooked vegetable and a salad item. Its use is widespread across Europe and southwestern Asia.

The recent growth in endive consumption in North America, Australia, and South Africa is in part due to the boom of prepared salads. The fact that endive is so well suited to fall and winter production in the open in mild-winter areas of temperate climate zones bodes well for it given the growth of interest in year-round local food production and regionally supported markets.

Most of the world's endive seed production is done in Europe, with a majority of it coming from the Netherlands, northern France, and a number of districts in Italy. It is a cool-season, dry-seeded species that needs more heat for ample seed production than what may be ideal for production of the market vegetable forms of the crop.

CROP CHARACTERISTICS

Reproductive Biology

Endive (*C. endivia*) has an annual life cycle, and it is highly self-pollinating in its reproductive habit. These two traits distinguish it from chicory (*C. intybus*), which is a biennial and is highly cross-pollinated. Endive flowers by elongation of the main stem and produces coarse, thick branches. Clusters of four to six sessile inflorescences are formed at the base of each branch, at each node, and at the end of each branch. Each inflorescence is made up of 15 to 25 single perfect flowers that are borne in an involucre of overlapping bracts, which functions much like the calyx of a simple flower. Each flower has five fused stamens that form a column enclosing the pistil. At maturity the stigma grows up through this column, coming in contact with the anthers and becoming covered with pollen. The ray florets form a ligulate corolla of pale to grayish blue (infrequently mauve) color that remains relatively closed through the time of receptivity. Endive is fully self-compatible, with a very high proportion of selfed seed, even in the presence of other *C. endivia* or *C. intybus* plants and insect pollinators. The combination of this self-compatibility and the semi-closed inflorescences results in very low rates of natural cross-pollination, probably far below 1% in most situations. However, cross-pollination can occur at higher rates in the right environmental conditions or the presence of certain species of pollinators, and adequate precautions should always be taken when growing a seed crop.

Climatic and Geographic Suitability

The crops of the *Cichorium* genus are quite cold-hardy, easily overwintering in most parts of their centers of origin (the Mediterranean basin and Southwest Asia). Genetic variability for tolerance to cold exists between crop types and specific varieties, but endive will usually come through winter temperatures between 20 and 25°F (–7 and –4°C) with little damage, and selected endive populations are able to survive winter lows of 17°F (–8°C). Endive is best suited to temperatures between 58 and 68°F (14 and 20°C) for optimal vegetative growth. During the reproductive phases of flowering and seed growth endive flourishes at higher temperatures, but they shouldn't regularly exceed 86°F (30°C) for high-quality seed. Cool springs that encourage good vegetative growth followed by seasonally dry summer conditions are desirable in order to obtain adequate yields and to harvest clean, disease-free seed. The growing season should be somewhat longer than that needed for many of the cool-season brassica and chenopod seed crops. Much of the commercial seed crop is produced in the Netherlands, northern France, and northern Italy. In North America the Willamette Valley of Oregon and a number of interior valleys of Northern California have proven to be well suited to producing high yields of high-quality endive seed.

Many seed growers have good success planting endive in the fall in order to establish a healthy plant that can take full advantage of all favorable spring weather, even during unsettled weather that makes it difficult to spring-plant. This ensures that the plants will have an adequate frame to produce a good seed set. Much of the commercial endive seed crop is planted in the early spring and must be given plenty of time to achieve a frame before the daylength and relative maturity of the plant promotes flowering.

Most endive varieties are fully annual and appear not to need a vernalization or cold treatment to initiate flowering. A number of authors claim that there are biennial endives. If a grower plants an endive variety in spring and it has little or no bolting, or if it has erratic bolting throughout the population, then it may be biennial, and it should be planted in the fall as described above. This should supply enough cold treatment in most temperate climates to fully promote flowering in any variety exhibiting biennial tendencies.

SEED PRODUCTION PRACTICES

Soil and Fertility Requirements

Endive seed production is suited to a range of soils that are well drained, yet have good water-holding capacity and moderate fertility. Soils with high levels of nitrogen will cause excessive vegetative growth and poor root development.

Growing the Seed Crop

Endive is more tolerant of hot temperatures than chicory or lettuce, but for seed production it prefers moderate temperatures that don't regularly exceed 86°F (30°C), especially during flowering and early, rapid endosperm development. In climates where temperatures regularly rise above this after the summer solstice, it is best to fall-plant the crop so that the majority of flowering will be completed by late spring the following year, before the heat of the summer.

Endive can be transplanted, with the advantages including an early organic weed control (cultivation) of the field and the opportunity to plant older plants at an early date to maximize mature yield. You can also perform an initial selection for type in the greenhouse. However, many growers believe that by direct-seeding *Cichorium* crops into the field, the resulting plants will have more fully developed root systems, which will ultimately result in healthier plants with higher seed yields.

Cichorium crops can be treated much like lettuce for seed production, though they are decidedly more robust and often taller in stature than lettuce when flowering. Endive can be planted at a spacing of 24 to 32 in (61 to 81 cm) between rows, with a final within-row spacing of 14 to 20 in (36 to 51 cm). Drilling the seed at six to eight seeds per foot (30 cm) and blocking with a sharp precision hoe to the width of the desired spacing will help ensure an adequate stand and allow you to select the most vigorous, healthy seedlings in the field. A second pass across the field on foot will be required several days later to eliminate any late-emerging seedlings and check for any doubles—two plants at or near the same spot; these won't produce satisfactory plants.

Weed control using standard mechanical cultivation practices can be employed, being careful not to injure the root system. *Cichorium* crops should not be hilled during cultivation. In arid seed production areas irrigation is indispensable. It is important to ensure the emergence of a uniform stand and to produce continuous, rapid growth, building a good plant frame for the reproductive phase of the life cycle, which in turn will produce good seed yields. However, overhead irrigation must be cut off from the seed crop once flowers emerge on the elongated flower stalk, as it can interfere with pollination and promote disease.

Seed Harvest

Proper timing of endive seed harvest is important to produce high-quality, mature seed without suffering any losses due to seed shattering. Seed of the *Cichorium* species is borne in fruits that are shaped like capsules or small barrels and can shatter relatively easily when fully ripe. As endive has an indeterminate flowering habit, it will often continue to set new flowers from onset of flowering until harvest. This means that you must make a decision on the time of harvest based on when the maximum quantity of fully mature seed is available, before an appreciable amount of mature seed has shattered. The most seed that you can hope to harvest from any one plant would be somewhere in the 60 to 80% range if you are growing the endive crop in a favorable climate. Having a uniformity of maturation between endive plants of a particular variety is common among endive varieties as they are largely self-pollinated and are thus often narrowly selected in the breeding process. This uniformity is important to ensure a crop of high-quality seed.

If seed is at an optimum maturity then you may not need to windrow plants (or can windrow for only a short time) before threshing. This is important to guard against shattering. Also, there exists some variation between different varieties and types in all *Cichorium* crops in terms of ease or reluctance in shattering, and this must be taken into consideration when harvesting their seed. For this reason plants are best cut for harvest early in the morning when the moisture from the morning dew decreases the rate of shattering. Conversely, threshing may be adequate to dislodge most of the seed from the capsules, but a percentage of these capsules that remain intact may require abrasion to break them apart. This can be accomplished through the use of a belt thresher.

GENETIC MAINTENANCE

Endive and escarole varieties are largely self-pollinated, and while they can cross with other members of *C. endivia*, these rare crosses can be largely avoided by following the isolation requirements specified below (see "Isolation Distances," below). Crosses can also occur with the chicories (*C. intybus*), but at a rate of less than 1% according to an elegant experiment conducted by Charlie Rick at the University of California–Davis in the 1950s.

Variant forms in terms of leaf shape, color, and stature will occur naturally in endive and escarole, and these plants should be rogued out. The leaf shape, degree of curl, and form and distance between lobes should be monitored. Selection for intensity of color is important in named varieties, and some varieties will also have a distinct glossiness, which may vary. Selection is possible in an early rosette stage and is advisable at the market maturity stage. Tipburn is common in endive, and mass selection to eliminate the worst individuals with this malady has proven successful in lessening its severity over several generations. Selection should also be done to eliminate early bolters in almost all varieties.

ISOLATION DISTANCES

As *C. endivia* is highly self-pollinated there are undoubtedly crossing events that occur in the presence of insects in the biologically diverse habitats found on many organic farms. The flowers remain closed until pollination and fertilization have been completed in almost all cases, although their form and color are much like those of the chicories, and they are visited by various species of bees and other pollinators. Precautions and recommendations given for lettuce are applicable for endive (see Lettuce, "Isolation Distances" on page 151). The minimum isolation distance between different crops of *C. endivia* should be 150 ft (46 m) in open terrain with no barriers but can be dropped to 50 ft (15 m) if there is a sufficient naturally occurring barrier between crops (see chapter 13, Isolation Distances for Maintaining Varietal Integrity, "Physical Barriers" on page 337).

LETTUCE

Lettuce (*Lactuca sativa*) is the most important salad vegetable. Its succulent foliage is used as the base for salads in many cultures around the world. Lettuce consumption has steadily increased with the advent of refrigerated shipping after World War II, slowly replacing other lesser-known salad crops in affluent societies.

Lettuce is an annual crop with leaves borne as a basal rosette with a wide range of leaf types and colors. It is indigenous to the Mediterranean basin, where there is evidence that the Egyptians used lettuce more than 5,000 years BCE. Several species of *Lactuca* exist in the wild, but *L. sativa* is not found outside of cultivation. Lettuce is probably derived from wild lettuce (*L. serriola*), a weedy species that grows in temperate zones around

the world and that is fully interfertile with the cultivated crop. The ancestral form of lettuce was undoubtedly much like wild lettuce, with narrow spiny leaves, prominent midribs, bitter flavor, an abundance of latex sap, and a tendency toward early flowering (bolting).

Lettuce has six predominant morphological types:

1. **Looseleaf** types have a dense rosette of leaves that can be smooth, savoyed, or ruffled and are arranged in a loose configuration; they have little or no compact heart.
2. **Crisphead** types have a majority of leaves that form in a tight, overlapping fashion, resulting in a spherical, cabbage-like

SEED PRODUCTION PARAMETERS: LETTUCE

Common names: lettuce, celtuce
Crop species: *Lactuca sativa* L.
Life cycle: annual
Mating system: highly self-pollinated
Mode of pollination: closed perfect flower; no need to stimulate pollen shed
Favorable temperature range for pollination/seed formation: 60–74°F (16–23°C)
Seasonal reproductive cycle: late spring through late summer or fall (4–5 months)
Within-row spacing: 1–1.5 in (2.5–4 cm)
Between-row spacing: 22–30 in (56–76 cm)
Species that will readily cross with crop: Wild lettuce (*L. serriola*) will cross with cultivated lettuce when in close proximity with lettuce seed crops.
Isolation distance between seed crops: 10–50 ft (3–15 m), depending on the crop type and barriers that may be present on the landscape

head with relatively few loose outer leaves. Crispheads are often erroneously called 'Iceberg' lettuce, which was an early successful variety of this type.

3. **Butterheads** also form a head, although it is less compact and smaller than the crisphead varieties. They also have broad, tender leaves that have a buttery texture and flavor and are easily bruised with handling and transporting.

4. **Romaine** or **cos** types have upright, oblong clusters of coarse, thick leaves with large midribs that overlap into loose head-like structures.

5. **Celtuce** types, also called stem lettuce or asparagus lettuce, are grown for their thick, erect stem, which is not unlike the bolting, flowering shoot of other lettuce types. This stem is peeled, and the succulent tender core is used as both a raw and cooked vegetable throughout China and in Egypt but is uncommon elsewhere.

6. **Latin** lettuce types have elongated leaves similar to butterhead types in their texture and may form loose, semi-closed heads. They are grown primarily in the Mediterranean region and to some extent in Argentina and Chile.

Seed production of this dry-seeded crop is best accomplished in a Mediterranean climate with a seasonal dry period for seed maturation and harvest preferred for high-quality seed. While the vegetable crop is thought of as a cool-season crop, lettuce requires warmer temperatures than many of the other cool-season, dry-seeded crops to fully realize its potential as a seed crop. For this reason lettuce seed is grown in the warmer reaches of climates where the seed of cool-season crops is traditionally grown, in places like the Mediterranean basin and the coastal valleys of the Pacific coast of North America. Conversely, excessively high

temperatures during the reproductive stages of lettuce's growth cycle may also cause lower seed yields and poorer seed quality.

CROP CHARACTERISTICS

Reproductive Biology

Lettuce is a self-pollinating annual that produces a dense rosette of leaves early in the season, followed by flower stalk initiation, whereby the central cylindrical stem elongates, and indeterminate flowering may last for up 2 months. The flowers are borne in a fashion characteristic of most species of the Asteraceae (as described under the Asteraceae family, "Reproductive Biology"). In lettuce the 10 to 25 florets in each capitulum are all simultaneously receptive to pollination for only a few hours on the same morning, longer if the weather is cool and cloudy. Each floret produces an ovary that produces only a single seed if successful. If the weather is excessively hot, any capitula sexually maturing on that particular day may abort all their ovules. Fortunately, lettuce plants have indeterminate flowering and can continuously produce new racemes from the main stem of the plant for up to 2 months (or longer in some favorable climates). The structure of the lettuce flower promotes self-pollination, with the style pushing pollen out of the anthers as it emerges through the pollen tube. This usually results in a self-fertilization, but the pollen is then also available to pollinating insects, and cross-pollination can happen during their pollen gathering. Sticky and relatively heavy, the pollen is not windblown.

Climatic and Geographic Suitability

Lettuces are grown across temperate climates worldwide. Moderate temperatures between

◀ A closed, developing lettuce seedhead (top) and an open, fully mature seedhead (below).

68°F (20°C) and 86°F (30°C) are best for production of both the vegetative and reproductive stages of lettuce. Many non-heading types have greater tolerance to temperature extremes outside of this range. Lettuce seed production is ideally suited to regions with a Mediterranean seasonal dry period during seed maturation and harvest, as rain can cause shattering of mature seedheads, sprouting of seed in the seedheads, and discoloration and diseases of the seed.

Higher temperatures and the longer daylengths of summer often accelerate flower initiation and bolting in many lettuce types. Celtuce and heirloom varieties are usually the first to bolt, therefore making it easy to mature a seed crop in most suitable temperate areas. Conversely, many modern crisphead, butterhead, and cos types have been bred to be daylength-neutral and are generally the most bolt-hardy lettuces, which can be problematic when trying to mature a seed crop in areas with shorter growing seasons.

Geographically, you must be sure that wild lettuce (*Lactuca serriola*) is not growing in your production area. This possible progenitor species of our cultivated crop has spread across the globe in temperate climates, and while it isn't often found flowering in open, cultivated ground, it can be found growing in less disturbed ground near the fencerows and on the edges of fields, successfully reproducing. All lettuce seed growers should familiarize themselves with the characteristic leaf shape and the profile of the flowering plant of wild lettuce so they can increase their chances of spotting this rather unassuming weed and eradicate it within a minimum of 500 ft (152 m) from any production field.

SEED PRODUCTION PRACTICES

Soil and Fertility Requirements

Lettuce can be grown on a wide range of soils for seed production, but soils must be well drained to minimize the chances of a number of lettuce diseases becoming a problem. Lettuce does well on lighter sandy or silt loams as a vegetable crop. It has rather shallow lateral feeder roots and requires a steady supply of water and nutrients throughout the season. When growing lettuce as a seed crop, many growers prefer somewhat heavier soils, like a clay loam or silt-clay loam that can more easily deliver nutrients and water to the crop over the course of the much longer season that is required to produce seed. These soils can be especially important for their ability to hold water during hot weather in late summer as the lettuce seed crop is maturing.

Adequate soil fertility that is sufficient to produce vigorous spring growth, to establish a good frame or basal rosette of leaves and support a bountiful seed crop, is mandatory. Soils that are rich in available phosphorus are desirable as P is essential for early plant development. This is especially important in areas with cool springs, as the mineralization of P in soils managed organically is low in cool weather. Nitrogen levels should be moderate; excessive foliar growth can be detrimental to producing a stout and sturdy flowering plant with a satisfactory seed yield.

Growing the Seed Crop

Lettuce requires a long growing season to mature a seed crop. For this reason it is imperative to get the crop in the field and growing vigorously as early as your local climate will allow. The lettuce seed crop can either be direct-seeded into the field or planted from transplants. Producing the crop from transplanting has several advantages: First, the crop can be given 3 to 4 weeks of time to develop in the greenhouse or cold frame, when field temperatures are still too cool for active growth of the young plant. When the transplants are well past the seedling stage and able to put on rapid growth they are put into the field—as long as the spring weather has warmed and conditions are advantageous to their growth. Second, young lettuce plants with a minimum of four to six true leaves have several characteristics that are very representative of the mature plant, and this gives you an opportunity to make an initial selection to type. The color and any colored spotting that may exist in the variety is usually clearly evident in the seedling tray at this point; any off-types for these traits that may exist in a particular seed lot for a specific variety can be identified in the greenhouse and discarded before the variety is transplanted. The third advantage of transplanting the crop is cultural, as you can avoid the blocking and thinning as well as early weeding of the direct-seeded crop when the lettuce seedlings are young and have to compete with the first flush of weeds. Transplanting eliminates much of this intense hand work, which is often done on hands and knees; for that reason alone transplanting has become a standard practice for lettuce, endive, and chicory crops (and other crops with very small seedlings) for many organic farmers. It should be noted, though, that direct seeding of these crops does have its advocates. Some farmers will argue passionately that direct-seeded crops always outperform transplants as they will establish better root systems, both primary and secondary roots, by growing through their entire life cycle without being disrupted.

Spacing: Whether you're direct seeding or transplanting, the final spacing of lettuce plants for seed production should be 12 to 16 in (30 to 41 cm) within the row and 18 to 30 in (46 to 76 cm) between rows. It is important for anyone growing lettuce seed organically to follow these guidelines and not be tempted to plant any closer, as the airflow through the standing crop is important to discourage the spread of foliar diseases. Cultivation should be shallow near the plants, as the lateral feeder roots lie close to the soil's surface and are easily disturbed.

Irrigation: Lettuce requires a steady, relatively large quantity of water to produce optimum seed yields, because its root system isn't very extensive. In conventional systems in the recent past this has often been accomplished through the use of overhead sprinklers during the vegetative stages of growth and then switching to furrow irrigation during flowering and seed maturation. Many organic growers are now using drip irrigation systems to good effect, but some still rely on overhead irrigation during the early stages of growth to establish a large, vigorous frame on their plants and ensure good yields, before switching to drip for the reproductive stage of the crop's growth.

Maturing the Seed Crop: There is considerable variation in the amount of time that it takes for lettuce to bolt, and in many regions a significant number of varieties will not fully

mature a seed crop before the season ends. It is wise for anyone new to growing lettuce seed to plant a number of prospective lettuce varieties as early as possible in the season to determine which will bolt and mature an appreciable seed crop in their climate. The longest-standing or slowest-bolting varieties are the crisphead types. They are the slowest to develop their characteristic heads of folded leaves and almost always are the slowest to bolt. Indeed, many crisphead varieties cannot be grown for seed in many of the shorter-season areas that can grow seed of other varieties. The next longest-standing types are many of the larger modern cos or romaine types, though some of the smaller varieties in this class can bolt much faster. The third longest-standing lettuce type is the newer, more highly bred butterhead from Europe, as older butterheads are usually faster to bolt. The looseleaf types are generally faster bolting than the first three, but there has been a lot more breeding of these types since the 1980s and many newer varieties are much slower to bolt than the older looseleaf varieties. The consistently fastest-bolting lettuces are the celtuce varieties. As the elongated stem of celtuce is the culinary item of interest, flower stem initiation has been selected for by the farmers who cultivated this crop for many centuries, causing it to occur early in the life cycle of the plant. Farmers growing a lettuce seed crop must determine which type of lettuce and which varieties within that type are suitable for seed production in their region.

Emergence of the Seed Stalk: In crisphead lettuces the seed stalk may not easily emerge, as the folded leaves forming the head are too firm and tight. This can result in delayed and distorted seed stalk development or elongated and possibly damaged flower shoots. Much modern conventional crisphead seed production involves applying synthetic growth regulators like gibberellic acid to the plant before it forms a head to promote early flowering. This is not an option for organic farmers and it is undesirable, as it doesn't allow you to evaluate the crop for its trueness-to-type as a particular crisphead variety. To assist the emerging seed stalk three tried-and-true methods do not involve applying growth regulators:

1. Lancing the head with a sharp knife in a broad X cut on top, sometimes called racing.
2. De-heading, which involves stripping away outer leaves of the head to the point where the seed stalk can easily push through the innermost leaves of the heart and emerge.
3. Giving the top of the head a sharp blow with the palm of your hand and thus cracking a number of the crisp folded leaves to the point that they will easily break away as the seed stalk attempts to push out into the open.

Each of these methods requires practice and is best learned from someone with experience. De-heading is time consuming and normally only practiced in breeding nurseries with relatively few plants.

Seed Harvest

As individual lettuce seedheads mature, the calyx of the flower expands and dries, forming a parachute-like structure called a **pappus** that aids in wind dispersal of the seed—much like lettuce's relatives the dandelions (*Taraxacum* spp.). The full expansion of the pappus is called feathering and signals the maturation of the seedhead. Lettuce has an indeterminate flowering habit, and as with other crops that mature seed sequentially, there is always the

risk that the earliest-setting seed will mature and shatter before later-setting seed fully matures. In lettuce grown as a seed crop in a favorable climate, the earliest-setting seed often has a higher germination rate, presumably because it develops before the hotter weather of late summer, which does not favor the formation of large, well-developed embryos. Therefore, both of the seed harvesting methods described here emphasize a timely harvest to get the first flush of seed.

A mature and feathered lettuce seedhead. ▶

The multiple-harvest method, as practiced on the small scale, has its pros and cons.

Multiple Harvest: Hand-harvesting lettuce seed is still widely practiced and lauded for its ability to produce the highest-quality seed. The reason for this is that only seed that is fully mature is easily shaken from the plants; therefore no seed that has not fully formed will easily dislodge from the plant using this method. While hand-harvesting varies from region to region based on climate and preferred practices, the basic steps are as follows. When between 35 and 50% of the seedheads have feathered, harvest crews with cloth sacks go through the field shaking the flowering portions of each plant vigorously into the sack, being careful not to damage the plants. This is then repeated in 7 to 14 days, depending on how fast the seed is maturing and showing another flush of feathered seedheads. In some climates a third or fourth harvest may

be possible, but you should evaluate later-setting seed for its quality on a number of occasions in order to make this judgment and not waste time harvesting poor-quality seed.

The presumed advantages of this method are: (1) You are only harvesting fully mature seed in stages—seed that is ready to shatter, or dislodge from the plant into your harvest sack; and (2) you lose less seed to early shattering or birds than you would with a more mechanized harvest approach, where growers often wait for a higher percentage of flowers to feather before harvesting in order to maximize yields. Unfortunately there are several disadvantages to consider with the multiple harvest method. First of all it requires more labor than other mechanized methods, with many hours of bending and shaking required over repeated harvests. Second, it is a dusty, dirty job. With each shake comes a cloud of dust from the surface of the plant as well as pieces of the pappus that are propelled into the air. Workers should always wear face masks during this process.

Single Harvest: The perception that machine-harvested lettuce seed is inferior is based primarily on the fact that it usually occurs when growers harvest seed using the single-harvest system often used by larger producers. When seed is harvested at any one moment in the long flowering period, many immature seedheads are threshed with the fully mature seed when the entire plant is threshed. With many other indeterminate flowering crops the less mature, poorly developed seed is easily eliminated during cleaning, as they are usually smaller and considerably lighter in weight than mature seed and can thus be removed by winnowing or screening. With lettuce this is not as easily done, as much of the undeveloped seed is not any smaller and is not appreciably lighter than normal seed and is not readily removed using these methods.

The basic method for harvesting lettuce seed currently used by most large-scale seed producers involves hand crews cutting the plants near their base when about 50% of the flowers on most of the plants in the field have feathered. Then they lay the plants into windrows and mechanically thresh the seed when the plants have dried sufficiently, before excessive shattering occurs. This method will often result in as much as a 10% drop in germination percentage, due largely to the immature seed content.

Frank Morton and his crew at Gathering Together Farm in Philomath, Oregon, have developed an improvement of this harvest method in their organic seed operation. Instead of cutting the lettuce plants off at the time of harvest, they pull the entire plant by the roots and then lay most of the plant onto a 6 ft (1.8 m) wide strip of landscaping fabric that has been unrolled along the edge of the patch. They are careful to not get the root-ball or any soil onto this fabric, as its main purpose is to catch shattering seed that may fall while the plants are laid down in a windrow or as the seed shatters over the next 7 to 10 days. The reason that the root-ball is kept intact is so the plant is kept barely alive for the first few days of this process, delivering all possible energy to mature the seed that is only partially mature. The plants are stacked three or four deep on the fabric from both sides with the tops of the plants just barely touching in the middle. This creates a mound of plants that is about 3 ft (0.9 m) high while the plants remain green with a mound of root-balls on each side about 1.5 ft (0.5 m) high. This mound of plants will keep most of the seed inaccessible from any birds that may show up to eat the seed off plants while they dry. The American goldfinch (*Carduelis tristis*) that is native to much of the North American continent is notorious for this and will get many of the seedheads near the top of the pile, but a finch will not venture down into the mounds if they are piled properly.

The use of a heavy grade of landscape cloth can be very important if there is rain while the crop is drying in the windrow. If there is only a short period of rainfall that delivers measurable precipitation but passes quickly, then the rain reaching the fabric will easily pass through it and the crop and any seed on the fabric will then be able to dry quickly. In several instances when a longer period of wet weather has threatened, Frank has taken a chain saw and cut the root-balls off the drying plants, then carefully rolled each strip of landscape fabric with the plants into a large roll that looks like a jelly roll. These rolls, which Frank calls "Morton's Jelly Rolls," can then be covered with tarps in the field; if wet weather persists for a longer period of time, they can be lifted onto a trailer or truck bed and be moved to shelter. They then can be held temporarily as a jelly roll and returned to the field as the weather clears or be unrolled and dried under cover. If kept under shelter you should ensure good airflow and possibly heat if cool, wet weather persists.

At the point when the crop in the windrows is almost completely dry, it is important to turn it to ensure even drying. This needs to be done gently to minimize shattering. To ensure that all of the seed is as mature as it can be when preparing to thresh, dry the crop to the point where the plant is thoroughly dry and leaves are crisp. However, the crop should be threshed in the mid- to late morning when plants are still pliable from the morning dew, but late enough in the morning so that the free moisture from the dew has dried. The pliability will help keep the small stems attached to the seedheads from breaking into many small pieces.

Threshing Lettuce Seed: Threshing lettuce manually can be done with a flail, pole, or rake. The ripened seedheads will shatter easily when the plants are sufficiently dry. Mature lettuce plants that are standing and are thoroughly dry, or plants that are dried in windrows, can also be threshed mechanically. However, care must be taken when using threshing equipment; adjust the intake reel so that the relatively fragile lettuce seed receives minimum impact in the threshing process. It is a widely accepted fact that lettuce seed that is mechanically threshed can have germination percentages that are as much as 10% lower than lettuce seed harvested manually.

Many seed people claim that the rotary bar on threshers can also damage lettuce seed, dropping the rate of germination unless you're skilled at properly adjusting the equipment for lettuce.

GENETIC MAINTENANCE

Maintaining the genetic integrity of lettuce is best accomplished at several different stages of its life cycle. As many people now start the lettuce seed crop in the greenhouse, it is possible to easily select for several obvious traits while the lettuce is in the seedling stage. Once the lettuce plant establishes its first four to six true leaves it is quite easy to distinguish whether the color of individual plants matches the varietal norm and whether any spotting, streaking, or other color demarcations that are present are consistent for that variety. All seed growers should also always be mindful of selecting for vigor in their genetic stocks at the seedling stage, as it has repeatedly proven to improve seedling vigor for a number of crops when practiced over several cycles of selection.

When the lettuce plants are at or near the transplanting stage (8 to 12 true leaves) it is possible to judge and select for the stature of the plant, whether it is upright or prostrate to the ground. It is also possible to distinguish whether a leaf is entire or lobed, if it is smooth (flat) or savoyed, if it is blistered or not, and to select for the overall shape of the leaves. As the plants approach their optimum size for harvest as a vegetable it becomes possible to select for the thickness and the glossiness of the leaves, as well as the degree of tightness of the head (for example, butterheads have smaller, looser heads than crisphead varieties)—or the absence of heading. It is also a good time to do a final selection for color, spotting/streaking, leaf shape or type, and stature before the plant bolts.

Selection for time of bolting is always important in seed crops. In lettuce there should always be selection pressure against early

Red and green lettuce varieties in the field. ▶

From left, the author, Frank Morton, and Micaela Colley identifying lettuce plants with resistance to downy mildew in Frank's disease nursery, affectionately known as Hell's Half Acre, Philomath, Oregon.

bolting, as the early bolters are very undesirable for anyone growing lettuce as a vegetable crop. However, as stated in the "Growing the Seed Crop" section, page 143, some lettuce varieties require such a long season to reproduce that they won't successfully make seed in some regions. Therefore, if you're selecting against early bolting you must be careful to not shift bolting too far in the opposite direction in some genetic backgrounds.

During the reproductive phase of the life cycle many organic seed growers will lose a significant percentage of their lettuce seed crop to *Sclerotinia sclerotiorium*. This fungal pathogen is rarely a problem during the vegetative phase of the plant's life cycle but can be particularly troublesome during the reproductive phase. It is characterized by a cottony growth that can often first be seen in older leaves after the floral shoot has emerged and will then often emanate from splits or cracks in the lower section of the flower stalk. Infected plants should be removed as soon as they appear to eliminate susceptible plants from the population. Frank Morton of Wild Garden Seeds has selected against this disease for a number of years in

his lettuce breeding work, and it has resulted in marked increases in horizontal resistance to the effects of this fungal pathogen in a number of his selected varieties.

ISOLATION DISTANCES

As *L. sativa* is highly self-pollinated there are undoubtedly crossing events that occur in the presence of insects in biologically diverse habitats found on many organic farms. The flowers remain closed until pollination and fertilization have been completed in almost all cases, although their form and color are much like those of the other members of the Asteraceae family and are visited by various species of bees and other insect pollinators. The minimum isolation distance between different crops of *L. sativa* should be 150 ft (46 m) in open terrain with no barriers, but can be reduced to 50 ft (15 m) if there is a sufficient naturally occurring barrier between crops (see chapter 13, Isolation Distances for Maintaining Varietal Integrity, "Physical Barriers" on page 337).

– 8 –

BRASSICACEAE

The Brassicaceae is one of the largest and most widespread families of flowering plants, with more than 330 genera and 3,700 species worldwide. Wild and cultivated species of this botanical family, often called the mustard family or the brassicas for short, can be found in almost every corner of the six arable continents of Earth. Most species are herbaceous and are found in temperate climates, though some can be found from tropical regions to the Arctic tundra. Members of the mustard family share several key characteristics:

1. Most contain glycosinolate compounds that produce the sulfurous odor or taste common to the family.
2. They all have the distinctive flower with four petals in the shape of a cross (historically this family was called Cruciferae, which means "cross bearer").
3. Most bear a fruit that is a long, narrow pod known as a **silique** with two carpels that are separated by a partition or septum.

The mustard family contains more important agricultural crops than any other plant family. The earliest use of the plants in this family was possibly for their medicinal properties.

▼ The Brassicaceae bear four-petaled flowers in the shape of a cross. This photo shows a radish flower (*Raphanus sativus*).

Mustard seed has been long used as a poultice or plaster for bronchitis and pleurisy and also to stimulate circulation to relieve muscular and skeletal pain. The culinary use of the pungent seed of early Brassicaceae domesticates was ground to make a condiment akin to the prepared mustard we use today. A number of these crops were also originally used for their oil-bearing seed for cooking and lighting, including radishes, turnips, mustards, and two species of rapeseed. Rape and its modern low-erucic-acid descendant, canola, are still among the most important oilseed crops in the world.

The use of the Brassicaceae seed as a condiment and oilseed predates the use of the leaves, roots, and flowering buds of these crop species, because many cycles of human selection were necessary in the ancestral forms of our familiar modern brassicas to lessen the concentration of bitter-tasting glycosinolate compounds in these other parts of the plants. Forms of mustard greens, kales, cabbages, radishes, and turnips have probably been grown since the beginnings of the cultivation of vegetables. All of these crops and the other crops in this family that followed were selected for their palatability and other important agronomic characteristics as they spread across many cultures for several thousand years of agricultural development.

Six Brassicaceae species constitute the majority of salad greens, potherbs, and root crops in this family that are economically significant around the world. Four of these six are within the *Brassica* genus: the cabbage clan (*B. oleracea*); Asian greens, turnips, and rapini (*B. rapa*); mustard greens (*B. juncea*); and rutabaga and Siberian kale (*B. napus*). The two other important Brassicaceae crop species are radishes (*Raphanus sativus*) and arugula (*Eruca sativa*). Of these six species, *Brassica oleracea* is the most important, with seven major crop botanical varieties or botanical groups" of what is commonly known as the cole crops: (1) cabbage, (2) kale and collards, (3) cauliflower, (4) Brussels sprouts, (5) sprouting broccoli, Calabrese, and heading broccoli, (6) kohlrabi, and (7) Chinese broccoli. When many authors list these crops by their Latin name they will also designate the botanical variety or botanical group to which it belongs; for example, cabbage is *B. oleracea* var. *capitata* or *B. oleracea* (*Capitata* group). While this may be helpful for some uses to delineate the difference between a cabbage and a cauliflower, there is little use for this level of botanical classification for anyone producing seed of these crops. The important fact is that all of the varied crops of *B. oleracea* are all the same species and are therefore fully sexually compatible and will easily cross-pollinate.

FAMILY CHARACTERISTICS

Reproductive Biology

The Brassicaceae was long known as the Cruciferae family based on the four-petal cross or cruciform-shaped **corolla** (a collective term for the petals of flowering plants) that is present in all members of this family. Their flowers are perfect or bisexual and also have four sepals and six stamens, usually in a configuration of four long stamens of a height equal to the pistil and two shorter stamens borne near the base of each flower. Petals are often bright yellow, as is common with many mustard types both wild and cultivated, but can also be white, pink, and occasionally lavender in color. The crops of this family are all prolific flowering plants, usually with an inflorescence that has many branches emanating from a single main stem. Flowers open in the morning, usually before the anthers shed pollen as the day warms. Prolific pollen production and productive nectaries are typical, which accounts

The fruit of most brassica plants is borne as a long, narrow pod known as a *silique*. This photo shows immature radish seedpods. ▶

for the attractiveness of these species to a diverse group of pollinators. Seed is borne in a fruit known as a silique, which has two carpels where the developing seed is enveloped in membranous tissue in each of the two "hulls" that make up the two sides of the pod-like silique.

The cultivated crops of the mustard family are universally cross-pollinated, with the vast majority of pollination being carried out by insects. Many of the crops in this family have a sexual "self-incompatibility system" that prevents individual plants from receiving their own pollen, thereby enforcing cross-pollination. This genetically controlled biochemical mechanism is an incompatibility between the pollen (male) and the stigma or style (female) of the flowers of a single plant. The species and crops that have this system are therefore considered obligate out-crossers, and they are well equipped to avoid inbreeding depression as long as you make sure to grow adequate populations of well-maintained plants.

Life Cycle

The vegetables of the Brassicaceae are almost entirely derived from ancestral plants with a biennial or winter annual life cycle. Most of the crops from this family probably had a strong vernalization requirement when first domesticated, which in many cases has been lessened over the course of human selection. Crops like spring radishes, mustard greens, and most of the Asian greens now require little or no vernalization to flower, while some crops like rutabaga, kale, and cabbage require a lengthy vernalization period to begin the reproductive process.

When growing the annual brassicas for seed it is important to plant and establish the crop early in spring to ensure good vegetative growth before the plant initiates flowering with the onset of warm weather. Planting too early can result in stunted plants with "checked growth" when they don't grow vigorously through prolonged cool weather. These plants may never reach their full potential as seed producers. Conversely, planting too late in the spring will also deprive the plants of lush vegetative growth that will support a large seed yield later in the season.

Commercial seed production of almost all biennial brassicas is accomplished by overwintering the first-year crop in the field. As many of these crops have a degree of cold hardiness, it is possible to grow them through the winter in a range of moderate temperate climates. Each of the biennial crops from the mustard family has an ideal size and stage of development when it is most cold-hardy and is most conducive to overwintering. This ideal size for overwintering is usually an earlier growth stage, before the crop has attained its full development as a vegetable. For this reason many of the biennial brassicas are planted in midsummer to shorten the length of the season so the plants do not grow past their ideal size for overwintering.

Climatic Adaptation

Climatically the brassicas are almost universally cool-season crops, although there are crop types and varieties within several of the species that have been adapted to produce well in hot weather. However, the bulk of the brassicas are grown in temperate climates during the cool seasons of spring and fall, and a number of theses crops are grown as winter crops in the milder temperate zones. It is also true that most of the crops in this family most successfully complete their life cycle, through flowering to producing mature seed, in cool, dry weather.

It follows that these are cool-season, dry-seeded crops. The large-scale commercial production of most seed crops from this family takes place in temperate regions around the globe with prolonged cool, often wet, spring weather and cool, dry summers. These regions include the Pacific Northwest of the United States and Canada, northwestern Europe, the Mediterranean basin, southeastern Brazil, parts of New Zealand and southeastern Australia, and coastal areas adjoining the Sea of Japan. The amount of dry summer weather and average temperatures can vary widely across relatively short distances in some of these regions, thus greatly influencing the quality of seed produced.

Seed Harvest

All of the vegetable crops of the Brassicaceae have an indeterminate flowering habit, producing a profuse number of flowers and maturing seed over a long period of time. The pod-like siliques that develop after fertilization are very compact and easily shatter in most brassicas when they mature. As with all indeterminate-flowering vegetables, the seed crop must be watched closely once a majority of the siliques start to turn a tannish color. If you harvest too early you risk getting a less-than-optimum yield and a high percentage of immature seed. However, when you wait too long, trying to get as many siliques to ripen as possible, there is a danger of a significant portion of the siliques shattering, thus sacrificing part of what is often considered the best

Mature radish seedpods (*siliques*).

portion of the most indeterminate seed crops. As with all indeterminate-flowering crops the secret is to harvest the seed crop when the maximum amount is very near maturity, but before any of the earliest seed to set is mature enough to shatter.

Brassica seed crops are cut when the shade of color of a majority of the siliques is somewhere between yellow and tan and they are becoming dry. The siliques should not yet be dry enough to readily split and release their seed. In most situations the plants of brassica seed crops (especially the biennials) should be cut at this point at, or above, the soil level and stacked into shallow windrows on dry, clean cultivated ground or onto some type of porous fabric to dry. This curing or drying may take at least 7 to 10 days in the preferred weather conditions (clear, warm, and dry). Hardware cloth, geotextile fabric, or other porous fabric will not hold moisture if precipitation occurs during this period. The windrows should be turned every few days to ensure even drying.

Seed of almost all members of the mustard family is easily threshed (the major exception is radish). By capitalizing on the natural tendency of the siliques to dehisce (split open), it is easy to separate the seed from the rest of the dried plant material. This is easily accomplished with either a stationary thresher or a combine that is fed manually. The settings used for either of these are quite simple, as the seed of almost all species in this family are small. Radishes require more expertise in the threshing process as their siliques are fibrous and do not dehisce readily. Radish seed is also much bigger than the seed of most other Brassicaceae and is easily damaged during threshing.

Seed Cleaning

The seed of most crops in this family is very easily cleaned. Winnowing of the empty silique hulls is quick and efficient since the hulls are much less dense than the small brassica seed and can be blown off. The seed is then readily cleaned of any chaff using a standard mill with a series of screens with round holes. Top screens with larger holes will catch larger pieces of debris while allowing round brassica seed to easily pass through. Bottom screens with holes that don't allow sound seed to pass through can easily sift out smaller pieces of chaff, soil, or small, poorly formed seed.

ISOLATION DISTANCES

Taxonomically the Brassicaceae have caused more confusion for seed growers isolating crops to avoid crossing between two seed crops than any other family of vegetables. Determining which crop is a member of which species based on form or any other manifestation of phenotype is difficult. To fathom that the very diverse crops of a species like *B. oleracea* all came from a simple leafy ancestral type that probably looked something like collards takes some imagination. When comparing the tall stems and well-spaced leafy buds of Brussels sprouts with the whitened, branched flower stems that form the curd of cauliflower, or with the swollen, bulbous stem of a kohlrabi, it is hard to believe that all these forms are members of the same species and will readily cross with one another. Just as surprising is that the very cabbage-like head of a Chinese cabbage, *B. rapa*, will not cross with the common cabbage (*B. oleracea*) but will cross with the root crop turnip, which is also *B. rapa*.

There are numerous horror stories among vegetable seed growers of seed crops being ruined through the unintended cross-pollination of two morphologically disparate brassica crops where no one bothered to check the species identity of the two crops before planting. Thus, farmers producing more than one brassica seed crop or growing seed in a seed production area must become diligent in learning the species designation of each brassica seed crop that is being grown.

Isolation for the brassicas follows the general rule for cross-pollinated crops that are primarily insect-pollinated. Two crops within a given species of the same crop type, the same form, the same maturity class, and the same color need only be isolated by 1 mi (1.6 km) if grown in open terrain or 0.5 mi (0.8 km) if there are substantial physical barriers between the crops (see chapter 13, Isolation Distances for Maintaining Varietal Integrity). An example of two seed crops of the same crop type that could use these isolation recommendations would be two red, round-headed, long-season-storage-type savoy-leaved cabbages. The isolation distance would definitely need to be increased to 2 mi (3.2 km) in open terrain if the second crop is an entirely different crop type—say, broccoli or cauliflower within *B. oleracea*—or if it is a different type of cabbage in any of the major morphological characteristics, such as if it is green, has a pointed head, is a short-season fresh market type, or is a smooth-leaved cabbage. This increased isolation requirement between two crops of the same species but different types can be cut to 1 mi (1.6 km) if there are substantial physical barriers between the two crops.

THE COLE CROPS

The vegetable crops of the species *Brassica oleracea* L. are often referred to as cole crops. This includes at least six major crop types: (1) cabbage, (2) cauliflower, (3) Brussels sprouts, (4) heading and sprouting broccoli, (5) European kale and collards, and (6) kohlrabi—all of which are often split into subspecies or botanical groups by many authors, though they are clearly members of this species and are fully sexually compatible. The word *cole*, which is used to describe the varied members of this species, is thought to trace back through Middle English and Anglo-Saxon dialects to the Latin word *caulis*, which translates as "stem" or "stalk." Indeed, all of these crops when grown through to seed maturity develop an impressive thick stem, which has even been used in some cultures as fuel for cooking fires after the plant has matured and dried. In some communities of the United Kingdom selection for tall-stalked types resulted in cabbage and kale varieties where the fully mature stems are harvested for walking sticks and also historically for roof rafters in traditional building designs.

The ancestor of this polymorphic species originally came from the Mediterranean basin.

SEED PRODUCTION PARAMETERS: THE COLE CROPS

Common name: cole crops, which comprise these six general types: (1) cabbage, (2) cauliflower, (3) Brussels sprouts, (4) heading and sprouting broccoli, (5) European kale and collards, and (6) kohlrabi

Crop species: *Brassica oleracea* L.

Life cycle: mostly biennial, a few "weak biennials" that perform as annuals

Mating system: cross-pollinated

Mode of pollination: insects

Favorable temperature range for pollination/seed formation: 60–77°F (16–25°C)

Seasonal reproductive cycle: heading broccoli and summer cauliflower, spring through late summer or fall (4–6 months); most other cole crops (true biennials), midsummer of Year 1 through to summer or late summer of Year 2 (12–14 months)

Within-row spacing: 18–36 in (46–91 cm), depending on crop type

Between-row spacing: 2–4 ft (0.6–1.2 m), depending on crop type

Species that will readily cross with crop: All of the cole crops are fully sexually compatible and will readily cross with one another.

Isolation distance between seed crops: 0.5–2 mi (0.8–3.2 km), depending on crop type and barriers that may be present on the landscape

The earliest, most primitive domesticates of *B. oleracea* were probably not unlike leafy, non-heading cabbages—kales or collards—with smaller leaves and varying amounts of leaf curl and leaf thickness. A rustic wild version of this type is still found along the coastal cliffs of Europe. Selection under cultivation undoubtedly produced larger leaf types similar to our modern kales and collards as they moved with the spread of agriculture many thousands of years ago before any written description of their use was recorded. Further selection occurred for a variety of unusual traits that appeared over time as these ancestral types spread geographically, were seized upon by different agricultural societies, and were developed into new crops. Selection for an enlarged terminal vegetative bud that doesn't open led to the development of cabbage. The selection for enlarged vegetative buds on the side of elongated stems led to Brussels sprouts. Selection of enlarged flower buds led first to sprouting broccoli and later to heading broccoli. Cauliflower was developed from plants with modified flowers consisting of undifferentiated shoot apices. Kohlrabi was selected from plants with excessively swollen stems.

Most of these cole crops are very important vegetables across many cultures around the world. They are cool-season crops, although there are versions of cabbage, cauliflower, and broccoli that are tolerant of hot weather when produced for the vegetative, economically important stage of their life cycle. The seed of all of these crops does require a moderate climate for commercial production. These cool-season, dry-seeded crops are traditionally grown in temperate regions around the globe with prolonged cool, often wet, spring weather and cool, dry summers. However, the amount of dry summer weather and average temperatures can vary widely across relatively short distances in many of these places, thus greatly influencing the quality of seed produced in each of these areas. This is especially important to consider when placing the cole crops for seed production. Also, the various types of cole crops can have significantly different environmental needs within these varied seed production areas. Crops such as heading broccoli and collards produce better yields and higher-quality seed in longer-season, warmer locations, while Brussels sprouts and cauliflower grow best in cooler areas, ones whose steady temperatures that don't usually exceed 74°F (23°C) during flowering and early seed development.

CROP CHARACTERISTICS

Reproductive Biology

The cole crops are cross-pollinated biennial species that produce a large reserve of food in leaves and/or stems by the end of their first year of growth, which is converted into a flowering structure during the spring of the second year. In the second season the flowering stem elongates, producing numerous branches with much smaller leaves and large numbers of yellow perfect or bisexual flowers with four petals, four sepals, six stamens, and a single stigma terminating in a style. The four petals are arranged in the cruciform pattern that is characteristic of all of the Brassicaceae species. The flowers have two nectaries that secrete nectar freely in most varieties of this species for the 3 days that the flowers are open. Coupled with the ample pollen produced by most members of this species, the *B. oleracea* crops are highly attractive to many species of pollinating insects. Seed is borne in a pod-shaped fruit called a silique, as in all brassica crops (see the Brassicaceae introduction, "Reproductive Biology" on page 154).

All members of the cole crops have a sexual self-incompatibility that is characteristic of most species of the Brassicaceae, thus preventing self-pollination and enforcing cross-pollination in these crops. Almost all of the pollination that occurs is due to insect activity. Organic farms that are adjacent to wildlands and that have little or no insecticide use usually have ample habitat for wild pollinator species. When growing *Brassica* seed crops that are smaller than 0.25 acre (0.14 ha) in these environments, it is often possible to have enough wild pollinators to handle pollination needs. Larger acreage may require supplemental honeybee hives.

Climatic and Geographic Suitability

Seed of the cole crops is grown in certain temperate coastal regions that often have prolonged mild springs, cool summers, and dry late-summer periods that are sufficient for seed harvest. These crops also require mild winters to overwinter in the field. However, there is a considerable range of cold temperatures that these crops can tolerate; many of the European kales will survive winter lows of 14 to 16°F (–10 to –9°C), whereas many of the heading broccoli and tropical cauliflower varieties can suffer damage when temperatures dip below 32°F (0°C). The other *B. oleracea* crops fall somewhere between these two extremes, and it is important to get to know the limits of the cold hardiness of both the crop type and the specific variety before you plant a seed crop.

The range of moderate summer temperatures that is best suited to *B. oleracea* seed crops is also dependent on the crop type and specific variety that is being grown. In general these crops require spring and early-summer temperatures that do not regularly exceed 78 to 80°F (26 to 27°C), especially during pollination and the early stages of seed development.

Specifically, both Brussels sprouts and cauliflower are notoriously sensitive to heat during these formative stages of seed development, and most varieties of these two crops can produce poor seed yields and low germination percentages when subjected to temperatures that exceed 74°F (23°C) for any extended length of time. Alternatively, many varieties of heading broccoli and European kale will tolerate temperatures above 80°F (27°C) during this critical stage of seed development. The other cole crops generally fall somewhere in between these extremes, but the good seed grower will seriously consider these environmental needs when placing a seed crop.

SEED PRODUCTION PRACTICES

Soil and Fertility Requirements

All of the cole crops require a reasonably high level of fertility to produce a robust foliar frame that will support a healthy plant through the biennial cycle and produce an optimal seed yield. Most rich, fertile soils—from heavier clays to sand and silt loams—can produce good seed crops for this group, especially when they are enriched with ample compost and are humus-rich. Excessive nitrogen should be avoided, especially during the first season of growth, as it can produce plants that are less cold-hardy during the overwintering process. A steady supply of moisture is also important throughout the season and is especially crucial when building a good plant frame during the first season of growth and to produce lush foliage leading up to flowering in the spring of the second season. It is, however, critical that the soils are well drained and do not hold any standing water, particularly for the overwintering crop. The pH should be between 6.0

and 6.6 for optimum nutrient utilization and growth. Soils that are more acidic than this also encourage the growth of the club root fungus (*Plasmodiophora bassicae*), with cabbage and Brussels sprouts being the most susceptible.

Growing the Seed Crop

A number of seed production methods are used for the cole crops, depending on whether they are treated as annuals or biennials for seed production and whether the particular seed crop needs to be grown to full vegetative maturity for trait selection before it flowers and produces a seed crop. The seed-growing methods for the cole crops will be explained for annual versus biennial production and will

consider whether to grow the crop to optimize selection or to produce the seed as easily as possible for the given climatic conditions.

Annual Seed Production of Cole Crops: The only cole crops routinely grown for seed production in one season are the modern heading broccoli types, which are weak biennials that only require a modicum of vernalization in order to initiate flowering. Some kohlrabi seed has also been produced as an annual crop using a treatment wherein the seed is vernalized by first having it pre-sprout and then holding it in cold treatment for more than a month before direct-seeding it in spring. This method works with kohlrabi varieties that have a low vernalization requirement and does not allow for selection of the fleshy tuber-like stem; therefore it can only be done if starting with high-quality stockseed.

Heading Broccoli. The annual heading broccoli types are often initially grown in nursery seedbeds before they are transplanted into the seed production field. Alternatively, broccoli starts can be started in flats in the greenhouse. Broccoli requires a long season to fully mature seed and needs to be transplanted early in the season to take full advantage of early to mid-spring to produce a substantial frame and head for maximum seed set. Cool spring temperatures below 50°F (10°C) also help to ensure that the plants will quickly satisfy the low vernalization requirement of this crop. However, there is a danger of the plants prematurely forming heads or "buttoning" if they are exposed to excessively cold temperatures or a number of other environmental challenges. Flowering shoots should be fully developed and

◄ When producing broccoli seed crops it is important to maintain heads with uniform development of the individual flower buds, or "beads," as seen on this heading broccoli type.

162

'January King' is a cold-hardy cabbage variety suitable for overwintering in the field with a fully-formed head in areas with mild winter conditions. ▶

flowering by late spring, as late-flowering plants can get overwhelmed by aphids if still flowering in midsummer. Heading broccoli is usually planted in rows that are 24 to 40 in (61 to 102 cm) apart; plants are spaced at 18 to 24 in (46 to 61 cm) apart within the row for seed production. Staking and tying of broccoli plants using a stake-and-weave trellising system is often done when the flowering shoots have developed sufficiently. Stout wooden stakes that are 5 to 6.5 ft (1.5 to 2 m) long are driven into the ground every 10 to 12 ft (3 to 3.7 m) with heavy cordage used to hold the plants up.

Biennial Cole Crops Started in Nursery Beds: Brussels sprouts, cabbage, cauliflower, collards, European kale, and sprouting broccoli are often initially grown in nursery seedbeds before they are transplanted into the seed production field. At that time they are transplanted at the desired spacing for either the seed-to-seed method if the climate is mild enough to allow overwintering of the crop, or the root-to-seed method (really the "plant-to-seed method") if the plants need to be dug and buried in pits or placed into a coldroom during the coldest months of the year. This latter step is rarely used for any commercial production of seed in the cole crops (with the possible exception of kohlrabi), which have large plants that require being stored with an intact root system.

Cabbage. Cabbage varieties vary considerably in the maturation of their heads and therefore you must carefully consider both their planting and transplanting dates during the first season of growth. In mild-winter seed production regions such as the Skagit Valley of Washington nursery beds are planted between mid-May and mid-July, depending on the relative maturity class of the cabbage variety. These planting dates are used to produce plants that are just beginning to appear to be making heads, but the heads are very loose with few leaves and are soft to the touch when pressed with your fingers. This stage of growth is desirable as it is more cold-hardy for overwintering in the field; however, it requires that you plant high-quality, well-selected stockseed to produce the crop, as it isn't possible to select based on the characteristics of the fully formed mature head (see chapter 17, Stockseed Basics).

Alternatively, in milder production areas, or when growing some of the more cold-hardy cabbage types like 'January King', it is possible to grow the cabbages to the vegetative full-head maturity in the first season and still successfully overwinter the plants for seed production in the field. This allows for selection of head characteristics such as size, shape, firmness, color, degree of savoy in the leaf, and relative maturity at the end of the first season or beginning of the second season. Both this method and the method of overwintering immature heads may allow for selection for cold hardiness at the respective stages of development, depending on the region of production.

For small-scale production of cabbage seed the mature plants can also be dug and placed in a coldroom with temperatures between 34 and 39°F (1 to 4°C) and with high humidity at the end of the first season. If the humidity is upward of 90 to 95% (see chapter 3, Understanding Biennial Seed Crops, "Overwintering the Crop" on page 19), bare-rooted plants with all of their loose leaves removed can be placed on shelves made from wood slats or wire mesh. To reduce the chances of rot don't pack the plants too tightly and place them only one layer deep on each shelf. This will also allow easy monitoring of their health when doing monthly checks to discard any rotting plants. These plants can be transplanted in early spring in many temperate climates, as most cabbage varieties are able to withstand considerable cold weather after storage. Cabbage plant spacing can vary considerably depending on the type, vigor, and size of the specific variety that you are planting. The spacing between rows can range from 30 to 48 in (76 to 122 cm) or even more, as some older winter varieties can produce very large flowering plants. The within-row spacing between plants can be anywhere from 18 to 36 in (46 to 91 cm), depending on the variety.

When cabbage is grown to a full vegetative head the first season, there is often a problem with the emerging seed stalk not being able to easily push through the many tightly folded leaves of the head when growth resumes in the spring. To some degree this problem is overcome due to the natural splitting of the cabbage head as it overmatures past the vegetable stage. (Any vegetable farmer knows how quickly this can occur in some cabbage varieties.) Unfortunately, when this process is left to happen naturally there is often a delay in the emergence of the seed stalk; even worse, there can be a reduced number of seed stalks emerging, and the ones that do emerge can be malformed or damaged and may not attain their full reproductive capacity.

The traditional method used to encourage easy emergence of the cabbage seed stalk is to make two incisions several inches long at right angles on the top of the mature head so that the seed stalk will emerge effortlessly as flowering is initiated. This cross-like cut on the top of the head should only be 0.5 to 1.5 in (1.3 to 4 cm) deep, depending on the stage of development of the seed stalk, to avoid damaging the growing point. While this method is still used, many growers now make four side cuts to the head, cutting from 1 to 2 in (2.5 to 5 cm) of the side of the head on all four sides of each cabbage. The top of the head often doesn't need to be cut or removed, as the emerging multiple seed stalks will readily push the remnants of the outer leaves off after the "cheeks" of the head have been cut off.

Cauliflower. Producing cauliflower seed is similar to producing cabbage seed in several key aspects, but the crop is more exacting in its climatic requirements; it is trickier to grow commercial-quality seed of this highly valued crop than any of the other brassica crops. Cauliflower heads are not easily overwintered in the field or in a coldroom like cabbage or the other biennial cole crops. Cauliflower is notorious for not performing well when exposed to temperature or moisture stress at any time during its life cycle. It will also produce a poorer grade of seed if exposed to temperatures above roughly 74°F (23°C) for any length of time during the period of primary seed set and seed formation. This is why commercial cauliflower seed production is limited to so few geographic regions on Earth, which include the Sequim-Dungeness Valley of Washington's Olympic Peninsula, the Lompoc Valley near Santa Barbara, California, coastal regions of Brittany and Normandy in France, and the coasts of Chile and South Africa, all of which remain cool and dry throughout the summer and have mild winters.

Cauliflower varieties are usually categorized based on the relative time needed for them to mature as a vegetable crop throughout the growing season. There are fast-growing, early-maturing types that are harvested from summer into early fall; intermediate-maturing types for the fall and early winter; and long-season, late-maturing types that are overwintered and harvested from late winter into mid-spring. There is also a unique class of cauliflower adapted to the tropics with variable maturity that can form heads in hot weather.

All of these cauliflower types also have varied vernalization requirements—from the little or no vernalization required by the tropical types to the relatively high amount of vernalization time needed by the overwintering types (see chapter 3, Understanding Biennial Seed Crops, "Vernalization" on page 17). Unfortunately, as the crop is generally more cold-sensitive than the other biennial cole crops, it cannot be overwintered in many of the climates that successfully produce cole crop seed. The tropical and most of the early-maturing summer varieties are very sensitive to cold temperatures and must be grown in the mildest winter climates mentioned previously (or in other comparable climates) that rarely get winter temperatures at or below freezing (32°F/0°C).

Hence, both the geographic placement of a seed crop based on the severity of winter and determining the time of season for planting for any of the specific maturity classes of cauliflower can be tricky. The tropical and summer cauliflowers are planted in the spring, developing heads with subsequent flower initiation in summer and harvestable seed in the fall, therefore producing seed as an annual. The intermediate and overwintering types are planted into nursery beds in early summer and transplanted in mid- to late summer depending on the region and growth rate of the variety. There is also a range in the degree of cold hardiness in these types, with the overwintering types usually being the most cold-hardy. For this reason regions that are suited to cauliflower seed production will often produce only one maturity class of cauliflower as a seed crop.

The goal with many of these types is to have a plant that forms a nice rosette of leaves in the fall, usually between 12 and 16 leaves before the onset of the coldest winter weather, but does not form a curd. In most cases where seed growers are attempting to grow these cauliflower types outside the traditional areas of production, they must experiment with each variety, transplanting seedlings or plugs anywhere from mid- to late summer depending on the variety and the length of their season to get to this stage of growth.

Cauliflower spacing in the production field is similar to cabbage plant spacing and can vary considerably depending on the type, vigor, and size of the specific variety that you are planting. But as a cauliflower plant in bloom is more compact than a flowering cabbage plant, they are generally grown at a tighter spacing. The spacing between rows can range from 30 to 40 in (76 to 102 cm), and the within-row spacing between plants can be anywhere from 18 to 30 in (46 to 76 cm), depending on the variety.

Brussels Sprouts. The seed of this biennial cole crop is also grown much like cabbage, and most open-pollinated varieties tolerate colder temperatures than even cold-hardy cabbage varieties. For this reason their seed can be produced in the same regions where cabbage seed is produced and sometimes in colder districts. In many of these climates it is possible to produce plants with sprouts that are nearly fully formed going into winter. This is desirable, as it allows you to select for the size, shape, color, flavor, and appropriate spacing of the sprouts on the stalk for the variety you're producing. Selection can also be done for cold hardiness in climates that are marginal for overwintering Brussels sprouts. Many varieties will survive temperatures down to approximately 10 to 14°F (−12 to −10°C) with fully formed sprouts but will experience freeze damage on sprouts near the base of the plant, especially where lower leaves have sloughed off over time. Brussels sprouts varieties do vary considerably in terms of their cold hardiness, and it is advisable to grow overwintering test plots if you are in a marginal climate before you attempt to produce a commercial Brussels sprouts seed contract.

Many of the other details for growing seed of Brussels sprouts are much like growing cabbage seed crops. Plants are often started in nursery beds in late spring to early summer and transplanted to the field by late summer. They are usually transplanted at a similar spacing as is used for cauliflower. It is also sometimes noted that the quality of Brussels sprout seed, much like those of cauliflower, can be reduced when temperatures go above 74 to 78°F (23 to 26°C) during the period when the seed is first forming and developing, though this is not well documented.

For smaller seed production acreage in cold climates Brussels sprouts plants can be dug with a sizable root mass, placed into bins with moist (but not saturated) soil, and put into a coldroom with temperatures between 34 and 39°F (1 to 4°C). These plants should be left in the field as long as possible before digging and then all of the older, lower leaves from each stalk should be removed before transplanting into the bins (washtubs can be used). Plants can be packed into the bins tightly so long as you provide good air circulation, but they should be checked every few weeks and discarded if rot develops. Plants should be transplanted as soon as danger of severe weather has passed in early spring.

As with all overwintering biennials, selection for health and vigor, as well as good floral characteristics, should be practiced as the plants start to actively grow at the beginning of the second year. In some cases seed growers will cut off the terminal bud at the top of the Brussels sprouts plants after overwintering to promote more uniform development of flowering shoots from the lateral buds. As with the other valuable cole seed crops, the plants are usually staked using the stake-and-weave trellising system as described with heading broccoli to prevent losses from lodging. This is especially important with Brussels sprouts as they are often taller than other cole crops.

Collards and European Kale. These two cole crops are very similar to each other in all seed production practices. The main difference is the superior cold hardiness of most European kales over collards. Seed of both crops is produced much like cabbage seed, and both have the benefit of being more cold-hardy than cabbage, thereby increasing the geographic range where their seed can be produced. Both crops can also be grown to their harvestable stage as vegetable crops during the first season of growth and then be overwintered successfully. As kale and collards are frequently harvested during the fall, winter, and early spring in many milder climates, this enables you to select plants for a number of traits during these optimum harvest periods before the crop flowers in the second season.

Many European kale varieties will survive temperatures down to at least 14 to 16°F (−10 to −9°C), although they will exhibit freeze damage. The less cold-hardy European kale varieties can be lost at temperatures in the range from 18 to 24°F (−8 to −4°C), depending on the duration of the freeze and the repetition of freeze–thaw events. The popular Italian kale 'Lacinato' is in this less cold-hardy category, and a number of growers have lost plants or sometimes apical growing points on plants at temperatures at the low end of this range. Accumulated knowledge on the cold hardiness of collards is less well known, though a number of common varieties seem to survive at least down to the temperatures quoted for the less cold-hardy kales: 18 to 24°F (−8 to −4°C). As with all of the cole crops, experiment by overwintering potential varieties of these crops in smaller plots on the farm before

'Lacinato' kale, also popularly known as Black Tuscan or "dinosaur" kale, has dark, narrow strap-like leaves but is somewhat less cold-hardy than other common European kales.

planting a commercial crop of the variety if you have any questions as to its cold hardiness.

European kale and collard seed fields are usually planted from nursery bed transplants or as greenhouse plugs in mid- to late summer, similar to early or midseason cabbage varieties, for seed production. Plant spacing for kale can be quite variable depending on the size of the particular variety when flowering. The spacing between rows can range from

30 to 40 in (76 to 102 cm) and the within-row spacing between plants can be anywhere from 18 to 36 in (46 to 91 cm). The within-row spacing for most of the kales is usually at the lower end of this range, while the collards would be similar to the larger cabbages at the upper end of this range.

Kohlrabi. Kohlrabi is the only cole crop that is regularly direct-seeded and thinned to a stand, then left in the same spot to overwinter and produce a seed crop the following spring. This is possible only in regions where there will be little appreciable loss of the crop from winter cold. A number of kohlrabi varieties will tolerate winter temperatures in the 16 to 22°F (–9 to –6°C) range if they are not subjected to too many freeze–thaw events during the winter. Growers will often throw soil up around the swollen stems to insulate the crop from extreme temperatures, though it is important to not cover the leaves of the plant. Alternatively, kohlrabi is the easiest of the cole crops to lift from the field and store in a coldroom over the winter for replanting in the spring. The bulb-like vegetable, which is often mistaken for a root crop, is actually stem tissue that is the food storage organ of the biennial plant. If it is lifted with much of the prominent taproot intact, it can be stored much like carrots or beets in a coldroom with high humidity and temperatures between 34 and 39°F (1 to 4°C). If a high-humidity environment in a storage facility is not available then the kohlrabi can be bagged like beets or carrots in plastic bags with airholes after the leaves and most of the soil removed from the plants (see chapter 6, Carrots, "Root-to-Seed Method" on page 87).

The final row spacing for a kohlrabi seed crop is often 22 to 36 in (56 to 91 cm), with plants spaced from 14 to 18 in (36 to 46 cm) on center within the row. Large, long-storage varieties such as 'Superschmeltz' may need up to 24 in (61 cm) between plants in the row to reach their full potential when flowering and producing seed. When using the seed-to-seed method it is possible to thin this crop in stages before getting to the desired density at the onset of flowering in the second season. Using this strategy, it is possible to keep two to three times the needed density of plants for over-wintering in the field before making the final selection of the healthiest plants in the spring.

There is also an alternative way to produce kohlrabi seed as an annual, by "vernalizing" the sprouted seed; this technique has been used in Europe and other regions for many years. It only works with kohlrabi varieties that are fast bolting, which indicates that they have a low vernalization requirement (see chapter 3, Understanding Biennial Seed Crops, "Vernalization" on page 17). Seed is soaked for 8 hours, then germinated quickly at 68 to 72°F (20 to 22°C). Within 24 hours more than 70% of the seed will be just starting to germinate; it must then be promptly stored at 30°F (–1°C) for 5 to 7 weeks. From this controlled, "gentle" freezing the seed is then allowed to thaw very slowly and planted into favorable spring conditions in the field. The crop will build some-what of a limited frame and then begin flower initiation as if it had been overwintered in the field or coldroom. The overall size of the plant and yield may be reduced with this shortcut to seed production. Most importantly you must use stockseed that is guaranteed to be true to type, because you'll have no chance to select the swollen stem of the vegetable, which will never fully develop using this method. There is also a danger of inadvertently selecting toward earlier-bolting plants if this method is used for breeding stock over multiple generations, as there is no opportunity to select against early bolters as would normally be done with a biennial (because they're all early bolters!).

Sprouting Broccoli. Sprouting broccoli is in essence the ancestral form of our modern heading broccoli and still has a loyal following in some regions of Europe with milder temperate climates. It has a multibranching habit that produces many small inflorescences that develop over the course of 1 to 2 months, usually in spring of the second season. True sprouting broccoli is a cole crop of the *B. oleracea* species and should not be confused with rapini or broccoli raab (*B. rapa*). Unlike heading broccoli it is a true biennial and requires a full vernalization period, probably at least 6 to 8 weeks, to begin flowering in the second season of growth (see chapter 3, Understanding Biennial Seed Crops, "Vernalization" on page 17).

As a seed crop, sprouting broccoli is essentially grown like cabbage, with plants started in nursery beds in early summer and transplanted into the field in mid- to late summer. While the many varieties of this crop have considerable fluctuation in the harvest window of their sprouts, from late winter to late spring, it seems that all varieties need to attain approximately two-thirds of their full vegetative size before their growth slows with the onset of winter. Most varieties then experience some regrowth with the first warmth and longer days of late winter before flower stalk initiation. Unlike cabbage, the different varieties of this crop seem to respond well when planted at much the same time in the first season of the biennial cycle. Depending on the region, sprouting broccoli is usually planted at some time in midsummer in a nursery and transplanted to the seed production plot 6 weeks later to be able to attain the two-thirds full vegetative size necessary to produce a satisfactory seed crop. Seed growers outside traditional production areas of Europe will have to experiment with planting dates to discover the optimum sowing and transplanting times in their climate.

Several of the commercial open-pollinated sprouting broccoli varieties sold widely are from the United Kingdom. Some of these varieties are quite cold-hardy and can rival some of the European kale varieties by surviving temperatures down to approximately 14°F (–10°C), although many will exhibit freeze damage. There are also some varieties that are less cold-hardy; these can suffer losses when exposed to temperatures in the 20 to 24°F (–7 to –4°C) range. Consider growing overwintering test plots of any varieties of interest before establishing larger commercial plantings.

Upon transplanting, sprouting broccoli plants are routinely planted into rows that can range from 30 to 40 in (76 to 102 cm) apart and have a within-row spacing anywhere from 18 to 36 in (46 to 91 cm). Older open-pollinated varieties tend to be larger, more sprawling plants that may require wider spacing, while newer varieties are often more compact and can be planted at a closer spacing.

SEED HARVEST

All of the cole crops have an indeterminate flowering habit, producing a profuse number of flowers and maturing seed over a long period of time, like the other species of the Brassicaceae. As with all indeterminate flowering vegetables, brassica seed crops must be watched closely once a majority of the pod-like siliques start to turn a tannish color. These siliques can easily shatter as they mature. For many indeterminate crops the oldest, first seedpods to form often contain the highest-quality seed and should not be lost to shattering if possible. On the other hand the seed crop should not be harvested too early or you will risk getting a less-than-optimum yield and a high percentage of immature seed. As with all indeterminate flowering crops, the secret is to harvest the seed crop when the maximum amount of seed is very near maturity, but before any of the earliest seed to set is mature enough to shatter.

Cole crops should be cut when the color of a majority of the siliques is somewhere between yellow and tan and they are becoming dry. The siliques should not yet be dry and brittle, at which point they will readily split open and release their seed. In most situations the cole crop plants should be cut at soil level, or slightly above, then stacked into shallow windrows on dry, clean cultivated ground or onto some type of porous fabric. Curing or drying may take at least 7 to 10 days in optimal conditions (clear, warm, and dry). Hardware cloth, geotextile fabric, or other porous fabric will not hold moisture if precipitation occurs during this period. The windrows should be turned every few days to ensure even drying.

Seed of the cole crops is easily threshed. As with other brassicas this is easily accomplished with either a stationary thresher or a combine that is fed manually. The settings used for either of these are quite simple; the seed of all of the cole crops is small and easily threshed. Indeed, the seed of all of the cole crops is virtually identical and is not easily distinguished by visual inspection. Therefore, all threshing and seed-cleaning equipment must be thoroughly and completely cleaned between crops if you are growing more than one brassica crop. Otherwise most brassica seed mixes are impossible to undo, and you will kick yourself for not taking the time to clean every facet of the machines between crops.

Seed Cleaning

The seed of most crops in this family are very easily cleaned. Winnowing of the empty silique hulls is quick and efficient as the hulls are much less dense than the small brassica seed and can be blown off. The seed is then readily cleaned of any chaff using a standard mill with a series of screens with round holes. Top screens with larger holes will catch larger pieces of debris while allowing round brassica seed to easily pass through. Bottom screens with holes that don't allow sound seed to pass through can easily sift out smaller pieces of chaff, soil, or small, poorly formed seed.

GENETIC MAINTENANCE

As with most of the cross-pollinated crop species, the open-pollinated (OP) varieties of the cole crops need to be maintained in generous heterogeneous populations of at least 70 to 90 individuals in order to retain their vigor and inherent genetic variability. This can be burdensome when simply growing one of these crops to increase your planting stock or making stockseed, as most of the cole crops are large plants that need a lot of room and two seasons to make seed. The other difficulty associated with taking on seed production of one of the cole crops is that many of the OP varieties of broccoli, Brussels sprouts, cabbage, or cauliflower have "slipped" in terms of their varietal integrity since the advent of hybrids in these crops. As skilled seed growers haven't selected many of these older cross-pollinated varieties regularly for a number of years, they seem to be rife with genetic variability outside of the realm of what is normally acceptable in a commercial crop variety. For this reason it is important to grow at least 30% more plants than the number of plants needed to get the expected seed yield for a particular contract. Hence it is advised that you plant at least 120 plants to have an adequate number of plants to select from to get the final 70-to-90 plant population. This allows you to rogue at least one-third of the plants that don't conform to the variety's characteristics. As most of these crops are biennials and may suffer winter losses, there is reason to start with at least twice as many plants in the first season as you will need in the final population.

Select during the crop's vegetative period for leaf type, leaf color, plant stature, and overall health and vigor, as these traits are often easily spotted during the early stages of growth. For the biennial cole crops it is very important to eliminate any plants that flower or show any signs of flower initiation (such as premature bud formation in cauliflower or sprouting broccoli) or premature bolting during the first season. Selection for cold tolerance is also always important for all biennials, as even plants that survive will frequently show some varying degrees of frost damage.

Broccoli and Cauliflower: As the cole crops have many different forms, the selection for the marketable vegetable part is different for each crop. All types of broccoli require selection for the variation found in the bead size of the individual flowers; the color, shape, and size of the heads; and timing of the heading. In both broccoli and cauliflower the branches of the inflorescence that form the head will grow apart as the head matures and cause the head to look "lumpy." To counter this, select for even, domed heads. Romanesco cauliflower types, of course, are an exception to this, as they have the triangular curds that stand out from the head quite prominently. Cauliflower also needs attention paid to the size and shape of the heads as well as the amount and cover of the wrapper leaves. The color of the curds can be white, yellow, orange, or green, and plants need constant selection to maintain a specific hue. Harvest maturity and uniformity of heading is important for all versions of these crops, although there are many diversified organic farmers using these crops for

vegetable production who question whether the harvest window needs to be as narrow as what the seed industry has pushed as the ideal in hybrid versions of these crops.

Cabbage: In cabbages, selection for shape, size, color, and relative maturity of the head are all very important. Also, the extent (or absence) of savoyedness in many varieties is a trait that requires constant vigilance of selection. Savoyed cabbage can also show varying shades of red or purple coloration in the wrapper leaves due to anthocyanin coloration. This needs to be monitored and selected for, just as the intensity of color in red cabbage, both exterior and interior, needs to be maintained. It is also very important to check the firmness of the head at full maturity whenever performing selection. It would appear that many plant breeders and seed growers have ignored the flavor of the cabbage varieties that they have been producing for a long time, as there are a lot of tasteless cabbage varieties out there. In many of the OP varieties available there is plant-to-plant variation for flavor within the variety. Taste-testing individual plants to identify good-tasting segregants is possible for a person with a developed palate.

Brussels Sprouts: The maintenance of traits for the sprouts of a Brussels sprout variety is very similar to the selection criteria used for cabbage. Although much smaller, the sprouts are routinely selected for their shape, size, color, firmness, and relative maturity, just as you would select cabbage heads. Flavor is also an often-ignored characteristic in Brussels sprouts that can be improved when variation exists in a population. The plants are selected for their stature, height, and the leaf size and placement on the plant. The sprouts should also not be too close together on the stalk, as crowding can impede harvest or promote insect or disease problems. As with the other biennial cole crops, it is also possible to select Brussels sprouts for cold hardiness.

Collards and European Kale: Collards and European kale are biennial leaf crops, and it is possible to select for all market traits in the first season of seed production. These traits include leaf color, leaf type, leaf curl, plant stature, and overall vigor. Collards usually have relatively flat leaves with some degree of undulation, which can be a distinguishing feature, while kale varieties exhibit a wide range of leaf curliness among the numerous varieties grown worldwide. The selection of this trait is often important to maintain the integrity of the variety. As with all biennials, it is very important to eliminate any plants that flower or show any signs of flower initiation (apical stem elongation) or premature bolting.

Kohlrabi: Kohlrabi is almost always grown as a seed crop in much the same way you would grow the vegetable crop during the first season of the biennial cycle. This affords you an opportunity to select for all of the traits that are important to kohlrabi during the first season before the crop is overwintered. These traits include size, shape, color, and smoothness of the bulb-like stem, as well as its relative maturity. The size, shape, color, and angle of growth of the leaves are also very important when selecting for uniformity in the kohlrabi crop. As this crop is usually overwintered

'Market Pride' (left) is a savoy-leaved green cabbage that requires an increased isolation distance from other, different types of cabbage, for example 'Red Drumhead' (right), when both are being grown as seed crops in the same region.

in the field, cold hardiness is critical to the seed grower. Selection for storability is also important, both for the vegetable farmer and for the seed grower, who stores the crop in a coldroom before replanting in the spring.

ISOLATION DISTANCES

It often takes people a bit of time to realize that all of the varied cole crops belong to the same species, *B. oleracea*. The visual difference between kohlrabi and cauliflower, for instance, seems very drastic, and most people are amazed that these two crops are both members of this species. Alternatively, the fact that European kale (*B. oleracea*) and Siberian kale (*B. napus*) are members of two different species is confusing to many farmers who don't have an intimate knowledge of these two crops. So if you're planning to produce seed of any of these crops, it is very important to make sure you have a firm understanding of which brassica crops belong to which species before planning seed-growing activities through any given biennial cycle.

Isolation for the brassicas follows the general rule for cross-pollinated crops that are primarily insect-pollinated. Two crops within a given species of the same crop type, the same form, the same maturity class, and the same color need only to be isolated by 1 mi (1.6 km) if grown in open terrain or 0.5 mi (0.8 km) if there are substantial physical barriers between the crops (see chapter 13, Isolation Distances for Maintaining Varietal Integrity). An example of two seed crops of the same crop type that could use these isolation recommendations would be two red, round-headed, long-season storage-type, savoy-leaved cabbages. The isolation distance would definitely need to be increased to 2 mi (3.2 km) in open terrain if the second crop is an entirely different crop type, such as broccoli or cauliflower within *B. oleracea*, or if it is a different type of cabbage in any of the major morphological characteristics—for instance, if it is green, has a pointed head, is a short-season fresh market type, or is a smooth-leaved cabbage. This increased isolation requirement between two crops of the same species but different types can be cut to 1 mi (1.6 km) if there are substantial physical barriers between the two crops.

MUSTARD GREENS

The crops that are commonly known as mustard belong to several species. The vegetable or leaf mustards, also known as Indian mustard, are from the species *Brassica juncea* (L.) Coss. This includes the large-leaved pungent greens used as potherbs and as salad greens in many cultures around the world today. This species also includes the mustard that is grown as an oilseed crop in Central Asia and brown mustard, which is used to make the hot, spicy European and Asian condiment mustards. In India and China *B. juncea* has long been used as a medicinal plant, most notably for its anti-inflammatory properties. The other mustards of commerce are black mustard (*B. nigra*), which was historically the main source of hot mustard flour or paste in Europe, and yellow or white mustard (*B. hirta*), the source of the bright yellow cream mustard used for hot dogs in North America. However, these species will not readily cross with *B. juncea*, and their culture and production will not be considered here.

Brassica juncea was probably first cultivated for its seed in Central Asia, in what is now northwest India. It must have migrated eastward early in its domestication to establish secondary centers of diversity in central and western China, eastern India, and Burma, as well as westward to the Near East. It has long

SEED PRODUCTION PARAMETERS: MUSTARD GREENS

Common names: mustard greens, Indian mustard, brown mustard
Crop species: *Brassica juncea* (L.) Coss.
Life cycle: most are "weak biennials" that perform as annuals
Mating system: mixed self-pollinated and cross-pollinated
Mode of pollination: insects
Favorable temperature range for pollination/seed formation: 60–77°F (16–25°C)
Seasonal reproductive cycle: spring through late summer or fall (4–6 months)
Within-row spacing: 18–36 in (46–91 cm), depending on crop type
Between-row spacing: 2–4 ft (0.6–1.2 m), depending on crop type
Species that will readily cross with crop: In addition to the variation of types in this species that need positive identification, including the brown mustard seed crops, there are also cases of feral *B. juncea* that may be growing as weedy mustard species.
Isolation distance between seed crops: 0.5–2 mi (0.8–3.2 km), depending on crop type and barriers that may be present on the landscape

'Osaka Purple' mustard (*Brassica juncea*).

much of the southern and southwestern United States, as well as up the Pacific coast to British Columbia, this crop will overwinter easily. Its popularity as a staple vegetable in the American South is equaled in parts of China, Indonesia, India, and other parts of the world where cool-season leafy green crops are prized during the off-season. The use of mustard greens in salad mix has been limited primarily to those growers supplying local markets, as the tender leaves do not have a very long shelf life.

CROP CHARACTERISTICS

Reproductive Biology

The reproductive characteristics of *B. juncea* are very similar to those for the other brassica vegetable crop species (see the "Reproductive Biology" section in the Brassicaceae introduction, page 154). The major difference is that the crops of *B. juncea* do not have the self-incompatibility reaction of most brassica crops and are therefore capable of self-pollination. This self-compatible reproductive state is similar to *B. napus*. According to several authors, the flowers of the leaf mustards can self-pollinate between 60 and 80% of the time in the field, with the remaining cross-pollinations due to insect activity. It is possible that, when these crops are grown in healthy agro-ecosystems with dynamic insect populations, the rate of cross-pollination may be higher than this.

Climatic and Geographic Requirements

The leaf mustards are cool-season vegetables. They are grown primarily as spring or fall crops in temperate climatic areas. For seed production they are best produced in regions with cool, dry summers, where summer high temperatures do not usually occur until later in the growing season after flowering and early seed

been used as a green leafy vegetable in many of these regions, grown primarily as a fall and early-spring crop. The leaf mustards are weakly biennial, requiring very little vernalization with lengthening days to initiate flowering. Therefore, spring-grown crops almost always bolt to seed by summer.

Mustard greens, as the vegetable crop is commonly called in North America, are a hardy, cool-season crop planted in the cool of spring or in the late summer to early fall and used until temperature extremes of the seasons that follow impact the luxuriant vegetative growth that is the hallmark of this crop. In

development. As with other brassicas, a mild, cool spring with ample, steady precipitation will produce mustard plants with a substantial root system and rosette of leaves that will in turn produce a robust flowering plant capable of producing a large complement of seed.

Mustard seed crops are usually produced from an early-spring planting in these cool yet relatively mild-spring areas. It is also possible to plant the crop in early to mid-fall of the previous year in mild-winter areas in order to have well-established crops to fully take advantage of the ideal spring growing conditions and flower earlier, which will result in an earlier maturation date of the seed crop. These overwintered crops can sometimes produce a heavier yield; however, the disadvantage of overwintering the crop is that temperatures in the range of 15 to 25°F (–9 to –4°C) can damage or severely destroy the crop, depending on its size and genetic background. Extended periods of cold, wet weather with little or no active growth can also result in more losses due to an accumulation of diseases.

Mustard leaf crops are able to mature seed crops in many of the seed production areas with Mediterranean climates, where brassica seed is successfully produced around the globe (see the Brassicaceae family introduction, page 153). In addition, much of the seed of these crops that is used locally, from the eastern Mediterranean to central Asia, is still grown and harvested within the agricultural communities that use these crops.

SEED PRODUCTION PRACTICES

Soil and Fertility Requirements

Mustards are more versatile than many other crops in their ability to tolerate a variety of soils and successfully produce a seed crop. Heavier soils should be avoided, and the soil should have good fertility and tilth to produce a healthy crop through to seed maturation. Soils must be well drained and have enough organic matter for good water- and nutrient-holding capacity, to support even spring growth for maximum flowering and seed production. Balanced soil nutrition similar to the recommendations for the cabbage clan is very important for *B. juncea* (see The Cole Crops, "Soil and Fertility Requirements" on page 161). A pH slightly above neutral is important for good growth, just as it is with other brassicas. Also, raising the soil pH to 7.2 or above is important in controlling the club root fungus.

Growing the Seed Crop

Most leaf mustard seed crops are planted in early spring and treated as annuals. As with most other members of the Brassicaceae that are successfully grown to seed as annuals, their ancestors were probably biennials or winter annuals that required a vernalization treatment of significant duration in order to flower in their second season of growth (see chapter 3, Understanding Biennial Seed Crops). Mustard crops undoubtedly still have a vernalization requirement, though it may be very low, possibly requiring a very low number of hours to complete vernalization and begin flowering. The cumulative amount of cold treatment necessary to trigger flowering could be as little as 20 to 30 hours below the vernalization threshold. While the vernalization threshold is usually considered to be 50°F (10°C) for most biennial vegetable crops, the threshold for many of the brassicas that easily flower and produce seed as annuals is probably higher, possibly between 50 and 62°F (10 to 17°C). Thus, producing some mustard seed crops from a spring planting in warmer areas with temperatures that don't routinely stay below 50°F (10°C) is possible with some types of mustard.

The crop is sown in rows that are between 24 and 36 in (61 to 91 cm) apart. Plants within the row are ultimately thinned to anywhere from 14 to 18 in (36 to 46 cm) apart. This depends in part on how large and robust a flowering plant is produced by the particular variety that you are growing. The initial spacing of the seedlings within the row can be three times this dense, with three to four plants in the 14- to 18-in (36- to 46-cm) segments, allowing for an initial selection for leaf color, shape, curl, and plant stature, as well as overall vigor and health. This can be done in the fall if the crop is to be overwintered or in the early spring if planted in spring. If spring-planted the window of time to do this without crowding the plants is relatively short, and the final appropriate spacing should be reached before any of the plants begin to physically touch in order to produce plants with unchecked growth.

Crops for overwintering are sown relatively late in the season, near the autumnal equinox in most temperate climates. Leaf mustard crops are more winter-hardy if they have not achieved their full vegetable crop size and seem to be hardiest when a single plant can fit under a standard 6 in (15 cm) clay pot. At the very beginning of the next growing season these plants will usually put on enough vegetative growth or frame to support a full flowering plant. The great advantage of overwintering the crop is that flowering occurs by mid-spring, thus avoiding the aphid pressure that can occur in late spring. Also, the seed matures by midsummer and is harvested when precipitation is least likely in most climates where brassica seed is grown commercially.

Seed Harvest

The mustard seed crop matures and is harvested very much like most of the crops in the Brassicaceae family (see "Seed Harvest" under the Brassicaceae family introduction, page 156). *B. juncea* seed crops that are produced from overwintered plants will often produce significantly higher seed yields than spring-planted crops. As with all indeterminate-flowering vegetables, the seed crop must be watched closely once a majority of the siliques start to turn a tannish color. When the earliest-setting siliques are dry and ready to shatter, the plants are ready to be cut and placed into windrows for the final maturation and curing.

Seed Cleaning

The seed of the *B. juncea* crops is very similar to most of the other brassica crop types and is cleaned in a similar fashion to all of the round-seeded types (see "Seed Cleaning" in the Brassicaceae family introduction, page 158).

GENETIC MAINTENANCE

Leaf mustard is a biennial that can be planted in fall for an extended vernalization period or in spring in areas with extended cool weather. In either case, it is possible to select for all of the vegetative characters that are important to vegetable growers while the crop is young and undergoing rapid leaf growth. Planting the crop in the fall affords you two seasons in which to select for these vegetative characters. In the first season of vegetative growth it is very important to eliminate any plants that flower or show any signs of flower initiation (apical stem elongation) or premature bolting with any biennial crops. As the overwintered mustard crop is planted so late, there is little chance of this occurring, but it is possible in milder temperate zones where the mustard crop may have experienced some cool fall weather followed by unseasonably warm temperatures late into the fall.

Selection should be done during the crop's vegetative period for leaf type, leaf color, leaf curl, plant stature, and overall vigor, as these traits are often more easily spotted during the early stages of growth.

If mustard leaf crops are overwintered in the field, there may be damage from the cold when temperatures reach between 15 and 25°F (–9 to –4°C). This offers you a good opportunity to select for the most cold-hardy plants, even if the damage doesn't kill them, as the plants receiving the most damage can be eliminated after the worst of the cold period is past.

Plants that exhibit disease symptoms, excessive insect damage, or poor overall health can be eliminated at any point during the biennial cycle. It is always a good idea to evaluate the plants during the flowering stage as well, making sure that all selected plants have a good, healthy complement of flowers with normal branching and timely flowering.

ISOLATION DISTANCES

The *B. juncea* crops along with the *B. napus* crops are unique among the *Brassica* species: Both see a high proportion of self-pollination events in each generation (see the "Reproductive Biology" section on page 175). However, there are still many cross-pollination events as the flowers of these species do open shortly after becoming receptive and releasing pollen. Their flowers are very typical of the Brassicaceae and are very attractive to a large number of insect pollinators, so they have many opportunities to cross within their respective species. For this reason the isolation distances that are necessary to maintain a high level of genetic purity are the same as those used for other brassica species (see "Isolation Distances" in the Brassicaceae introduction, page 158) and follow the general rule for cross-pollinated species that are primarily insect-pollinated.

In summary, when growing two *B. juncea* crops of the same crop type, form, color, and maturity class, they need to be isolated by 1 mi (1.6 km) if grown in open terrain or 0.5 mi (0.8 km) if there are substantial physical barriers between the crops (see chapter 13, Isolation Distances for Maintaining Varietal Integrity).

When producing two different crop types of *B. juncea*, the isolation distance would definitely need to be increased to 2 mi (3.2 km) in open terrain with no barriers. This increased isolation requirement between two *B. juncea* crops of different types can be cut to 1 mi (1.6 km) if there are substantial physical barriers between the two crops.

RADISH

Radish (*Raphanus sativus*) is one of the oldest vegetable domesticates and was originally cultivated in China. From there it spread to Japan and other parts of Asia; it also moved westward to Asia Minor and the Mediterranean basin, where it established a second center of diversity and spread to many cultures. The ancestral form of radish was probably closely related to the modern green- and red-fleshed winter radishes that are still grown extensively across Asia today. The great geneticist and ethnobotanist Edgar Anderson declared radishes to be "one of our most ancient cultivated plants."

The fleshy edible part of the radish plant that we call the root is actually both the hypocotyl and the root. These "roots" vary widely in shape, size, and color, and many more types have come onto the seed market in current years thanks to the specialty market. Radishes also vary in the length of time the roots remain edible after harvest. The small, fast-maturing spring radishes (also called summer radishes) only remain edible for 10 days to 2 weeks under proper storage. The larger, slower-growing winter radishes are traditionally summer-planted, grown through late summer and fall, and used as a winter storage vegetable

SEED PRODUCTION PARAMETERS: RADISH

Common names: radish, spring radish, winter radish, daikon
Crop species: *Raphanus sativus* L.
Life cycle: both "weak biennials" that perform as annuals and biennials
Mating system: cross-pollinated
Mode of pollination: insects
Favorable temperature range for pollination/seed formation: 60–77°F (16–25°C)
Seasonal reproductive cycle: spring and summer radishes: spring through late summer or fall (4–6 months); winter radishes and daikons (true biennials): midsummer of Year 1 to summer or late summer of Year 2 (12–14 months)
Within-row spacing: 18–36 in (46–91 cm), depending on crop type
Between-row spacing: 2–4 ft (0.6–1.2 m), depending on crop type
Species that will readily cross with crop: All radish types are fully compatible and will cross. There is also a wild radish (*Raphanus* spp.) that may cross with cultivated radish.
Isolation distance between seed crops: 0.5–2 mi (0.8–3.2 km), depending on crop type and barriers that may be present on the landscape

'French Breakfast' radish that has been selected for uniform root shape.

Radishes have also been cultivated for their edible leaves and seedpods. In parts of Asia specific varieties have been bred for their mildly pungent, edible seedpods, which are either lightly cooked or made into traditional quick pickles. The leaf radish is also grown in Asia as an attractive edible green leafy vegetable with red-tinged stems and hairless leaves.

CROP CHARACTERISTICS

Reproductive Biology

Radish is a cross-pollinated species with the cruciform perfect flowers common to all brassicas. As with most of the other cultivated members of this family, radish is genetically self-incompatible (see Brassicaceae, "Family Characteristics" on page 154) and will rarely self-pollinate. While the small spring radishes and the much larger winter radishes may seem vastly different in form, size, and seasonal adaptation, they both are *R. sativus* and are therefore sexually compatible and will readily cross if flowering in the same vicinity at the same time.

that can remain edible for 4 to 6 months in storage. As a storage crop, winter types with their firmer, denser flesh have been an important winter vegetable in Asia for thousands of years. Winter radishes include the round and tapered Chinese types of various colors, the daikon and large round white radishes of Japan, and the 'Black Spanish' and 'Munchener Bier' types from Europe.

In terms of their life cycle the spring radishes and winter radishes are quite distinct. The older, more ancestral winter radishes are true biennials requiring a period of cold treatment or vernalization in order to bolt (initiate flowering) and produce seed (see chapter 3, Understanding Biennial Seed Crops, "Vernalization" on page 17). They are usually planted in mid- to late summer to produce a fully grown storage root by fall. They are then either held in the field in mild-temperate climates or stored over the winter and replanted in the spring for the reproductive phase of their life cycle in the second season.

Spring radishes are small, usually round or with a tapered "icicle" shape, and come in shades of red, pink, white, or violet. These types can be easily grown for seed as annuals as they are quick to produce a rather small, fleshy root in spring and have been selected to require very little vernalization to initiate flowering. All of the spring varieties vernalize easily with probably no more than 100 hours of cool nights and vernalization temperatures that may be as warm as 54 to 64°F (12 to 18°C) during the short 30 to 40 days that it takes to produce edible radishes. They then flower by early to midsummer and produce a seed crop in the first season of growth.

▼ Radish flower.

Climatic and Geographic Suitability

As with other brassicas, radishes are best adapted to a cool climate for the rapid growth of their storage roots. However, radishes are more tolerant of heat when flowering than some of the more heat-sensitive crops in this family. For this reason radishes are successfully grown in the hotter inland valleys of Idaho, Washington, and Oregon in the Pacific Northwest, in southern France, and in some of the warmer reaches of the Mediterranean, as well as parts of China and Japan, where temperatures easily get to 80 to 90°F (27 to 32°C) and above during flowering and seed set. Historically, in the warm Upper Midwest of the United States, Michigan was an important radish-seed-growing area during the mid-20th century. In fact, radish seed production does not produce adequate yields of seed in some of the cooler coastal valleys where many of the

other brassica seed crops thrive. However, the drier Mediterranean climate, with dry weather during maturation and harvest, is desirable for growing high-quality seed.

SEED PRODUCTION PRACTICES

Soil and Fertility Requirements

Radishes are more particular in their need for good soil tilth and fertility than most other vegetable crops. To ensure the development of well-shaped roots that are a true reflection of a variety's potential it is best to grow the roots on a lighter soil, preferably a sandy loam, a silty loam, or a soil that is somewhat high in organic matter. These soils, along with adequate moisture and fertility, will promote even, unchecked growth. In heavier soils with uneven watering and poor fertility the roots will often have rougher exteriors with many off-shapes, making selection difficult.

Fertility of the soil should be adequate but not excessive. Radishes need a steady supply of nutrients through both root and flower production, but too much nitrogen will cause excessive top growth when you're growing the roots for selection. It will also cause excessive leaf growth when maximum flower production is necessary for a good seed set later in the season. All flowering crops require plenty of available phosphorus to maximize their reproductive capacity and subsequent seed yield.

Growing the Seed Crop

The basic difference between growing a seed crop of spring radishes and winter radishes is that the winter radishes have traditionally been grown as a biennial, requiring two seasons to complete their cycle, whereas spring radishes have become an annual seed crop due to many generations of selection to greatly reduce their need for vernalization.

Spring radishes need to be planted in early to mid-spring, depending on location, when temperatures and daylength are sufficient to produce rapid root growth and high-quality roots. Once a majority of the roots have reached market maturity you have the option of using the root-to-seed method or the seed-to-seed method as described on the following pages. In the root-to-seed method you lift or pull the plants from the soil and select for several root traits (see "Genetic Maintenance," page 188). In the seed-to-seed method you thin to an optimum spacing and allow the plants to produce seed in place, without a selection of the roots.

Winter radishes are traditionally grown over the course of two seasons with the roots either stored in the ground where they are produced or pulled and put into a coldroom till spring. In the root-to-seed method you select the roots you have stored or pull the roots in the spring to evaluate them and select the best for replanting in the second season. In either situation it is a good practice to pull and select the roots for the important root traits that characterize the variety you are producing. For the seed-to-seed method there are two distinct methods used for winter radishes. The first, more traditional method produces the roots in the first season, overwintering them and growing the seed crop in the second season without lifting them for selection. The second method produces the seed as an annual, vernalizing the crop during an extended spring season so the crop flowers and produces seed in one season.

Root-to-Seed Method for Spring Radishes: In the root-to-seed method for radishes you sow the crop into a root nursery and produce a crop of roots just as if you were producing them as a vegetable crop. Rows should be at least 12 to 14 in (30 to 36 cm) apart with plants thinned to six to eight roots every 12 in (30 cm) within the row, so roots can attain their true shape rapidly. When they have achieved their marketable size as a vegetable they should be harvested promptly for transplanting before they have a chance to become pithy. Pulling the roots is best done on a cool, overcast day. Before pulling the crop to evaluate the roots it is important to look at the tops, which should be reasonably uniform in their vigor, shape, size, and color. Off-types for foliar characteristics can be discarded. The radish plants are then pulled, laid on the soil, and inspected for shape, color, smoothness, crown size, and refined taproot (see "Genetic Maintenance," page 188). As you select the best roots, twist off the tops above the growing point, making sure that the apical growing point is not damaged and, if possible, keeping two or three young leaves intact to start photosynthesizing as soon as the crop is transplanted. Removing the bulk of the leaves before transplanting is important to slow the transpiration stream, which carries moisture from the roots to the leaves. This selection and removal of leaves should occur swiftly to keep the roots in optimum condition for regrowth after transplanting.

The selected roots can then be replanted immediately or stored in cool, high-humidity conditions until the time of planting. Because spring radishes require so little vernalization time—which is met in the 4 to 5 weeks that it takes to produce the roots in spring in temperate climates—they are almost always planted within 24 hours of selection. However, some commercial seed growers will store the roots in a coldroom at 35 to 42°F (2 to 6°C) for 7 to 10 days to ensure that all of the roots in the population have met their complete vernalization requirement and will bolt swiftly

and uniformly soon after transplanting. The drawback of this trick is that it can delay the eventual flowering and maturity of the crop if the plants are already fully vernalized at the time of root selection. This can be a problem in marginal seed production areas with a shorter growing season. In most cooler, shorter-season areas there is usually no need for the extra vernalization to begin with, as these are typically cooler temperate zones.

The roots are then transplanted into the seed production field in rows that are spaced between 24 and 36 in (61 to 91 cm) apart. The within-row spacing can be anywhere from 4 to 12 in (10 to 30 cm) between roots, with a lot of debate among radish seed growers as to the optimum spacing for seed yield. Radishes will produce a bigger plant when given extra room, but much of the additional flowering that occurs doesn't often mature in many temperate climates.

Root-to-Seed Method for Winter Radishes: The root-to-seed method is fundamentally different for winter radishes than for spring radishes due to the fact that winter radishes are true biennials and the roots need to be produced as a fall crop and overwintered if you are to maintain a good varietal selection. This means that planting the crop needs to be timed so the roots mature as a vegetable in the fall of the first year and are then held in the ground or in a coldroom, to be transplanted in spring and produce a seed crop in the second season. In contrast with spring radishes, winter radishes take at least 60 days to size up to vegetable harvest size for selection. Therefore, they need to be planted in a nursery in mid- to late summer in most temperate zones, depending on the warmth and duration of your fall. Spacing for the nursery would be similar to growing the crop as a vegetable, with rows 18 to 24 in (46 to 61 cm) apart and plants spaced at 3 to 4 in (8 to 10 cm) apart within the row.

When winter radishes have achieved their marketable size and before they are subjected to hard freezes they should be pulled and evaluated. Selection of the most representative, disease-free roots is best done under cool, cloudy conditions. Traits unique to the various types of winter radish should be considered (see "Genetic Maintenance," page 188). Selected roots can be replanted into a seed production field if winter temperatures do not drop below approximately 22°F (–6°C), depending on the cold tolerance of the variety that you are growing (though some varieties will suffer from freeze damage under some conditions above this temperature). Otherwise, the roots can be stored in a coldroom under the same conditions recommended for beets or carrots (see chapter 6, Carrots, "Root-to-Seed Method" on page 87). Winter radish roots are then replanted in spring after danger of hard freezes is past. Transplant the roots into the seed production field with rows spaced 24 to 36 in (61 to 91 cm) apart. As with spring radishes, the within-row spacing can be anywhere from 4 to 12 in (10 to 30 cm) between roots, though some types of winter radishes may require the wider spacing to accommodate the mass of the plant that is produced.

Seed-to-Seed Method for Spring Radishes:

The seed-to-seed method for spring radish types is relatively straightforward. Stockseed is sown directly into the seed production field after the spring weather has settled at a rate of six to eight seeds per 12 in (30 cm) of row, with row centers between 24 and 36 in (61 to 91 cm) apart. Sowing seed any thicker than this is pointless unless your stockseed has an exceptionally low germination rate. Coupled with an initial within-row weeding 7 to 10 days after emergence, it is important to thin

the most robust plants to two or three per 12 in (30 cm) of row. In the seed-to-seed method there is no root selection, other than what bit of root may be showing above the soil, though it is still possible to select for foliar or reproductive characteristics. Radish plants are allowed to go straight from harvestable root to flowering plant in situ, producing seed at the end of that season.

Seed-to-Seed Method for Winter Radishes:

There are two basic versions of the seed-to-seed method for biennial winter radishes. Both differ from the root-to-seed method in that they don't allow you to inspect and select the roots that are going to make seed. Hence, these methods should only be used when the crop is planted from well-selected stockseed (see chapter 17, Stockseed Basics).

In the first method, the winter radish seed crop is grown as a biennial in a similar fashion to the root-to-seed method. The winter climate must be mild enough for the roots to be overwintered in the ground. The seed is planted several weeks later during the summer so that the radish roots only grow to an inch (2.5 cm) in diameter before they go through the winter. At this size they will be more cold-hardy than if they were grown to their optimum vegetable size, and they will still make a full-sized seed-bearing plant in the second year. The same crop spacing that is used for the spring radish seed-to-seed method can be used for this method, but wait to thin the crop until after it comes through the winter, as you will probably lose some plants due to cold or rot.

In the second seed-to-seed method winter radishes are grown as annuals, with seed being produced in the first season of growth. This is possible in temperate regions with extended cool spring weather if they are sown as early

An immature radish silique, or seedpod. ▶

◄ Mature radish seedpods.

as possible and can accumulate enough vernalization hours by mid- to late spring. This method is commonly used for daikon and other Asian radishes in the Willamette Valley of Oregon, where nighttime temperatures regularly dip below 50°F (10°C) until late spring, thus supplying plenty of cold treatment hours to initiate flowering in most Asian winter radish varieties. However, there are some of these varieties that will not reliably flower using this method; they require a longer vernalization period than they can get in a normal spring. These are the late-maturing, high-solids daikon varieties that are the best storage varieties. This makes sense, as the selection for the denser, longer-storing roots that would be kept in cold conditions would result in varieties that require a longer cold vernalization period. These varieties must be grown as biennials. If you wish to use this seed-to-seed method on any unproven winter radish varieties in your area, it would be wise to grow an early spring-planted observation plot to see if the particular variety will reliably bolt in your region. The same crop spacing that is used for the spring radish seed-to-seed method can be used for this method, though some winter radish varieties may require the wider row and plant spacing.

Seed Harvest

As with all of the crops of the Brassicaceae, radishes have an indeterminate flowering habit, producing a profuse number of flowers and maturing seed over a long period of time. Radish seed forms in pod-like siliques that are somewhat different from the other common cultivated brassica crops. Radish siliques are quite a bit fatter, almost bulbous, and are more fleshy and fibrous when compared with other brassica seedpods. Fortunately, at harvest and

in post-harvest handling these pods do not shatter readily at maturity, and therefore seed is not easily lost during these steps as often happens with other brassica seed crops.

As the radish pods are maturing, they turn from a greenish yellow color to a tannish brown and lose their fleshy appearance, becoming lighter and more papery to the touch. As with the other indeterminate-flowering brassicas, the radish seed crop is ready to harvest when approximately 60 to 70% of the pods have reached maturity. The plants are then cut aboveground mechanically by swathing them or by hand with a machete. Don Tipping of Seven Seeds Farm in Oregon has used a chain saw to cut the thick stems at the base of the plant.

The plants are then dried by stacking them into shallow windrows on dry, clean cultivated ground or onto some type of porous fabric like hardware cloth, geotextile fabric, or other porous fabric that will not hold moisture if precipitation occurs during the drying period. This curing or drying will usually take at least 7 to 10 days with the preferred weather conditions (clear, warm, and dry). As the radish pods don't crack or shatter readily, they should be allowed to dry thoroughly, becoming brittle for the threshing process.

Seed Cleaning

Radishes require more expertise in the threshing process as their siliques are fibrous and do not dehisce readily. Radish seed is much bigger than the seed of most other brassica crops and is easily damaged during threshing. Combines and stationary threshers with rollers that can be set to effectively crack the pods, without crushing them, are the most efficient way to recover most of the seed successfully. After the seed has been threshed it can be separated from most of the pods, chaff, stems, and other debris by screening and winnowing.

The seed can then be thoroughly cleaned using a standard mill with a series of screens with round holes. Radish seed requires larger screen sizes (with round holes) than are used for other brassica crops.

GENETIC MAINTENANCE

As with all cross-pollinated species, having a good knowledge of what the ideotype is for any variety that you are selecting for makes life easier. Radishes can be quite variable, and their phenotype or appearance can change considerably when grown in one environment as opposed to another. That's why it is important to remember that the ideotype is often hard to define in any cross-pollinated species; it's really just an average of all of the plants that are hovering near the elusive ideal representation of that variety. This is very true of radishes, as they seem to be shape-shifters across different environments and soil types, even more than some of the other root crops like carrots or parsnips, which also epitomize this syndrome. For this reason it is important to select radish roots to type when they are grown under favorable conditions: a fertile soil with good tilth; an even, steady source of soil moisture; and moderate weather conditions that promote even growth (see "Soil and Fertility Requirements," page 182). If you have less-than-ideal conditions when producing roots for selection, it is often best to relax how strictly you practice root selection and concentrate

The author training farmers in evaluating and selecting radish root characteristics in the field. ▶

instead on negative selection using the plant breeders' credo: "Eliminate the uglies."

For the best results, roots need to be selected at a mature vegetable stage. As all of the roots will not mature at the same time, it is best to pull the crop when approximately 75 to 80% of them are a marketable size. Pull the crop and place the plants neatly on the beds for evaluation. Keep the roots out of direct sunlight if possible until they are replanted. Root selection on a cloudy, cool day is ideal. Many open-pollinated radish varieties are frequently not highly uniform due to lack of careful selection. Expect to discard at least 20% of the roots, which can often be up to 50% (or more!) if the variety hasn't been selected for several generations. It is important to grow large populations of roots to ensure that there will be plenty for selection.

The basic traits for selection in all radish varieties include root shape, color, and size. The size and shape of the leaves are also important traits for many fresh market varieties, as are seedling vigor and disease resistance for any crop used in organic farming.

Root Size and Shape: Each radish variety has a fairly distinct ideal shape that is often distinguishable from other similar varieties. Negative selection against off-type shapes and roots that are too big and/or too small is always a good place to start, homing in on more exacting selection pressure if you are familiar with the variety.

Root Color: There are two reasons for color variation in radish. First, as with all pigmented root crops, there is always inherent genetic variation for color. Second, color variation can be caused by previous outcrossing with a different-colored variety and can be limited if judicious selection is practiced. Genetically recessive colors are harder to eliminate as they can hide in the heterozygous state for generations. The genetic dominance/recessive hierarchy for color is dark purple is dominant to red; red is dominant to pink; pink is dominant to white. Each color down the hierarchy is recessive to all colors above it, with white recessive to all pigmented roots. Progeny selection can speed the elimination of off-type colors should they appear (see chapter 17, Stockseed Basics).

Leaf Size and Shape: While many radish seed growers pay little or no attention to the leaves of radish, the height, shape, and uniformity of the leaf type can play an important role in radishes that are sold in bunches on the fresh market. In spring radishes the tops can be anywhere from 4 to 8 in (10 to 20 cm) tall. Historically, the taller-topped varieties were used almost exclusively for bunching spring radishes as they were easier to bunch and made a fuller, more bountiful-looking bunch. Alternatively, many winter radishes may have tops that are as tall as 12 in (30 cm) or more. Also, all radish types can have leaf shape variation. Leaves may range from deeply lobed (as in many daikons), to moderately lobed (in many spring radishes), to strap leaves with no lobing. While uniformity of height or degree of lobing in the foliage is important in some fresh market applications, it may not be crucial to select for absolute uniformity of this trait for all radishes.

Seedling Vigor: Seedling vigor and early robust growth are always important to organic growers, as these traits affect the plants' ability to compete with weeds, outgrow the damping-off complex, and contribute to earlier yields in the case of fast-growing crops like radishes. Selecting for the earliest-emerging seedlings and eliminating all later-emerging seedlings will improve seedling vigor over several cycles of selection. As radishes are one of the quickest crops to emerge after planting, it is quite easy to pick a point after a satisfactory subset of plants emerge and eliminate all late-emerging plants. This can be coupled with an early hoeing and thinning of the plot.

ISOLATION DISTANCES

Isolation for radishes follows the general rule for cross-pollinated crops that are primarily insect-pollinated. Two crops within a given species of the same crop type, the same form, the same maturity class, and the same color need only to be isolated by 1 mi (1.6 km) if grown in open terrain or 0.5 mi (0.8 km) if there are substantial physical barriers between the crops (see chapter 13, Isolation Distances for Maintaining Varietal Integrity). An example of two seed crops of the same crop type that could use these isolation recommendations would be two red, round, short-top spring radishes. The isolation distance between two crops would definitely need to be increased to 2 mi (3.2 km) in open terrain if the second crop is a different-colored or different-shaped spring radish (for instance, a white or an elongated icicle or French Breakfast type).

Also, if there were two winter radish seed crops of different types, such as a 'Black Spanish' type and a snowy white cylindrical daikon, you would also need this increased distance between crops. Alternatively, two similar crops of a white daikon could be isolated at the lesser 1 mi (1.6 km) distance in open terrain and 0.5 mi (0.8 km) with barriers. Perhaps it goes without saying that any spring radish type needs be isolated from any winter radish by the increased distance if both are grown as seed crops.

RUTABAGA & SIBERIAN KALE

There is much speculation as to whether the three crops of *Brassica napus*—rutabaga, Siberian kale, and canola—originated from a naturally occurring cross between turnip (*B. rapa*) and one of the crops of the *B. oleracea* clan. The genetic evidence is quite convincing: *B. napus* has two distinctly different sets of chromosomes (a disomic polyploidy or allopolyploid in genetic terminology), with one set of chromosomes very similar to those found in turnip and another set of chromosomes very similar to those found in *B. oleracea* crops. From this exotic pairing across species boundaries must

have come a wealth of genetic variation that gave the farmer-breeders of the day much to work with in developing these three diverse crops. From the scant knowledge we have of the evolution of these crops we know that they were probably developed in Northern Europe; they first appear in the written record during the Middle Ages. As with many other *Brassica* species, the original domesticate of *B. napus* was probably an oilseed crop. This species remains one of the most important oilseed crops in the modern world in the form of rape and canola. These oilseed crops can

SEED PRODUCTION PARAMETERS: RUTABAGA & SIBERIAN KALE

Common names: Siberian kale, Russian kale, rutabaga, Swede turnip
Crop species: *Brassica napus* L.
Life cycle: biennial
Mating system: mixed self-pollinated and cross-pollinated
Mode of pollination: insects
Favorable temperature range for pollination/seed formation: 60–77°F (16–25°C)
Seasonal reproductive cycle: midsummer of Year 1 through summer or late summer of Year 2 (12–14 months).
Within-row spacing: 18–36 in (46–91 cm), depending on crop type
Between-row spacing: 2–4 ft (0.6–1.2 m), depending on crop type
Species that will readily cross with crop: Siberian kale and rutabagas are fully sexually compatible and will readily cross with each other. Be conscious of the species of kale that you are growing for seed, as well as whether the "turnip" you may be growing is truly a rutabaga or turnip (*B. rapa*).
Isolation distance between seed crops: 0.5–2 mi (0.8–3.2 km), depending on crop type and barriers that may be present on the landscape

belong to either *B. napus* or *B. rapa* or may be one of the newly developed *B. juncea* types. The majority of the canola that is grown for oil in North America today is from *B. napus* varieties, which are known as Argentine canola types. This poses a risk for anyone growing seed crops of rutabagas or Siberian kale, as a canola crop being grown for oilseed can easily contaminate other *B. napus* crops being grown for seed.

The rutabaga has traditionally been most popular in places with cool, marginal conditions for many of the storage vegetables and has often been associated with subsistence agriculture. The name *rutabaga* comes from a dialect of Swedish. However, there is little consistency among English speakers as to the name of this vegetable. Much of England and Canada calls them Swedes or Swedish turnips, while in the north of England, Scotland, Ireland, and Atlantic Canada they call them turnips. Adding to this confusion are the forage

▼ Flower of Siberian kale (*Brassica napus*).

types that are usually called fodder turnips across these areas.

Siberian kale, as its name implies, has a long association with Northern Europe. It has long been in the shadow of the more widely grown *B. oleracea* crop known as kale, which has been a standard vegetable in Western Europe (and in the more knowledgeable North American markets) for a long time. The advent of prepared salad mixes for restaurants and produce markets has thrust the little-known Siberian kale into a much more prominent role among vegetables. With its more tender texture, milder flavor, and range of leaf shapes and colors, Siberian kale has become an important ingredient in this market, thus increasing its importance as a seed crop.

CROP CHARACTERISTICS

Reproductive Biology

The reproductive characteristics of *B. napus* are very similar to those for the other brassica vegetable crop species (see Brassicaceae, "Reproductive Biology" on page 154). The major difference is that the crops of *B. napus* do not have the self-incompatibility reaction and are therefore capable of self-pollination. According to several authors the flowers of these crops can self-pollinate between 60 and 80% of the time in the field, with the remaining cross-pollinations due to insect activity. It is possible that when these crops are grown in healthy agro-ecosystems with dynamic insect populations, the rate of cross-pollination may be higher than this.

Climatic and Geographic Requirements

The two major vegetable crops of this species are not generally thought of as having many similarities, since rutabagas are a root crop

and Siberian kale is a leafy green vegetable. However, these crops have several marked similarities in their culture, both as vegetables and as seed crops. As vegetable crops both are routinely planted in mid- to late July for fall harvest. In milder climates like the maritime areas of the mid-Atlantic states, most of the southeastern states, and the Pacific coast of the United States and Canada, it is possible to harvest these crops into winter and even overwinter them in many locations. Both are true biennials requiring at least 8 to 10 weeks of vernalization to initiate flowering, which can be accomplished either in cold storage or by overwintering the crop in the field. The vernalization requirement is easily satisfied by overwintering in most temperate climates, where about half of the hours of each day are below 50°F (10°C) for at least 4 months (see chapter 3, Understanding Biennial Seed Crops, "Vernalization" on page 17).

Commercial seed of rutabaga and Siberian kale is grown in several of the major brassica seed production areas around the world (see the Brassicaceae introduction, "Climatic Adaptation" on page 156). However, both of these crops certainly still have regional seed production in places where these crops are traditionally grown. As with all brassica seed production there is the danger of several devastating seedborne diseases when these crops are grown in climates conducive to disease. Plants must be monitored closely and tested for seedborne diseases if the seed is to be sold commercially.

SEED PRODUCTION PRACTICES

Soil and Fertility Requirements

Successful seed crops of both of these crop types can be grown on a variety of soil types ranging from light sandy soils to fairly heavy loams. Heavier soils should be avoided when growing the first-year rutabaga root crop as clay soil can cause branching of the roots, making it harder to select if you're using the root-to-seed method. Soils must be well drained, especially for overwintered crops, yet have enough organic matter for good water and nutrient-holding capacity to support even spring growth for maximum flowering and seed production. Balanced soil nutrition similar to the recommendations for the cabbage clan is very important for these crops (see The Cole Crops, "Soil and Fertility Requirements" on page 161). Nutrients that are biologically available from soil humus are always critical for biennial crops like rutabagas and Siberian kale, which often grow in the same spot in the field for more than a year. A pH that is slightly above neutral is helpful for good growth, as it is with other brassicas, and raising the pH to 7.2 or above is important in controlling the club root fungus, especially for rutabagas.

Growing the Seed Crop

Siberian Kale: These cool-season crops require a mild growing season for lush vegetative growth before vernalization and subsequent flowering when grown for seed. This is achieved by growing these crops in cool coastal valleys, where they are easily established during mid- to late-summer weather and can easily produce the vegetative growth during the long, cool fall season characteristic of these areas.

Siberian kale is always overwintered for seed production. Most varieties can generally withstand temperatures between 14 and 20°F (–10 to –7°C), with several varieties, including 'White Russian,' surviving temperatures at or below 14°F (–10°C) if there are not too many repeated periods of freezing and thawing during the winter. When grown as a seed crop

The broad, lobed leaves of 'Red Russian' kale have a distinctive magenta red midrib, and the plants are typical of Siberian kale. ▶

Siberian kale is often planted in late summer, early to mid-August in the Pacific Northwest and coastal British Columbia, which results in a smaller, hardier plant than when it is planted earlier for vegetable cropping. Siberian kale is also known to have better resistance to the root rot complex of organisms in waterlogged soils in comparison with European kale and other *B. oleracea* crops overwintered in wet climates.

Rutabaga: Rutabagas are usually overwintered for commercial seed production using the seed-to-seed method in a similar fashion to the method used for carrots (see chapter 6, Carrot, "Seed-to-Seed Method" on page 86). For seed-to-seed production rutabagas should be sown between the last week of July and mid-August. The plants may then be selected over the course of the fall for vigor, plant stature, and foliar characteristics. When selecting in the fall it is a good idea to leave at least two to three times as many plants as you'll need for your final stand. After the challenges of winter, as the plants put out new growth, you can select to a final stand for seed production. Alternatively, rutabagas may be lifted, evaluated, and then stored or replanted for seed production the following year (see chater 6, Carrot, "Root-to-Seed Method" on page 87). As with all biennial root crops grown for seed, the evaluation and selection of the roots afforded by using the root-to-seed method is crucial to maintaining uniform root-quality characteristics, unless your crop is grown from well-maintained stockseed.

Siberian kale and rutabagas are routinely planted in rows 24 to 26 in (61 to 66 cm) apart for the seed crop. Both crops are often direct-seeded at about 1 lb (0.5 kg) of seed per acre; plants are subsequently thinned to 14 to 18 in (36 to 46 cm) between plants

for the final stand, depending on the size of the variety. Kale is often transplanted for seed crops at the same row spacing, and some growers will transplant two to three times the necessary plants within the row (every 4 to 9 in/10 to 23 cm) in the first year so that they can select to the best survivors the following spring. They will usually initiate flowering in March or April after going through winter and will often be ready to harvest in early to mid-July in most locations.

Seed Harvest

The seed crop that is produced by either of these two crops matures and is harvested very much like most of the crops in the Brassicaceae family (see "Seed Harvest" in the Brassicaceae family introduction on page 156). The *B. napus* crops can produce prodigious seed yields and seem to mature very quickly once a critical number of siliques have been set. As with all indeterminate flowering vegetables, the seed crop must be watched closely once a majority of the siliques start to turn a tannish color.

Don't harvest the crop too early or you will risk getting a less-than-optimum yield and a high percentage of immature seed. However, if you wait too long, trying to get as many siliques to ripen as possible, you risk seeing a significant portion of the siliques shattering, thus losing a quantity of what is often considered the best-quality seed. As with all indeterminate flowering crops, the secret is to harvest the seed crop when the maximum amount of seed is very near maturity, just as the earliest seed to set is mature enough to shatter.

Seed Cleaning

The seed of the *B. napus* crops is very similar to most of the other brassica crop types and is cleaned in a similar fashion to all of the round-seeded types (see the Brassicaceae introduction, "Seed Cleaning" on page 158).

GENETIC MAINTENANCE

Both rutabagas and Siberian kale are biennials, which affords you two seasons in which to select. In the first season of vegetative growth it is very important to eliminate any plants that flower or show any signs of flower initiation (apical stem elongation) or premature bolting. Selection should also be done in the first season for leaf type, leaf color, plant stature, and overall vigor, as these traits are often more easily spotted during the early period of growth. It is also possible to select Siberian kale for the extent of undulation and dentation of the leaf margins.

When rutabagas are grown using the root-to-seed method and the roots are pulled at the end of the first season of growth, it is an excellent time to evaluate the root characteristics. Selection should be for uniform root shape, smooth root surface, uniform exterior coloration, and short necks. Selection against too prominent a taproot or the branching of the taproot is also important. Breeders may also select for interior color by cutting into the "cheek" of the root with the same method that is used for beets (see chapter 5, Garden Beet, "Genetic Maintenance" on page 60). The piece of root that is cut off can also be used for tasting, either raw or cooked, to select for a sweeter or milder root. This piece of the root can be analyzed further for its dry matter content. Lastly, roots can be selected for their ability to withstand long-term storage.

If either of these *B. napus* vegetable crops are overwintered in the field there may be damage from the cold if temperatures dip below 20°F (−7°C). This gives you a good opportunity to select for the most cold-hardy plants, even if the damage doesn't kill them, as the plants receiving the most damage can be eliminated after the worst of the cold period is past.

Plants that exhibit disease symptoms, excessive insect damage, and poor overall health can be eliminated at any point during the biennial cycle. It is always a good idea to evaluate during the flowering stage as well, since all selected plants should have a good, healthy complement of flowers with normal branching and timely flowering.

ISOLATION DISTANCES

The *B. napus* crops, along with the *B. juncea* crops, are unique among the brassica species as they both have a high proportion of self-pollination events that occur in each generation (see "Reproductive Biology," page 193). There are still many cross-pollination events, however, as the flowers of these species do open shortly after becoming receptive and releasing pollen. Their flowers are very typical of the Brassicaceae and are very attractive to a large number of insect pollinators, so they have many opportunities to cross within their respective species. For this reason the isolation distances that are necessary to maintain a high

level of genetic purity are the same as those used for other brassica species (see the Brassicaceae introduction, "Isolation Distances" on page 158) and follow the general rule for cross-pollinated species that are primarily insect-pollinated.

In summary, when you're growing two *B. napus* crops of the same crop type, form, color, and maturity class, they need to be isolated by 1 mi (1.6 km) if grown in open terrain or 0.5 mi (0.8 km) if there are substantial physical barriers between the crops (see chapter 13, Isolation Distances for Maintaining Varietal Integrity).

When producing two different crop types of this species, the isolation distance would definitely need to be increased to 2 mi (3.2 km) in open terrain with no barriers. The most obvious example of this in *B. napus* would be between a rutabaga variety and a Siberian kale variety, but it would also include the case when you are growing a red and a green Siberian kale in proximity to each other. This increased isolation requirement between two *B. napus* crops of different types can be cut to 1 mi (1.6 km) if there are substantial physical barriers between the two crops.

TURNIP & ASIAN GREENS

Brassica rapa is one of the oldest crop species used by human societies, and like many of the brassicas it was probably first cultivated as an oilseed crop. It is also one of the most diverse cultivated species, with at least seven major subspecies used today in far-reaching agricultural systems around the globe. This species was surely widely dispersed at an early stage in the history of agriculture, as it has deep roots in the agricultural systems of three diverse regions of the world: the Mediterranean basin, Central and South Asia, and East Asia. The original center of diversity of this species is often thought to be the Mediterranean basin, and a wild type of *B. rapa* is still found in Western Europe today. The Mediterranean versions of *B. rapa* are turnip, rapini, and turnip rape, which is probably the oldest version of rapeseed cultivated by humans. Probable secondary centers of origin for this species are thought to include the Central Asiatic or Indian center for the tendergreen mustards (also called mustard spinaches) and the Chinese center in East Asia for Chinese cabbage, Chinese mustards, and most of the other *B. rapa* green leafy vegetables that are now often

SEED PRODUCTION PARAMETERS: TURNIP & ASIAN GREENS

Common names: Asian greens, Chinese cabbage, turnips, broccoli raab, rapini
Crop species: *Brassica rapa* L.
Life cycle: mostly biennial; a few "weak biennials" that perform as annuals
Mating system: cross-pollinated
Mode of pollination: insects
Favorable temperature range for pollination/seed formation: 60–77°F (16–25°C)
Seasonal reproductive cycle: Asian greens and some rapini: spring through late summer or fall (4–6 months), most turnips (true biennials): midsummer of Year 1 through summer or late summer of Year 2 (12–14 months)
Within-row spacing: 18–36 in (46–91 cm), depending on crop type
Between-row spacing: 2–4 ft (0.6–1.2 m), depending on crop type
Species that will readily cross with crop: The range of all of Asian vegetables that are *B. rapa* is broad and ever changing, and they will all readily cross with one another.
Isolation distance between seed crops: 0.5–2 mi (0.8–3.2 km), depending on crop type and barriers that may be present on the landscape

categorized as Asian greens. As all of these crops predate the recorded history of agriculture, it remains very difficult to understand the earliest dissemination of these crops through agricultural trade and human settlement.

The turnip and turnip rape crops are very old and important to human civilization in the Mediterranean and Europe. By the Middle Ages the turnip had become a staple of the diet of the common folk throughout Europe. It was especially important as a winter storage crop in many places with long, cold winters, as it was one of the most nutritious and easily stored crops of the time, only requiring "pitting" in the ground for storage. In more than one instance in England the humble turnip was depicted on a coat-of-arms to indicate that the family members were benefactors of the poor. In Ireland they were used for carving jack-o'-lanterns on All Hallow's Eve to ward off evil spirits before the tradition was brought to North America, where the more easily carved pumpkin took their place. Rapini, also called broccoli raab, broccoli de rape, or Italian turnip, has an immature flower bud and shoot that resembles biennial sprouting broccoli (*B. oleracea*), but with somewhat smaller, less compact flower buds. In Italian *rapini* means "little turnip," and this vegetable crop is undoubtedly derived from turnips that were selected for their prolific and tender sprouts.

Tendergreen or mustard spinach types appear to represent a separate, unique group within *B. rapa* that emerged in a secondary center of origin in Central or South Asia, where they are still widely grown today.

These mustards are milder than the pungent *B. juncea* mustard green types and are characterized by large, dark green foliage borne on cold-hardy, fast-growing plants. The Asian greens or Chinese mustards are a very diverse group of crops with many different forms that have spread very widely across East Asia. These *B. rapa* crops have been selected and adapted

▼ Broccoli rapini, a.k.a. raab or Italian turnip, is harvested for its leaves, shoots, and immature flower buds.

by the agricultural societies of eastern Asia for annuals like mizuna, tat soi, and pak choi and biennials like Chinese cabbage and komatsuma. Many of these crops have become increasingly important around the world as Asian cuisine has spread in popularity. Chinese cabbage, also called napa in North America, is probably the most widely grown vegetable of this group.

▼ *Brassica rapa* flower.

CROP CHARACTERISTICS

Reproductive Biology

The reproductive biology of the *B. rapa* crops is very consistent with most of the species of the Brassicaceae (see the Brassicaceae introduction, "Reproductive Biology" on page 154). It is a highly outcrossing species with a strong self-incompatibility reaction that

eliminates most self-pollination and therefore avoids inbreeding within populations.

Almost all of the annual crops of this species are non-heading leafy crops that are probably mild biennials with a very low vernalization requirement to initiate flowering. This minimum vernalization is easily satisfied when the plant goes through relatively few cool nights during the early stages of growth. Biennial crops of *B. rapa*, like the densely heading Chinese cabbage or turnips, require a longer vernalization of anywhere from a few weeks to the full 8- to 10-week duration necessary for most biennial vegetable crops (see chapter 3, Understanding Biennial Seed Crops). Of course the amount of vernalization that is required to initiate flowering for a crop like turnips can vary greatly among varieties, with some bolting readily if grown under cool spring conditions, while other storage types can require a more lengthy period of cold treatment to flower.

Climatic and Geographic Suitability

The *B. rapa* crops are akin to other members of the Brassicaceae in climatic adaptation. While these crops all favor mild yet cool-weather conditions to grow optimally during all stages of their life cycle, there are some varieties of several of the Asian greens crops that exhibit some tolerance to hot weather as vegetables. As with the bulk of the brassicas, however, they are best adapted to flowering and producing mature seed in cool, dry weather.

As with all of the other seed crops of the Brassicaceae, commercial production of most seed crops from this family takes place in temperate regions around the globe with prolonged cool, often wet, spring weather and cool, dry summers. These regions include the Pacific Northwest of the United States and Canada, northwestern Europe,

the Mediterranean basin, southeastern Brazil, parts of New Zealand and southeastern Australia, and coastal areas adjoining the Sea of Japan. In the case of the *B. rapa* crops, though, varieties or local strains of many of the various types (turnips in Eastern Europe, rapini in the Mediterranean basin, mustards in the Indian subcontinent, and many Asian greens throughout East Asia) are still grown as seed crops for local consumption. While many of these places may not be ideal for commercial seed production, the local farmers have learned to adapt the growing of their local strains to their seasonal fluctuations, producing successful seed crops for local use. There must also certainly be some adaptation of the germplasm to these less-than-optimum seed-growing conditions.

SEED PRODUCTION PRACTICES

Soil and Fertility Requirements

Successful seed crops of all of the *B. rapa* crops can be grown on a variety of soil types ranging from light sandy soils to fairly heavy loams. However, most of these crops do best on loose friable loams, especially when growing the first-year turnip root crop, since a loam with a higher clay content can cause branching of the roots, making it harder to select if you're using the root-to-seed method. Soils must be well drained, especially for overwintered biennial crops, yet have enough organic matter for good water- and nutrient-holding capacity to support even spring growth for maximum flowering and seed production. Balanced soil nutrition that is similar to the recommendations for the cabbage clan is very important for these crops (see The Cole Crops, "Soil and Fertility Requirements" on page 161).

Nutrients that are biologically available from soil humus are always very important for biennial crops like turnips and Chinese cabbage, which often grow in the same spot in the field for close to a year.

Growing the Seed Crop

Most of the non-heading, leafy green *B. rapa* crops, including most of the Asian greens, the tendergreen mustards, and rapini, are planted in early spring and treated as annuals for seed production. As with most other members of the Brassicaceae that are successfully grown to seed as annuals, their ancestors were probably biennials or winter annuals that required a vernalization treatment of a significant duration in order to flower in their second season of growth (see chapter 3, Understanding Biennial Seed Crops). These non-heading, leafy green crops undoubtedly still have a vernalization requirement, though it may be very low, possibly requiring a very low number of hours to complete vernalization and begin flowering. The cumulative amount of cold treatment necessary to trigger flowering could be as little as 20 to 30 hours below the vernalization threshold. While that threshold is usually considered to be 50°F (10°C) for most biennial vegetable crops, the threshold for many of the brassicas that easily flower and produce seed as annuals is probably higher, possibly between 50 and 62°F (10–17°C). Thus, producing any number of the Asian greens seed crops or the rapinis from a spring planting in warmer areas with temperatures that don't routinely stay below 50°F (10°C) is still possible. The tendergreen mustards probably require a longer vernalization period, but there is little research investigating vernalization requirements in most vegetables.

The two most prominent *B. rapa* biennials, turnip and Chinese cabbage, probably need what is often considered the standard vernalization period for biennial vegetable crops to flower and produce seed. The standard period is a cold treatment of from 6 to 10 weeks with temperatures under 50°F (10°C) (see chapter 3, Understanding Biennial Seed Crops, "Vernalization" on page 17). This is easily accomplished in many mild, temperate climates when the crops are left in the field through the winter. As with all biennials, each variety of these *B. rapa* crops has a different threshold temperature near or below 27°F (−3°C) where the freezing events, especially when repeated over the course of an extended cold spell, will damage the crop to the point that it won't regrow when conditions get favorable in late winter or spring. Research on turnip seed production in Oregon in the early 1950s found that it is possible to reliably initiate flowering and produce turnip seed when the crop is spring-planted in the Willamette Valley. The researchers found that turnips seeded by late April had plenty of time to vernalize, flower fully, and produce high seed yields in the cool, extended springs of western Oregon.

Most of the *B. rapa* crops can be sown in rows that are between 24 and 36 in (61 to 91 cm) apart, and plants within the row are ultimately thinned to a spacing that can be anywhere from 14 to 18 in (36 to 46 cm) apart. This depends in part on how robust and large a flowering plant is produced by the particular variety that you are growing. The initial spacing of the seedlings within the row can be three times this dense, with three or four plants in the 14 to 18 in (36 to 46 cm) segments, allowing for an initial selection for leaf color, shape, curl, and plant stature, as well as overall vigor and health. This can be done in the fall if the crop is to be overwintered or in the early spring if planted in spring. For spring-planted crops, the window of time to do this without crowding the plants is relatively short,

◄ *Brassica rapa* siliques, or seedpods.

and the final appropriate spacing should be reached before any of the plants begin to physically touch in order to produce plants with unchecked growth.

Seed Harvest

The *B. rapa* seed crop matures and is harvested like most of the crops in the Brassicaceae family (see the Brassicaceae introduction, "Seed Harvest" on page 156). *B. rapa* seed crops that are produced from overwintered plants will often produce significantly higher seed yields than spring-planted crops. As with all indeterminate flowering vegetables, the seed crop must be watched closely once a majority of the siliques start to turn a tannish color. When the earliest-setting siliques are dry and ready to shatter, the plants are ready to be cut and placed into windrows for the final maturation and curing.

Seed Cleaning

The seed of the *B. rapa* crops is very similar to most of the other brassica crop types and is cleaned in a similar fashion to all of the round-seeded types (see the Brassicaceae introduction, "Seed Cleaning," page 158).

GENETIC MAINTENANCE

The leafy, non-heading *B. rapa* crops, the Asian greens and tendergreen mustards that are annuals and require early planting in spring, afford a window of selection during the vegetative growth period of early to mid-spring in most temperate zones. Once these crops initiate flower formation with the onset of hotter weather and increasing daylength, it becomes more difficult to evaluate the vegetative traits such as leaf type, leaf color, plant stature, and vigor, as well as flavor and texture. Selection for these traits must be done before flower initiation. Rapini, which is also grown for seed as an annual, needs special attention paid to it for selection during the stages surrounding bud formation. Traits such as the number of shoots, size and color of buds, harvest period, and height of the plants are scrutinized during selection. Tightness of the buds is also important, although rapini has a less compact inflorescence than broccoli or sprouting broccoli.

Both Chinese cabbage and turnips are biennials, which usually affords you at least part of two seasons in which to perform selection. In the first season of vegetative growth it is very important to eliminate any plants that flower or show any signs of flower initiation (which typically starts with apical stem elongation) or premature bolting. Selection should also be done for Chinese cabbage in the first season for leaf type, leaf color, plant stature, and vigor. While it is desirable to select this crop for the ability to produce a uniform, tight head, it isn't always practical, since Chinese cabbage is often grown to a loose, pre-heading stage for overwintering in the field in many temperate seed production areas because the crop is more cold-hardy at this stage of development.

Turnips are evaluated and selected initially in the fall for foliar traits. The foliage of many turnip varieties is very important for many people who relish eating turnip greens. The tops will certainly exhibit variation for color, gloss, stature, and height, as well as for the extent of undulation and dentation of the leaf margins. To routinely select for root characteristics, turnips should be grown using the root-to-seed method. Selection should be for uniform root shape, smooth root surface, smooth crowns, uniform exterior coloration, and a fine taproot with minimal branching. Selection for flavor, texture, and the internal color of the turnip's root can be accomplished by cutting into the "cheek" of the root to check for these quality traits using the same method used for beets (see chapter 5, Garden Beet, "Root Color" on page 61) while being careful to not cut into the taproot as with beets. However, because roots are less dense and they generally don't store as well as other biennial root crops, they will often rot quickly when using this method. To attempt to use this cut-cheek method with turnips it is best to experiment with a small number of roots at first; the density of a turnip's flesh can vary greatly from one variety to another. Cut several roots to see if the cut surface will suberize within a few hours (see chapter 5, Garden Beet, "Genetic Maintenance" on page 60) and plant them into a test plot in the field to see if they will actively grow without appreciable rot within a few days. If they do, then it is possible to select for eating quality; if not, then you may be forced to go without direct selection for these qualities.

Turnip roots can be lifted, evaluated, and selected at the end of the first season of growth and the roots stored in a coldroom for replanting in the spring of the second season. This root-to-seed method is somewhat limited in larger-scale production of turnip seed as the roots of many turnip varieties do not store as successfully as most biennial root crops, which are much easier to store for long periods of time (see chapter 6, Carrot, "Root-to-Seed Method" on page 87). Alternatively, commercial turnip seed crops are often grown in milder temperate zones where the crop is left in the ground over the winter to produce a crop the next season in situ, without evaluation and root selection. This is the seed-to-seed method and requires that you plant well-maintained stockseed that will reliably produce a crop with uniform roots (see chapter 6, Carrot, "Seed-to-Seed Method" on page 86). The seed-to-seed method is also the preferred when you're spring-planting a turnip seed crop, as any lifting, evaluating, and replanting of the crop in late spring would postpone the flowering and maturation of the crop to an unacceptable degree in some seed production areas.

Plants of any of the *B. rapa* crops that exhibit any disease symptoms, excessive insect damage, or poor overall health can be eliminated at any point during the annual or biennial life cycle. It is always a good idea to also evaluate the plants during the flowering stage, as all selected plants should have a good, healthy complement of flowers with a normal flowering architecture that is common to the variety and the crop species. If any of the *B. rapa* vegetable crops are overwintered in the field, there may be damage from the cold if temperatures dip below 25°F (–4°C). This gives you a good opportunity to select for the most cold-hardy plants, even if the damage doesn't kill them, as the plants receiving the most damage can be eliminated after the worst of the cold period is past.

ISOLATION DISTANCES

As with all of the Brassicaceae family, there is much confusion among the *B. rapa* crops when it comes to determining isolation distances between crops. Crops that may seem very similar, such as turnips (*B. rapa*) and rutabagas (*B. napus*), are indeed from two different species and do not normally cross.

This is confounded by the use of common names, as rutabagas in some English-speaking countries are sometimes called Swede turnips. Rapini (*B. rapa*), which is often called broccoli raab in North America, is quite similar to both sprouting broccoli and the so-called Chinese broccoli, both of which are members of *B. oleracea* and are harvested for their immature flowering shoots. The possible mistakes when isolating crops can take two forms:

1. You may not properly isolate two seed crops of the same species. For instance, a turnip and a rapini crop may seem so dissimilar in their form that you don't realize they are both of the *B. rapa* species and need to be isolated accordingly.
2. You may wrongly assume that sprouting broccoli and rapini are the same species and isolate them from each other when it isn't necessary.

The *B. rapa* crops are cross-pollinated largely by insects, as are most of the Brassicaceae, and their isolation requirements are therefore the same as for all of the other crops in this family. When growing two *B. rapa* crops of the same crop type, form, color, and maturity class, they need to be isolated by 1 mi (1.6 km) if grown in open terrain or 0.5 mi (0.8 km) if there are substantial physical barriers between the crops (see chapter 13, Isolation Distances for Maintaining Varietal Integrity).

When producing two different crop types of *B. rapa*, the isolation distance would definitely need to be increased to 2 mi (3.2 km) in open terrain with no barriers. This increased isolation requirement between two *B. rapa* crops of different types can be cut to 1 mi (1.6 km) if there are substantial physical barriers between the two crops.

– 9 –
CUCURBITACEAE

The Cucurbitaceae family includes several of the most economically important vegetable crops worldwide. It is alternatively called the cucumber family, squash family, or gourd family. This family has close to 120 genera with more than 800 species, including three economically important Old World vegetables: cucumber (*Cucumis sativus*), watermelon (*Citrullus lanatus*), and melon (*Cucumis melo*). The important New World vegetable contribution to this family is a wide array of squash types from four species of the *Cucurbita* genus. All of the cultivated genera of the Cucurbitaceae family have trailing, climbing "vines," with fleshy, tender stems that almost always have tendrils. Cucumbers, watermelons, melons, and winter squash are often called vine crops in North America, while most summer squash varieties have fruit that are borne on plants with a bush habit. The other important crops of the Cucurbitaceae include several species of gourds, a number of ornamental vines, several minor melon species, and chayote, an herbaceous perennial vine that yields both an edible root and a rich textured fruit similar to summer squash. All of these crops are frost-sensitive.

Botanically speaking, all of the cucurbit crops covered in this manual are classified as fruit. Certainly, melons and watermelons are treated as fruit in culinary terms, as they are generally eaten as dessert. Yet all of the important cucurbits are considered vegetable crops, starting with the production and distribution of their seed, the cultivation methods that are employed to grow them, and the marketing and storage used in their distribution. Most of the cucurbits have a monoecious flowering habit, with unisexual flowers that are morphologically similar from species to species. The main differences are the size and the shades of yellow of the flowers, which is distinct between these crops.

The crops of the Cucurbitaceae are very important in almost all agricultural cultures worldwide. While there are limits on which cucurbit crops can be grown successfully in many temperate climates, there is ever-increasing fresh market production of cucumbers, summer squash, and some types of melons under cover in regions and seasons where it isn't normally possible to get commercial production. In tropical and subtropical climates circling the globe cucumbers remain the number one consumed vegetable on a fresh weight basis. Melons and watermelons are also widely grown and held in high regard in these hot climates.

Seed production for the various cucurbit crops is done across a fairly large geographic swath of the Earth's surface. In general, these crops are heat lovers and require warm springs followed by appreciable heat and warm nighttime temperatures during the summer months. Most of the cucurbit seed crops require summer temperatures of at least 77 to 86°F (25 to 30°C) during the day and produce the best seed crops when temperatures don't dip below 68°F (20°C) during the night. The exception to this would be most summer squash types and cucumbers, which can thrive with temperatures between 68 and 77°F (20 to 25°C) during the day and temperatures that don't usually drop below 59 to 62°F (15 to 17°C) at night. Cucurbit seed is produced commercially in many parts of the Middle East, South Asia, Southern Europe, North Africa, Israel, Indonesia, Taiwan, Korea, China, and Japan. In North America most commercial cucurbit seed is currently grown in the Arkansas River Valley of Colorado, the Central Valley of California, and in central and west Texas. Cucurbit seed is also still produced regionally in many agricultural societies that still hold value in culturally significant varieties of these crops.

FAMILY CHARACTERISTICS

Reproductive Biology

All of the cultivated vegetables of the Cucurbitaceae family are annuals and must produce a substantial vine, flower, and then produce some of the largest fruit on Earth, all in one growing season. Indeed, the giant show pumpkins of the *Cucurbita maxima* species are the largest fruit produced by any angiosperm species grown worldwide.

Most of the cucurbit vegetables have monoecious flowers, though there are some notable cases of andromonoecy across several species of this family. While the pistillate and staminate flowers of these various species can vary greatly in their size and color, from the various shades of yellow of the *Cucurbita* spp. to the greenish yellow of the watermelons, they are anatomically very similar across species. Cucurbit flowers are sensitive to temperature and daylength, with the ratio of staminate to pistillate flowers going up under the longer days near the summer solstice and more pistillate flowers appearing as the daylengths shorten.

The monoecious flowering habit is an effective biological way to ensure a relatively high degree of cross-pollination. The cross-pollination in all of the vining types of cucurbits is probably enhanced by the fact that the vines of one plant are often physically interwoven with the vines of adjacent plants as they reach the reproductive phase of their life cycle. This means that bees that are methodically working the flowers of one area of the crop are repeatedly moving from one plant to another. Cucurbits seem to be largely pollinated by honeybees and wild bee species in North America, as other pollinators do not seem to be as interested in the cucurbit flowers. Farmers growing cucurbits (which need pollinators for both fresh market and seed crops) are often concerned that they will not get adequate pollination from native populations of bees. This concern is heightened for many diversified organic growers, who often have flowering crops that they believe are more desirable, both in the quality and density of flowers, to the bees than their cucurbit crops. Though the number of flowers per unit area of land for the cucurbit crops is relatively low in comparison with many of the other flowering crops like clover or buckwheat, the fact is that if bees are present, then the cucurbits always seems to have good visitation from a number of bee species.

Life Cycle

All of the vegetable crops of the Cucurbitaceae are tender, warm-season annuals. They are all sensitive to frost and require a reasonably hot climate to mature their fruit. As annuals that have to put on an appreciable amount of vegetative growth early in the growing season to support a good fruit set, these crops need to be planted under favorably warm, settled weather conditions that will result in vigorous, unchecked growth. Traditionally these crops, when grown in regions most favorable to seed production, have been direct-seeded at the time of planting. Direct-seeding the cucurbits has long been regarded as superior to transplanting by many growers, as it seems to produce a more productive plant. But increasingly, organic growers are now transplanting their cucurbit crops for several sound reasons:

1. Less predictable spring weather in recent years means that "settled conditions" aren't always as likely at the time of planting.
2. Transplanting enables the farmers to avoid 10 to 14 days of weed growth that would normally be growing along with the crop seedlings.
3. In areas where either the striped or spotted cucumber beetle is a problem, transplanting is a way to grow plants past the most susceptible stage (the cotyledon stage) before transplanting them into the field.

The cucurbit crops must become established, producing flowers relatively early in summer to produce a good fruit that will mature safely ahead of the first killing frost. Always choose the crop and the particular variety of it based on the knowledge that it will produce a healthy, mature seed crop in most years at your location. This is especially important when you're growing a variety or a crop species that you have not grown before in your area. The crop-to-crop differences and even the variety-to-variety differences in adaptation and relative maturation across regions can be significant and can mean the difference between a successful seed crop or none at all. For this reason I strongly urge all seed growers considering seed crops of these species to grow a seed suitability trial, to evaluate the seed maturity and seed yield of any varieties that you may be considering growing commercially.

Climatic Adaptation

The cucurbits all produce their best seed crops with warm nighttime temperatures, which support pollen tube growth and help ensure a high percentage of fertilization of the embryos in each fruit that is set. Adequate heat units during the day are also important to stimulate rapid fruit growth and maturity. In summary, the best regions to produce seed of these crops need to have: (1) a relatively long, frost-free growing season; (2) enough hot weather for good fruit production; and (3) warm nighttime temperatures for maximum seed set.

Seed Harvest

The most important aspect of seed harvest for the cucurbit crops is making sure that the fruit that is harvested for seed is fully ripe, even overripe, but making sure as well that it isn't rotting and moldy to the point where the seed is damaged. In fact, allowing the fruit of all of these crops to go a bit beyond full ripeness may ensure that the seed is fully mature. As with other wet-seeded crops, like the Solanaceae, the overripe fruit also usually release their seed from the flesh or placental tissue much more easily than is the case with slightly underripe or even perfectly ripe fruit.

Fruit of the Cucurbitaceae crops is usually harvested by hand for seed production in order to have human eyes make a final assessment of whether: (1) the fruit is fully ripe; (2) there are marked levels of disease or rot on a particular fruit or plant; and (3) the fruit from a particular plant is true to type. Harvesters can reject individual fruit if they are either not yet ripe or rotting, and they can make a final act of selection if the plant has excessive disease or off-type fruit. At harvest, fruit are sometimes windrowed in the field or hauled out of the field before seed is extracted. Depending on the crop and climate, growers are sometimes able to make multiple harvests of ripe fruit from the same field. (See the "Seed Harvest" section of each Cucurbitaceae crop for information on post-harvest fruit handling.)

Seed Extraction, Fermentation, and Cleaning

Seed Extraction: Mechanical extraction of larger quantities of seed from all of the cucurbit crops is done using what is alternately called a vine harvester, cucurbit machine, or cucurbit seed extractor. This machine is pulled through the field while workers toss the ripe fruit into it, or the ripe fruit is gathered and run through the extractor while it is running in a stationary position. All of the cucurbits are run through these machines in much the same fashion, though each crop requires different settings. You must gain experience extracting the seed of each particular type of crop, as it can be damaged easily if the seed extractor is not set properly.

Seed Fermentation: Cucumber seed is borne in a gelatinous sac that adheres tightly to the seed. This sac needs to be removed from the seed either through physical abrasion or through fermentation (see Cucumber, "Seed Extraction, Fermentation, and Cleaning" on page 216). While cucumbers are the only cucurbit crop where the seed is routinely fermented, there are cases in which all of the other cucurbits are fermented, either to either loosen the seed from the placental tissue or to make it easier to clean. Importantly, the fermentation used for the cucurbits other than cucumbers is usually done for less time than the 3 to 4 days needed to adequately ferment cucumbers. (See the "Seed Extraction, Fermentation, and Cleaning" section for each individual cucurbit crop for specific recommendations.)

Seed Cleaning: The final cleaning of all cucurbit seed, through the use of a sluicebox or with a simpler decanting method, is essentially identical to that described for tomato (see chapter 12, Tomato, "Washing Seed" on page 323). The main difference is that for each crop with its variable seed size and weight, the amount and force of the water that is released down the flue will need to be adjusted to clean seed of the different species and different crop types. The drying of the seed needs immediate attention after this wet seed processing.

Fermentation and Seedborne Diseases: Fermentation of cucumber seed is advantageous in helping to control at least one of the most important diseases of cucumber, gummy stem blight (*Didymella bryoniae*). As the environment that exists during the fermentation process can vary considerably from one location to another, it is important to remember that any disease that is affected may not be completely controlled by this process. Gummy stem blight has become one of the most devastating diseases of cucumber in North America in the past 20 years, and organic research needs to be conducted to determine the optimum fermentation conditions to reliably control the transmission of this disease via cucumber seed.

GENETIC MAINTENANCE

All of the cucurbit crops have plant characteristics such as leaf size, shape, and color, as well as vine length, that can be evaluated before flowering, thus affording growers an opportunity to eliminate off-type plants before they flower and spread their genes throughout the population. When the cucurbit plants start to flower it is possible to evaluate their time of flowering and the proportion of male to female flowers, which can be quite variable in these mostly monoecious species.

At the time of flowering and within the first couple of weeks of fruit development it is possible in many of the cucurbits to evaluate the basic shape and color of the fruit, which enables you to eliminate plants at an early stage in the flowering of the crop. The earlier that you can eliminate any off-type plants, the less time an off-type individual has to spread its pollen (and thus its genes) throughout the population.

As cucurbit fruit reach a marketable size it becomes easier to select them for their trueness to type for size, shape, and the color of their skin. Occasionally, there will be a single off-type fruit on a plant that otherwise has fruit typical of the variety. When this occurs it is undoubtedly due to environmental conditions during the formation or early growth of that particular fruit, and, if the other fruit are normal, then the plant should not be eliminated. As with all seed crops, routine elimination of the plants that are most susceptible to any endemic diseases is always worthwhile.

ISOLATION DISTANCES

Isolating a crop of the Cucurbitaceae to avoid cross-pollination is relatively straightforward as long as you are willing to learn the species name of the crop you intend to grow for seed in this large diverse family. Cross-pollination across species boundaries that has resulted in viable offspring has occurred between varieties from two different squash species under controlled conditions, but is very rare in the field and most long time squash seed growers have not observed it.

In general, all of the members of the Cucurbitaceae family generally need 1 mi (1.6 km) of isolation distance when you're growing seed crops of two different varieties of the same species in open terrain. When barriers on the landscape are present it is possible to separate two crops of the same species by as little as 0.5 mi (0.8 km), especially when these adjoining crops are relatively small (see chapter 13, Isolation Distances for Maintaining Varietal Integrity).

As with all of the cross-pollinated crops, you need to distinguish between the distances that are necessary when the two adjacent cucurbit crops are of the same horticultural type versus when you're growing two different types of the same species. Perhaps the best example of this among the Cucurbitaceae are the widely varied crops within the species *Cucurbita pepo,* which includes everything from zucchinis to patty pans, acorns, delicatas, jack-o'-lantern (true) pumpkins, and even a subspecies of bottle gourds—all of which will cross easily. Hence, the minimum distances needed to isolate varieties from two of these disparate groups needs to be doubled over the basic isolation distances stated above, with 2 mi (3.2 km) between two crops when grown in open terrain and 1 mi (1.6 km) when significant barriers exist on the landscape between the two crops. For examples of the different horticultural types within each crop species, see the "Isolation Distances" section for each specific crop.

CUCUMBER

Cucumber (*Cucumis sativus* L.) probably originated in Africa but was likely carried to South Asia and the Middle East, its primary center of diversity, at an early stage of its domestication more than 3,000 years ago. Cucumbers quickly spread across equatorial regions of the Eastern Hemisphere. Cucumber cultivation became so prominent in the agriculture of southern China in the early stages of domestication that this region became a secondary center of diversity for this important vegetable crop. The Romans introduced cucumbers to Europe. Columbus carried cucumber seed to the Western Hemisphere on his first voyage in 1492, and by 1539 the conquistador Hernando de Soto found "cucumbers better than those of Spain" being cultivated in southern Florida.

Cucumbers occupy two basic market classes: the fresh market types usually sold as whole fruit, and processed types that are pickled using fermentation or short-brine methods. The major fresh market types include the Middle Eastern Beit Alpha type, the Asian trellis type, the North American slicer, and the European or Dutch greenhouse type (still called English cucumbers in some markets). The major processing types are the North American and the European pickling types, which vary in the extent of their spines. It should be remembered that that these are only

SEED PRODUCTION PARAMETERS: CUCUMBER

Common name: cucumber
Crop species: *Cucumis sativus* L.
Life cycle: annual
Mating system: largely cross-pollinated, with some self-pollination
Mode of pollination: insect
Favorable temperature range for pollination/seed formation: 75–86°F (24–30°C)
Seasonal cycle: late spring through late summer or early fall (~4–5 months)
Within-row spacing: regular spacing 8–16 in (20–41 cm); hill spacing 4–6 ft (1.2–1.8 m)
Between-row spacing: bush type 5–6 ft (1.5–1.8 m) for regular and hill spacing
Species that will readily cross with crop: *Cucumis sativus* var. *hardwickii* is a wild relative of cultivated cucumber that is only found in the foothills of the Himalayas of northern India.
Isolation distance between seed crops: 0.5–2 mi (0.8–3.2 km), depending on crop type and barriers that may be present on the landscape

the major types in commerce; many regional types and varieties outside of these classes are grown for local fresh market and pickles across many agricultural regions where cucumbers have been grown for thousands of years.

The cucumber is one of the most widely grown vegetable crops across much of the tropical and temperate regions of the world. It is especially valued in tropical climates, where it is often the number one vegetable eaten by volume. In traditional Chinese medicine cucumbers are noted for their "cooling effect" on the body, which is recognized as more than the simple effect gained from eating a fruit largely made up of water. The expression "cool as a cucumber" may have more to it than just describing a person who remains calm under pressure.

Much of the best commercial cucumber seed production is done in relatively hot, arid climates. However, all of the climates that are best suited to cucumber seed aren't excessively hot during the early parts of the season. This includes parts of the Middle East, Southern Europe, North Africa, Israel, Indonesia, Taiwan, and southern China. In North America most of the commercial cucumber seed is currently grown in the Sacramento Valley of California. Before the advent of hybrids, however, significant seed was produced in southern Ontario, southern Quebec, and Michigan.

CROP CHARACTERISTICS

Reproductive Biology

Cucumbers are typically monoecious, with separate staminate (male) and pistillate (female) flowers on the same plant. The staminate flowers usually form first during the longer days near the summer solstice, with a higher concentration of pistillate flowers emerging as the days get shorter. The proportion of male and female flowers is roughly equal during the peak season for fruit production. Cucumbers are usually highly cross-pollinated due to the monoecious condition, though they are self-compatible and will frequently self-pollinate.

There are also cucumber types that are hermaphrodites, with all perfect flowers, and andromonoecious varieties with a combination of staminate and perfect flowers, much like most melons (*C. melo*). The popular North American round-shaped specialty type 'Lemon' is an example of an andromonoecious variety. There is also a gynoecious flowering type that has predominantly female flowers and is used to produce hybrid cucumbers. This gynoecious type was grown in Korea as an open-pollinated variety with a low incidence of monoecious plants when an American plant breeder, Elwyn Meader, first collected it in the late 1940s while working as an agronomist under the Marshall Plan for the US Army. He recognized its potential value as a female parent in producing hybrids and brought it back to the United States, where it was selected for all-female flowering (see "Growing the Seed Crop," page 215).

The male flowers of cucumber are often formed in clusters of three to five flowers, with each attached to the plant on a thin peduncle at a leaf axil. Each of these staminate flowers has five petals that are fused at their base and five stamens tightly packed within each staminal collar. The female flowers are easily identified, as they have the prominent ovary at the base of the flower. These pistillate flowers are usually borne as either a solitary flower or as two or three flowers per node. Any spines that will appear on the fruit will be apparent upon close inspection of the ovary. There are three broad stigmatic lobes on top of a short, thick style. Both the stigmatic surface of the pistillate flower and the anthers of the staminate flower turn a deep golden yellow when sexually mature in most cucumber varieties.

Climatic and Geographic Suitability

Large-scale seed production of cucumbers is done in warm to hot seasonal conditions. Optimum growth and fruit set occur between 75 and 90°F (24 to 32°C), with nighttime temperatures that don't usually dip below 68°F (20°C). Nighttime temperatures below this can cause erratic pollen tube growth, resulting in lower fertilization rates and ultimately lower seed yields. As with all crops, temperatures that are too high at the time of the pollination will also retard pollen tube growth and prevent fertilization and subsequent seed formation. In cucumbers this can happen at temperatures above 95°F (35°C).

▼ A female cucumber flower (*Cucumis sativus*).

A developing cucumber fruit after fertilization of the ovary.

As with many of the cucurbits, the best seed production areas are regions with a reasonable amount of heat during the day, warm temperatures at night, and relatively low humidity. Too high a level of humidity during the growing season should be avoided due to the potential for increased levels of foliar disease.

SEED PRODUCTION PRACTICES

Soil and Fertility Requirements

Fertile loam soils that are well drained are desirable for cucumber seed production. Lighter, sandy loams that are able to warm up easily are especially useful in shorter-season areas with cool springs. Silt and clay loams with good tilth are desirable in longer-season areas for their increased ability to deliver a steady supply of water and nutrients. Heavier soils that are well drained will usually translate to greater seed yields over the course of the season.

Cucumbers are heavy feeders that thrive when ample amounts of well-decayed organic matter or compost are incorporated into the soil. There is a long tradition of using well-decayed horse or cow manure for growing cucumbers, which favors the kind of unchecked growth that produces high yields of fruit and seed. If using the traditional method of planting the crop into hills, then two to three shovelfuls of stable manure are worked into soil at the spot of each hill. A soil pH of between 6.5 and 7.0 is desirable.

Growing the Seed Crop

The cucumber crop is well adapted to producing a full complement of fruit even in many shorter-season temperate climates. In fact, it is

215

often noted that cucumbers require the shortest span of time from planting to production of fruit of any of the common fruit-bearing vegetables. This means that the crop can be direct-seeded in most temperate regions and still mature a reasonable seed crop. Seed can be planted when spring weather has settled and the danger of frost has past. Soil temperatures should be at least 68°F (20°C) at the time of planting, and temperatures between 75 and 95°F (24 to 35°C) will ensure emergence in 3 to 5 days. If planted in rows, seed should be sown liberally, six to eight seeds per foot (30 cm), and then thinned to anywhere from 8 to 16 in (20 to 41 cm) apart, depending on the growth habit of the plants.

Alternately, a more traditional method of growing the crop is to plant the seed into "hills" that are spaced equidistant from each other in the field. The spacing used for cucumber hills is often 5 × 5 ft (1.5 × 1.5 m) or 6 × 6 ft (1.8 × 1.8 m) apart. Placing the hills equidistant from one another allows for "check row" cultivation, where it is possible to run cultivation equipment through the field in both directions. This is a very good option for more thorough mechanical weed control right up until the point where the cucumber vines start to trail out into the open space between the hills. Seed is often planted into the hills at a rate of 6 to 12 seeds per hill, with seed sown in an area that is at least 6 in (15 cm) in diameter. Each hill is then thinned to three or four plants once it is clear which are the most vigorous and healthiest individuals. The idea of a "hill" is also often misunderstood, as this area where the seed is sown may not be a noticeable hill, although sometimes there is an actual raised mound when manure is worked into the spot before planting.

Seed Harvest

Cucumber seed is fully mature when the fruit turns either to a pale ghostly yellow (for white-spined varieties) or to a deep golden yellow (for black-spined types). In order to have seed with fully developed endosperm, good vigor, and the highest possible germination rate, it is very important to allow the fruit to achieve these respective colors and not rush the harvest. In drier temperate climates it isn't unusual for cucumber seed growers to leave the fruit on the plants until the first frost. The mature fruit can then be located, harvested, and tossed directly into a mobile seed extractor, if one is available for larger commercial plants. If the fruit is harvested for hand extraction after frost has killed the vines, it is usually best to get it out of the field in a timely fashion, as repeated frosts and saprophytic organisms can quickly cause excessive rot that may damage some portion of the seed crop. In general, it is never a good idea to allow the fruit to get moldy in the field.

When the fully ripened fruit that is harvested is sound, it is sometimes held for a number of days in a warm, well-aerated place before extracting the seed. A number of growers believe that such an after-ripening period produces an even higher-quality seed than that extracted immediately after harvest. This, of course, is only practical when the seed crop is relatively small.

Seed Extraction, Fermentation, and Cleaning

Mechanical extraction is used for larger commercial seed lots and is usually done with what is often called a cucurbit machine or cucurbit seed extractor. At harvest the fruit is put into a hopper that leads to a series of rollers or rotating blades; these macerate the fruit to free the seed as much as possible from the pulp. The pieces of rind, the pulp, and the seed then drop into a large rotating cylinder made of screen material. Much of the small pieces of the rind, pulp, and liquid go through the

▼ A cucumber fruit that has become too mature for eating, but that is perfect for leaving on the vine to fully ripen its seed.

screen material with the seed into a trough at the bottom of the extractor. Seed that adheres to the pulp is loosened through the spinning motion of the cylinder, passing through the screen. Because the cylinder is at a slight angle, the larger pieces of fruit and pulp slowly work their way to the end of the cylinder and into another trough. Examine the contents of this waste trough before composting them for appreciable amounts of seed that may not have been captured.

Cucumber seed shares an unusual morphological trait with tomato. Both crops have seed that is enclosed within a placental sac that is difficult to remove mechanically. As with tomato, there are basically two options for separating the seed from this gelatinous, tightly adhering sac. The traditional method to rid the seed of this sac is to ferment the freshly extracted seed, thereby loosening the sac by way of the yeast consuming the sugars within this gelatinous membrane. Alternatively, chemical extraction is commonly used in conventional production of cucumber seed. It uses diluted hydrochloric acid and is still very controversial with a number of organic certification agencies. The concentration of hydrochloric acid that is necessary to strip away this membrane make this practice environmentally unacceptable to most proponents of organic agriculture. Therefore, the fermentation seed extraction method will be the only method covered here.

Seed Fermentation: The seed that successfully passes through the screen with smaller pieces of pulp and the peel of the cucumber fruit is then put into barrels or other suitable containers to ferment in a process similar to that used for tomatoes (see chapter 12, Tomato, "Seed Fermentation," on page 323). The fermentation is activated from native, naturally occurring yeasts that are

Cucumber seed can be extracted from the fruit by hand (left) and placed in a container prior to fermentation (above).

The temperature range, duration of fermentation time, and separation of the cucumber seed from the pulp is essentially the same as for tomato.

Seed Cleaning: The final cleaning of cucumber seed through the use of a sluiceway or with the simpler decanting method is identical to that described for tomato. The main difference may be when using the sluice to clean cucumber seed, where the amount and force of the water may need to be increased to clean the heavier seed (see chapter 12, Tomato, "Washing Seed" on page 323). The drying of the seed requires immediate attention, just as with tomato seed, after this wet seed processing.

Fermentation and Seedborne Diseases: Fermentation of cucumber seed is advantageous in helping to control at least one of the most important diseases of cucumber, gummy stem blight (*Didymella bryoniae*). As the environment that exists during the fermentation process can vary considerably from one location to another, it is important to remember that any disease affected by this process may not be completely controlled by it. Gummy stem blight has become one of the most devastating diseases of cucumber in North America over the past 20 years, and organic research needs to be conducted to determine the optimum fermentation conditions to reliably control the transmission of this disease via cucumber seed.

present on the exterior of the cucumbers. The fermentation in cucumber starts quickly and in my experience may even be stronger than what occurs in tomatoes. Certainly, there are very few instances when it is necessary to add baking yeast and sugar to boost the process so long as the seed mixture is kept within an appropriate temperature range. As with tomato fermentation, there is some debate as to whether adding water to the seed mash hinders the fermentation process. Sometimes it is necessary so that all of the seed is submersed in liquid; however there is often enough liquid present, especially from overripe cucumbers, to easily ferment the batch. Also, like tomatoes, there is some debate as to whether the addition of water encourages sprouting. While there isn't any published research on this question, there is speculation that naturally occurring sprout inhibitors may be diluted when water is added to the mix.

GENETIC MAINTENANCE

Several important traits can be checked before the plant begins flowering. As with all crops, seedling vigor and early robust growth are always important traits to select for in crops under organic production systems. Evaluate the cucumber plants for the length of the internodes before flowering; determinate varieties have shorter internodes than the indeterminate types. There are also substantial differences in leaf characteristics like size, shape, and color that can often be detected.

When the plants start to flower it is possible to evaluate and select those of a given variety for uniformity in the timing of flowering and the production of their first female flowers. There is also variation for the number of male or female flowers per node. In most cases it is easy to distinguish the spine color of the fruit, before the ovary has been fertilized and starts to expand. As the fruit develops, the basic shape and the relative length-to-diameter ratio of the fruit are sometimes evident even before the cucumbers have attained a marketable size. The earlier that you can eliminate any off-type plants, the less time that those off-type individuals have to spread their pollen (and hence their genes) around.

As the fruit reach a marketable size it becomes possible to select them for their trueness to type for skin color, the color of mottling on the skin (if present), and the intensity of mottling. Shape and size of the fruit and spine color should also be rechecked at vegetable maturity. Some cucumber varieties may also have raised ribs, which may be distinct for their width and color and are usually distinct for a particular variety if present. Then at fruit maturity it is possible to check for the skin color, which does relate to spine color (see "Seed Harvest," page 216). As with all seed crops, routine elimination of the plants that are most susceptible to any endemic diseases is always worthwhile.

ISOLATION DISTANCES

For cucumbers the basic isolation distance between different cucumber seed crops without barriers is 1 mi (1.6 km), especially when these adjoining crops are relatively small (see chapter 13, Isolation Distances for Maintaining Varietal Integrity).

As with all of the cross-pollinated crops, you should distinguish the distances that are necessary when the two adjacent cucumber crops are of the same horticultural type versus different types of cucumbers. When growing two varieties of the same type of cucumber—for instance, two Beit Alpha types—the minimum isolation distance between them should be 1 mi (1.6 km) in open terrain. When there are larger commercial seed crops of two different types of varieties, if one is a Beit Alpha and the other is an Asian trellis type, then the minimum isolation distance separating the two crops in open terrain should be at least 2 mi (3.2 km). If the seed crops are being grown in terrain where there are natural barriers on the landscape that hinder the movement of insect pollinators, then you can lessen the isolation distance between two cucumber varieties of the same type to 0.5 mi (0.8 km); 1 mi (1.6 km) between two different types of varieties.

Please note that the Armenian cucumber, snake cucumber, and serpent cucumber are actually melons (*Cucumis melo*, Flexuosus group) and not true cucumbers, and therefore they will not cross with cucumbers (*C. sativus*).

MELON

The center of origin for melon (*Cucumis melo* L.) is Africa, but the plant was likely carried to South Asia, its primary center of diversity, at an early stage of its domestication more than 3,000 years ago. Melons quickly spread across equatorial regions of the Eastern Hemisphere, reaching China and East Asia and undergoing further diversification, to a degree that the melons originating from this part of the world are generally considered a different subspecies, *C. melo* subsp. *agrestis,* that is distinct from the melons most frequently grown in the West, *C. melo* subsp. *melo.*

A further division of the prevalent types of melons based on their ancestry and horticultural characteristics is found in the botanical groups that were originally created by Naudin in 1859. While his original groups have been added to and changed often depending on the author, melons are usually divided into between 7 and 15 botanical groups. The five botanical groups of melon that are the most important on a worldwide basis, both commercially and in the scope of their geographic use, are these:

1. **Reticulatus** group includes muskmelons and cantaloupes of North America and Israeli 'Galia' types; their fruit has a rind with a fine reticulated mesh, is aromatic as it matures, and it easily slips from the stalk at full maturity.

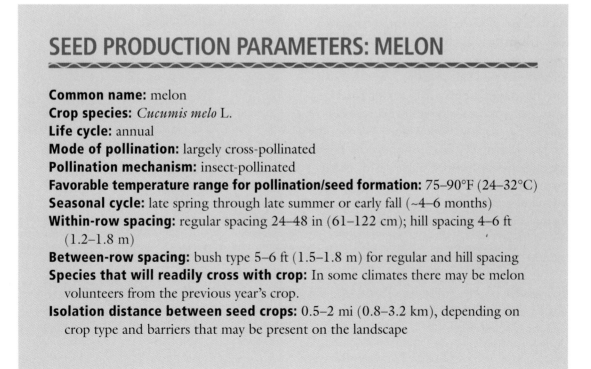

SEED PRODUCTION PARAMETERS: MELON

Common name: melon
Crop species: *Cucumis melo* L.
Life cycle: annual
Mode of pollination: largely cross-pollinated
Pollination mechanism: insect-pollinated
Favorable temperature range for pollination/seed formation: 75–90°F (24–32°C)
Seasonal cycle: late spring through late summer or early fall (~4–6 months)
Within-row spacing: regular spacing 24–48 in (61–122 cm); hill spacing 4–6 ft (1.2–1.8 m)
Between-row spacing: bush type 5–6 ft (1.5–1.8 m) for regular and hill spacing
Species that will readily cross with crop: In some climates there may be melon volunteers from the previous year's crop.
Isolation distance between seed crops: 0.5–2 mi (0.8–3.2 km), depending on crop type and barriers that may be present on the landscape

2. **Cantalupensis** group includes the true European cantaloupes and Charentais types; their fruit has a thick, scaly, or warted rind; slips; and is aromatic.

3. **Inodorus** group includes winter melons, casaba, canary, and honeydew types; their fruit has a fine, smooth rind; does not slip; and has no odor at maturity.

4. **Conomon** group, widely grown across Eastern Asia, includes Asian pickling melons that are not sweet and are used like cucumbers; there also a few types that become sweet at maturity, with a smooth, thin rind and crisp flesh; their fruit does not slip; and it is not aromatic.

5. **Flexuosus** group from the Middle East includes the snake melon, serpent cucumber, and Armenian cucumber, with long curved fruit that is harvested immature and eaten raw or cooked; fruit is ribbed with a wrinkled rind, does not slip; and is not sweet or aromatic.

All of these types are fully sexually compatible and will cross easily when grown too close to one another (see the "Isolation Distances" section, page 227).

While these groups give us a cultural context in which to categorize the most prevalent melon types grown worldwide, it is important to know that much of the melon breeding that has been done since 1960 is based on crosses between varieties from different botanical groups in order to increase storability and disease resistance. This means that in the modern era there are many varieties that are intermediate between these older groups. At present, many seed companies and seed growers are more interested in classifying melons by the practical characters that are most frequently seen as differentiating the most prevalent types and varieties. These groups are distinguished by: (1) fruit color, which can be orange, green, or white; (2) aromatic versus non-aromatic; (3) fruit that slips versus fruit that doesn't slip; (4) fruit that has a netted rind versus a smooth rind; and (5) fruit that softens at maturity versus fruit that remains firm at maturity.

Melons are heat lovers that are best adapted to tropical, subtropical, and some long-season temperate zones. Heat and sun are important in developing the sugars that make many of the melon types so popular across the globe. Much of the best commercial melon seed production is done in hot, arid climates. This includes parts of the Middle East, Southern Europe, North Africa, Israel, Indonesia, Taiwan, Korea, China, and Japan. In North America most of the commercial melon seed is currently grown

'Sharlyn,' is a 'Galia'-type melon (*Cucumis melo*, Reticulatus group) with firm flesh and flavor that is somewhere between a muskmelon and a honeydew. ▶

in the Arkansas River Valley of Colorado, the Central Valley of California, and west Texas.

CROP CHARACTERISTICS

Reproductive Biology

Most melons are andromonoecious, with a combination of hermaphrodite flowers and male flowers on the same plant. The hermaphrodite or perfect flowers have both stamens (male flower parts) and pistils (female flower parts). The hermaphrodite flower is borne singularly in leaf axils and has a short, thick style and a broad three-lobed stigma with nectaries at its base. The anthers in these hermaphroditic flowers are small and do not produce much pollen; thus they contribute only minimally to the pollination of these flowers. The staminate flowers are formed in the leaf axils in clusters of three to five. The form and structure of the staminate flowers is almost identical to that of cucumbers (see Cucumber, "Reproductive Biology" on page 213).

There are also some melon varieties that are monoecious with separate staminate and pistillate (female) flowers on the same plant, much like true cucumbers, *C. sativus*. These monoecious-flowering types are especially found in the *C. melo,* Flexuosus group, where this trait is nearly universal. Under both monoecy and andromonoecy, flowering melons are usually highly cross-pollinated, though they are self-compatible and will frequently self-pollinate. As with some of the other cucurbit species, the first or crown set fruit often does not produce as well in terms of quality or yield as the subsequent two or three fruit to set. Therefore, growers will sometimes pull off the crown set fruit and pinch off the apical growth of the primary leader after they are assured that there are two or three fruit set. In the language of the seed growers, the plants have then been "stopped" to produce higher quality and yield in the fruit that are favored.

Climatic and Geographic Suitability

Large-scale seed production of melons is done in hot seasonal conditions. Optimum growth and fruit set occur between 80 and 95°F (27 to 35°C), with nighttime temperatures that don't usually dip below 70°F (21°C). Nighttime temperatures below this can cause erratic pollen tube growth, resulting in lower fertilization rates and ultimately lower seed yields. As with all crops, temperatures that are too high at the time of the pollination will also retard pollen tube growth and prevent fertilization and subsequent seed formation. In melons this can happen at temperatures above 100°F (38°C). For melons the best seed production areas are regions with a significant amount of heat during the day, warm temperatures at night, and relatively low humidity. Too high a level of humidity during the growing season should be avoided due to the potential for increased levels of foliar disease.

SEED PRODUCTION PRACTICES

Soil and Fertility Requirements

Fertile loam soils that are well drained are desirable for melon seed production. Melons form a large, deep taproot that is at least 3 ft (0.9 m) deep. Heavier clayey loam soils are desirable for their nutrient- and water-holding capacity but must be free of waterlogging. Lighter, sandy loams that are able to warm up easily are especially useful in establishing early crops to avoid the hottest seasonal temperatures during the period of fruit set (see "Climatic and Geographic Suitability," above).

Melons, like all cucurbits, are heavy feeders, and they thrive when ample amounts of well-decayed organic matter or compost are incorporated into the soil. There is a long tradition of using well-decayed horse or cow manure for growing all the cucurbits, which favors the kind of unchecked growth that produces high yields of fruit and seed. If you're using the traditional method of planting the crop into hills, then work two to three shovelfuls of stable manure into soil at the spot of each hill. Turning in leguminous cover crops well in advance of planting has also long been used to grow robust melon crops. Melons and other cucurbits seem to perform best when at least half of their nitrogen is derived from organic sources. This gives good organic farmers a high likelihood of producing good yields if the climate is well suited to melon production. A soil pH of between 6.5 and 7.5 is desirable. Melons grown in arid regions are often grown successfully on slightly alkaline soils.

Growing the Seed Crop

Melons are traditionally direct-seeded in climates that are well suited to seed production. Seed can be planted when warm weather has truly arrived, all danger of frost has passed, and soil temperatures are at least 75°F (24°C); it is even better when the soil temperature is well above 80°F (27°C), with 90°F (32°C) being the optimum. This enables the unchecked growth that will produce healthy, vigorous plants if the soil, nutrition, and soil moisture are adequate.

If planted in rows melon seed crops need to be given more spacing than the 18 in (46 cm) that is often used when growing for fresh harvest. Melon plants for seed should be planted at least 24 to 48 in (61 to 122 cm) apart within the row, depending on the vigor and size of the variety. Several seeds can be planted at the appropriate interval for the variety and then thinned once you are assured that the selected plants are well established and healthy. The rows should be at least 6 to 7 ft (1.8 to 2 m) apart for optimum production.

Alternatively, a more traditional method of growing the crop is to plant the seed into hills within the rows in the field. The spacing used for melon hills can be anywhere between 4 ft (1.2 m) apart within the row by 6 ft (1.8 m) between the rows to the comparable 6 × 8 ft (1.8 × 2.4 m) apart. Placing the hills equidistant from one another at 6 × 6 ft (1.8 × 1.8 m) spacing allows for "check row" cultivation, where it is possible to run cultivation equipment through the field in the same fashion in both directions. This is a very good option for more thorough mechanical weed control right up to the point where the melon vines start to trail out into the open space between the hills. Seed is often planted into the hills at a rate of six to eight per hill, with seed sown in an area that is at least 6 in (15 cm) in diameter. Plants are then thinned to two or three once it is clear which are the most vigorous and healthiest individuals in each hill; make sure, however, that the plants aren't crowded during this pre-thinning period. The idea of a "hill" is also often misunderstood, as this area where the seed is sown may not be a noticeable hill, although sometimes there is an actual raised mound when manure is worked into the spot before planting.

Seed Harvest

For optimum seed quality and ease of harvest it is best to allow the melon fruit to remain on the plant for 1 to 2 weeks after the majority of favorable plants have fully matured. This after-ripening period serves multiple purposes:

1. It allows the ripe fruit to fully mature, as cucurbit fruit often produces a somewhat better seed quality if fruit is allowed to

"after-ripen" after the fruit has reached the ripe edible stage.

2. It allows any laggard high-quality fruit to finish ripening.

3. It softens the placental tissue in the seed cavity to the point where it easily separates from the seed upon cleaning.

Fruit maturity can be detected in multiple ways depending on the type of melon. In many cases the "ground color" or major color of the rind will have either a major or minor change in the color, from some shade of green to a warmer shade of color with a hue of yellow or gold. For many varieties from the North American types, most 'Galia' types, and many of the 'Ananas' and true cantaloupes the fruit will slip from the plant, whereas the honeydew/casaba varieties (Inodorous group), the Charentais varieties, and the Armenian cucumber (Flexuosus group) do not slip from the plants and need an experienced grower to know when the plants are ripe.

If the seed crop is relatively small and the fully ripened fruit is sound, the fruit can be gathered and held for a number of days in a warm, well-aerated place to complete the after-ripening, particularly if outdoor conditions are unfavorable. Otherwise, the seed extraction is done in the field, either by hand or through the use of a vine seed extractor as specified below.

Seed cavity of a ripe North American cantaloupe. Unlike cucumber seed, melon seed is not enclosed within a placental sac. Despite this fact, most seed growers still put the seed through a 1- to 2-day fermentation step during prior to final cleaning and drying.

Seed Extraction, Fermentation, and Cleaning

Mechanical extraction is used for larger commercial seed lots and is usually done with what is often called a cucurbit machine or vine seed extractor. At harvest the ripe fruit is put into a hopper that leads to a series of rollers or rotating blades, which macerate the fruit to free the seed as much as possible from the pulp. The pieces of rind, the pulp, and the seed then drop into a large rotating cylinder made of screen material. Much of the small pieces of the rind, pulp, and liquid go through the screen material with the seed into a trough at the bottom of the extractor. Seed that adheres to the pulp is loosened through the spinning motion of the cylinder, passing through the screen. Because the cylinder is at a slight angle, the larger pieces of fruit and pulp slowly work their way to the end of the cylinder and into another trough. Examine the contents of this waste trough before composting them for any appreciable amounts of seed that may have not been captured.

Melon seed, despite its resemblance to cucumber seed, does not have a placental sac and technically does not require the same kind of fermentation step that frees the cucumber seed from the sac. After the extraction described above, the seed is often cleaned under a high-pressure spray of water that removes any of the orange or green placental material from the seed. While this is often enough to clean the seed, some growers give the melon seed a 1- to 2-day fermentation period, which helps free it from all placental material and also helps avoid any staining that can occur in orange-fleshed types from the beta-carotene in the placenta. The problem with fermentation in all of the cucurbits other than cucumber is that it can easily damage the seed and lower the germination percentage if it is done for too long. If the conditions are right, only about 24 hours of fermentation is necessary. It is the same process described for tomato (see chapter 12, Tomato, "Seed Fermentation" on page 323). Watch the process closely, and separate the seed from the fermentation bath as soon as the desired results are reached.

Seed Cleaning: The final cleaning of melon seed through the use of a sluiceway or with the simpler decanting method is identical to that described for tomato. The main difference may be that when using the sluice, the amount and force of the water may need to be increased to clean the heavier melon seed (see chapter 12, Tomato, "Washing Seed" on page 323). The drying of the seed needs immediate attention, just as with tomato and cucumber seed, after this wet seed processing.

GENETIC MAINTENANCE

For melon, as with all of the cucurbit crops, there are several important traits that can be evaluated before the plant begins flowering. Seedling vigor and early robust growth will lead to the establishment of healthy seedlings. Because melons are more exacting than many other crops in their need for favorable environmental conditions to produce a healthy seedling, it is definitely important to select and maintain a level of vigor and health in all of your seedstocks. This is especially true for seed produced for organic production systems. It is also important to evaluate melon plants for the length of the internodes before flowering: Determinate varieties have shorter internodes than the indeterminate types. There are also substantial differences in leaf characteristics such as size, shape, and color that can be detected and rogued out if any off-types appear before flowering.

Once the melon crop starts to flower, it is possible to evaluate and select the plants of a given variety for uniformity in the timing of flowering and the production of their first hermaphrodite flowers, which can be easily recognized by the presence of the attached ovary. Evaluation and selection for the number of pistillate flowers per node and the proportion of pistillate to hermaphroditic flowers can be considered, as with other cucurbit crops.

As the fruit develop, their basic shapes are sometimes evident long before they have attained a marketable size. Melon fruit shape can vary greatly from type to type, from the globe or almost round shape of many cantaloupes to the long serpent shapes of Armenian cucumbers. If you are monitoring the crop from an early stage and watching both the flowering habit and growth of the first hermaphroditic flowers and the setting of the fruit, then you can monitor the shape and type of fruit right from the formation of the ovaries. Indeed, as with all cucurbits, you can select for fruit shape even before the ovary has been fertilized and the seed begins to form. (See the "Genetic Maintenance" section in the Cucurbitaceae family introduction on page 210.) As with all seed crops where the fruit is the main agricultural crop of interest, the earlier you determine that a fruit is not true to type, the earlier you can rogue out the plant and eliminate it as a source of pollen that might contaminate the rest of the seed crop.

With the melons, as with other cucurbits, it is often possible to observe and select for traits other than just fruit shape, including the major or ground color of the fruit, the depth of the color, and whether the fruit is ribbed, has warts, or has prominent sutures. You can also see the extent of the mesh or netting as the fruit matures to get a final idea of the true ground color and secondary color of the mesh, ribs, and sutures. As stated for all of the cucurbits, these traits change over time, and a good grower familiar with a particular variety can often recognize the presence or absence of an important trait at the earliest juncture and quickly eliminate off-type plants from the population. At maturity, it is possible to see all of the external characters already mentioned in their full glory for evaluation. You can also then check individual plants for internal traits such as color, intensity of color, and seed cavity size as well as sensory characteristics like flavor, sweetness, texture, and juiciness.

ISOLATION DISTANCES

For melons the basic isolation distance that should be used between different melon seed crops without barriers is 1 mi (1.6 km), especially when these adjoining crops are relatively small (see chapter 13, Isolation Distances for Maintaining Varietal Integrity).

As with all of the cross-pollinated crops, you should distinguish between the distances necessary when the two adjacent melon crops are of the same horticultural type versus when there are two different types of melons being grown. When growing two melon varieties that are of the same type—for instance, two honeydews with pale green flesh—then the minimum isolation distance between them needs to be 1 mi (1.6 km) in open terrain. When there are seed crops of two different types of varieties—for instance, if one is a North American muskmelon with orange flesh and the second is a honeydew type with pale green flesh—then the minimum isolation distance separating the two crops in open terrain should be 2 mi (3.2 km). Indeed, even if both varieties were honeydew types, but they were distinctly different in their days to maturity, size, disease resistance profile, or flesh color (there are white, pale green, and orange-fleshed honeydews), then they would also require this increased isolation distance in open terrain.

If two melon seed crops are being grown in terrain where there are significant natural barriers on the landscape that hinder the movement of insect pollinators, then you can lessen the isolation distance between two varieties of the same type to 0.5 mi (0.8 km) and to 1 mi (1.6 km) between two varieties of distinctly different types.

Note: It is important to remember that melon (*C. melo*) plants *will not cross* with true cucumber plants (*C. sativus*). But all growers of *any* seed crop in the *Cucumis* genus should be thoroughly certain whether the crop they are growing is one of the many variants of a melon (*C. melo*) or a true cucumber (*C. sativus*). There are a number of *C. melo* types within either the Flexuosus or Conomon botanical groups that are either called cucumber in English or have a form more akin to a cucumber, and that are often eaten before full maturity like a cucumber. As an example, there have been seed growers who have contaminated their seedstocks when growing a melon crop near an Armenian cucumber crop, thinking that the latter was a true cucumber and that the two would not cross. All members of the *C. melo* genus, no matter how diverse in phenotype, will cross readily if proper minimum isolation distances are not observed.

SQUASH

Squash is the general term used to refer to the five species of *Cucurbita* that bear large edible fruit. These species are indigenous to the New World and have long been important cultivated crops in the Western Hemisphere. All five of these species produce fruit that are sometimes eaten when immature (known as summer squash) as well as in the mature stage of growth (referred to as winter squash). All of the squash types bear edible seed that is an important part of the diet in some cultures. Though less common, flowers and leaves of squash are also eaten in some instances. The fruit of the wild progenitors of all of the modern squash types are extremely bitter, and it is believed that the first agricultural societies to domesticate these species did so primarily for their edible seed. As genetic variants with non-bitter fruit flesh were discovered, farmers selected and bred these non-bitter types for culinary use. These selected plants with their edible fruit became the ancestors of our modern-day squash types.

The five domesticated species of the *Cucurbita* genus produce a wide array of squash types that are used around the world as fresh vegetables, storage vegetables, seed crops, and ornamental and ceremonial icons in some cultures. Any list of important squash types will leave out many significant regional types and

SEED PRODUCTION PARAMETERS: SQUASH

Common names: winter squash, pumpkin, summer squash, marrow, gourd, zucchini

Crop species: *Cucurbita pepo* L., *C. maxima* Duchesne, *C. moschata* Duchesne, *C. argyrosperma* Huber, *C. ficifolia* Bouché

Life cycle: annual

Mating system: largely cross-pollinated, with some self-pollination

Mode of pollination: insect

Favorable temperature range for pollination/seed formation: 72–90°F (22–32°C)

Seasonal reproductive cycle: mid- to late spring through fall (4–5 months)

Within-row spacing: bush types 1–4 ft (0.3–1.2 m); bush types in hills 3–4 ft (0.9–1.2 m); vine types 1.5–3 ft (0.45–0.9 m); vine types in hills 6–8 ft (1.8–2.4 m)

Between-row spacing: bush types 3–5 ft (0.9–1.5 m); vine types 6–12 ft (1.8–3.6 m)

Species that can cross-contaminate: Many small-fruited ornamental gourds are *Cucurbita pepo* var. *ovifera* and will readily cross with all *C. pepo* vegetable crops.

Isolation distance between seed crops: 0.5–2 mi (0.8–3.2 km), depending on crop type and barriers that may be present on the landscape

Acorn (left) and Yellow Crookneck (right) are two of the most popular commercial types of *C. pepo*.

sub-types, but an overview of these five species may at least catch a glimpse of the major types and their uses.

***Cucurbita pepo* L.** is by far the most widely grown and economically important species worldwide. There were two centers of origin for this species that occurred independently of each other. The first (*C. pepo* subsp. *pepo*) occurred in southern Mexico about 10,000 years ago and the second (*C. pepo* subsp. *ovifera*) in what is now the southeastern United States at least 4,000 years ago. *C. pepo* can be grown across many temperate zones and is somewhat tolerant of cool weather in maturing fruit. This species is extremely polymorphic and includes most of the commercial varieties that are grown and eaten as immature fruit or summer squash. The major *C. pepo* summer squash cultivar groups include zucchini, cocozelle, crookneck, straightneck, vegetable marrow, and scallop. Also in this species are most of the pumpkins used as Halloween jack-o'-lanterns and varieties grown for edible seed, including variants known as the hull-less seeded types. Winter squash types include the acorn cultivar group and unusual varieties such as 'Spaghetti' and 'Delicata'. Winter squash of *C. pepo* usually do not store as well as those of *C. maxima* or *C. moschata*, and as a consequence, they were often known as autumn squash by squash devotees of an earlier era in North America. Also in this species are small gourds used as ornaments and that occur in a wide range of colors (often striped) and shapes. The wild forms of *C. pepo* (known variously as var. *ozarkana*, var. *texana*, and subsp. *fraterna*) produce bitter gourds. These grow in the southeastern US and northeastern Mexico and cross readily with their cultivated relatives, thus contributing genes to crop populations of *C. pepo* from time to time.

C. maxima Duchesne is probably the second most widely grown squash species. It is primarily used as a winter squash, although there are some types that are used as a fresh vegetable when harvested at an immature stage. Like *C. pepo*, some varieties are well adapted to cooler climatic zones—notably

some of the giant-fruited *C. maxima* varieties grown high in the Andes Mountains in Bolivia and Peru and neighboring countries. *Cucurbita maxima* is typically cultivated in cooler areas than *C. moschata* or *C. argyrosperma*, although there are *C. maxima* types adapted to hot areas ranging from the humid Amazon River basin to the arid southwestern United States. The center of diversity for *C. maxima*

▼ The maturity of *Cucurbita maxima* squash, like this 'Red Kuri' (a.k.a. 'Uchiki Kuri'), can be judged by the color and texture of the fruit's stem, which will turn tan and corky as it matures.

is the temperate zone of the Andean region of northern South America. Its wild progenitor, *C. maxima* subsp. *andreana,* is still found in much of this region and readily crosses with the cultivated types, imparting the bitter fruit flesh trait found in wild *Cucurbita* species and reinvigorating the domesticated types with genetic variability. Unlike *C. pepo* and *C. moschata*, which both spread widely through trade before the era of European conquest, it appears that *C. maxima* remained close to its center of diversity until Spanish and Portuguese sailors carried seed of it along their trading routes. From that point onward it spread quickly, and selection for the varied types and sizes of fruit has resulted in a wide diversity across the many cultures that grow them. Many of these squash have very high-quality flesh, both in terms of the sugars and dry matter that confer good eating quality and in terms of the carotenoid pigments that give the flesh beautiful shades of orange and deep yellow color and are nutritionally significant for their vitamin A and nutraceutical content. In fact, of all the squash species, *C. maxima* fruit are the most nutritionally significant based on our present knowledge of nutrition.

A number of *C. maxima* varieties hold the distinction of having the largest fruit of any plants on Earth. Some of the show pumpkin varieties bred in the recent past and their counterparts in the Andes can weigh more than 1,000 lbs (454 k) when grown by a skilled grower. Other *C. maxima* types that have been important over the past century are in the buttercup, banana, Hubbard, marrow, delicious, turban, and Kabocha classes. The Kabocha type, derived from crosses of 'Buttercup' with various Japanese long-storage varieties, has become commercially important across a number of temperate regions around the world. Most *C. maxima* squash are harvested and eaten at full maturity when they have developed

their maximum sweetness, solids, and carotenoid content. However, people also harvest immature fruit to eat as summer squash.

C. moschata **Duchesne** is an important species of squash in many tropical, subtropical, and warm temperate zones, as this species needs more heat than *C. pepo* and *C. maxima* to mature fruit and produce a commercially acceptable seed crop. There is some evidence that either there are two independent locations that served as centers of origin for this species or that one followed the other after trade moved the seed from one place to another. One of the problems in identifying a likely region of domestication is that the progenitor wild species for *C. moschata* is unknown and may be extinct. One center of diversity for *C. moschata* is the coastal lowlands of Peru, where there is evidence that *C. moschata* was first cultivated at least 5,000 years ago. A second putative area of domestication was in the border region between southern Mexico and Guatemala, at least 4,000 years ago.

When Columbus arrived in the Caribbean he found *C. moschata* widely cultivated by indigenous peoples. It was also found to be cultivated in parts of what is now the southeastern United States long before the arrival of Columbus in the New World. A variety now called 'Seminole Pumpkin' was grown by the Seminole people of Florida. And though *C. moschata* is especially well adapted to hot, humid climates, farmers eventually developed varieties through selection that could be grown as far north as the Great Lakes region of North America.

The fruit are usually either smooth and tan-colored with a short bell shape or a longer crookneck, or they are the short, often furrowed, oblate-shaped pumpkin types that have a tan- to buff-colored rind and flesh that is yellow to pale orange, or sometimes blackish green. The short bell types include the popular, high-quality butternut varieties, while the traditional crookneck varieties and the short, furrowed Cheese pumpkins do not have such a high-quality flesh and are used both for table stock and as fodder for livestock. The carotenoid content of most of the varieties in this species does not approach the levels of the best *C. maxima* varieties. Varieties of *C. moschata* are sometimes grown for production and sale of their nutritious seed. Widely grown throughout the Americas, *C. moschata* also has centers of diversity in Asia and Africa.

▼ The 'Tahitian' squash (*C. moschata*), seen here in the foreground, is quite similar in appearance to many *C. argyrosperma* varieties. Determining the species of the seed crop is always important to know if isolation is necessary.

C. argyrosperma **Huber** (formerly *C. mixta* Pang.) varieties were long thought to be part of the *C. moschata* species, as they share key morphological features. One distinctive trait of certain varieties of this species is their distinct silver seed, hence the Latin species name *argyrosperma*. The center of diversity for this species is southern Mexico, probably around 8,000 years ago. Its wild progenitor, *C. argyrosperma* subsp. *sororia*, produces bitter gourds (as do all other wild *Cucurbita* species) and grows as a dominant weed throughout much of lowland Mexico, where it often hybridizes with its cultivated counterpart in farmers' fields. Historically, the geographic spread of *C. argyrosperma* was slow compared with *C. pepo*, *C. maxima*, and *C. moschata*, though it has spread to other tropical climatic regions, including Africa.

The cultivated varieties of *C. argyrosperma* have smooth rinds that are striped or solid with pale greens or earth tones and either are shaped as oblate spheres or have long necks attached to a small bell that contains the seed cavity. The amount of phenotypic diversity in this species is limited, as are the number of commercial varieties; two examples of the latter include the 'Silver Seeded Gourd' and the 'White Cushaw' or 'Green Striped Cushaw'. Many varieties of *C. argyrosperma* were selected almost exclusively for their large nutritious seed, rather than for the fruit flesh; the culinary quality of the flesh is generally poor, pale, and rather low in carotenoid content. There are a number of regions throughout Mexico, Central America, and the Caribbean where landraces or farmer varieties are still grown and continue to adapt to the needs of the farmers.

Cucurbita ficifolia **Bouché** is the fifth domesticated species of *Cucurbita,* but the least well known because of its limited geographic distribution. It is sometimes called

233

the Malabar gourd because it was incorrectly thought to have originated in India. Like the other cultivated *Cucurbita, C. ficifolia* originated in the Americas—probably at high altitude in southern or central Mexico, although its wild progenitor is still unknown. *C. ficifolia* fruit look somewhat like watermelons in shape, size, and color pattern and typically have white fruit flesh (not found in any other squash species!) and flat black (less commonly tan) seed. It is cold-tolerant relative to other cultivated *Cucurbita,* and because of this trait it has at times been used as a rootstock for grafting cucumbers. Perhaps due to its capacity for long, sprawling, vigorous growth, *Cucurbita ficifolia* is erroneously said to be a perennial, although like all of the other four cultivated squash species it is an annual. One of the most popular ways to consume *C. ficifolia* in Latin America is prepared with raw sugar as a candied fruit, or at times it is used as a food for livestock.

Attempts have been made in some arid climates (Arizona in the United States and in Australia, in particular) to domesticate one of the wild, xerophytic wild species, *C. foetidissima* HBK (called the buffalo gourd), as either a starch crop or an oilseed crop. The very large storage root produced by this species is a good source of carbohydrates, and the large quantity of seed per fruit is rich in oils and proteins. The commercial ventures targeting development of these new crops, however, were not widely successful and have been largely abandoned.

CROP CHARACTERISTICS

Reproductive Biology

All squash are monoecious, meaning that there are separate staminate (male) and pistillate (female) flowers on the same plant. The staminate flowers usually form first during the longer days near the summer solstice with a higher concentration of pistillate flowers that emerge, as the days get shorter. The proportion of male and female flowers is typically 4:1 to 5:1 but may be roughly equal during the peak season for fruit production. Squash are usually highly cross-pollinated due to the monoecious condition, though they are self-compatible and will frequently self-pollinate.

The flowers of squash are among the largest of all vegetable crops. The petal color ranges from pale yellow to a deep orange-yellow, depending on the species, type, and variety. The male flowers are formed on a stout peduncle at a leaf axil. Each of these staminate flowers has fused petals with staminal collars that contain three stamens flanked by ample nectaries. The female flowers also produce ample amounts of nectar and are easily identified as in all cucurbits by the prominent ovary at the base of the flower petals. These pistillate flowers have a broad stigma with two lobes on top of a short thick style. The ovaries may have either three or five carpel chambers. The bountiful nectar produced in both the staminate and pistillate flowers is the primary attractant to most pollinating species that visit squash flowers. The sticky, large pollen grains produced in staminate flowers are easily picked up by pollinating insects entering the flowers to get nectar. Squash are typically pollinated by bees—either the common honeybee, *Apis mellifera,* or a range of native bees, including most notably the so-called squash or gourd bees of the genera *Peponapis* and *Xenoglossa.* The latter are solitary, ground-nesting native bees thought to have co-evolved with *Cucurbita* throughout much of the Americas. In places where they are numerous, they are often active well before dawn, beating out the later-rising honeybees to the floral resources in the wild or cultivated squash flowers.

Climatic and Geographic Suitability

Large-scale seed production of squash is done in warm to hot seasonal conditions. Optimum growth and fruit set occur between 72 and 90°F (22 to 32°C), with nighttime temperatures that don't usually dip below 68°F (20°C). Nighttime temperatures below this can cause erratic pollen tube growth, resulting in lower fertilization rates and ultimately lower seed yields. As with all crops, temperatures that are too high at the time of the pollination will also retard pollen tube growth and prevent fertilization and subsequent seed formation. In most squash varieties this can happen at temperatures above 95°F (35°C). As with many of the cucurbits, the best seed production areas feature a reasonable amount of heat during the day and warm nighttime temperatures.

▼ A male squash flower (left) and a female flower (right), with the developing fruit at its base.

SEED PRODUCTION PRACTICES

Soil and Fertility Requirements

Fertile loam soils that are well drained are desirable for squash seed production. Lighter, sandy loams that are able to warm up easily are especially useful in shorter-season areas with cool springs. Silt and clay loams with good soil tilth are desirable in longer-season areas for their increased ability to deliver a steady supply of water and nutrients. Heavier soils that are well drained will usually translate to greater seed yields over the course of the season for this reason.

Squash plants are heavy nitrogen feeders, and they thrive when ample amounts of well-decayed organic matter or compost are incorporated into the soil. There is a long tradition of using well-decayed horse or cow manure for growing squash and other cucurbits such as cucumbers, which favors the kind of unchecked growth that produces high yields of fruit and seed. If you are using the traditional method of planting the crop into hills, then work two to three shovelfuls of weed-free stable manure into the soil at the location of each hill. A soil pH between 6.5 and 7.0 is desirable.

Growing the Seed Crop

The diverse types of squash have a wide range of climates to which they are adapted for producing a full complement of mature fruit. In fact, it is often noted that *C. pepo* squash require a relatively short span of time from planting to fruit production for what is one of the largest of the common fruit-bearing vegetables. This feature means that the crop can be direct-seeded in most temperate regions and still mature a reasonable seed

crop. In contrast, *C. moschata* and *C. argyrosperma* generally need relatively long seasons to produce good yields of a high-germinating seed crop. Seed for all of the species can be planted when spring weather has settled and the danger of frost has passed. Soil temperatures should be at least 68°F (20°C) at the time of planting, and temperatures between 75 and 95°F (24 to 35°C) will ensure emergence in 3 to 5 days.

The modern spacing for seed production on bush-type squashes (that is, varieties selected genetically to have short vine internode lengths) can be as little as 12 in (30 cm) between plants within the row and 18 in (46 cm) between plants for vining types. The modern spacing used between rows for bush types is sometimes as close as 3 ft (0.9 m) to as much as 6 ft (1.8 m) for vining types. This tight spacing, especially on the within-row spacing for bush-type cultivars, is generally much closer than the spacing that was traditionally used in squash seed production before the late 20th century. This shift in cultural practice is attributable, in large part, to the following: First, much of the conventional squash seed production is now done using high external inputs, essentially delivering higher nutrients to the smaller root volume per plant that results when the plants are crowded. Second, much modern squash seed that is produced is F_1 hybrid seed. Since the plants of the inbred lines that are used as parents in hybrids are sometimes less vigorous than the plants of the OP varieties, this tight spacing in hybrid seed production is adequate for these less vigorous inbred lines.

The more traditional method of growing the crop, as a vegetable or as a seed crop, is to plant the seed into hills placed equidistant from one another in the field. The spacing used for hills in bush types is usually 3 to 4 ft (0.9 to 1.2 m) apart within the rows and 4 to 5 ft (1.2 to 1.5 m) between rows. This spacing can vary based on the size and bushiness of the typical plant of the variety. (See Cucumber, "Growing the Seed Crop" on page 215 for a more general discussion on seeding the hills.) Throughout much of Latin America and elsewhere in the tropics, squash is commonly planted (in hills) as an intercrop with corn, even in large-scale commercial plantings.

The spacing for the vining types of squash when planted into hills can vary even more than the bush types due to the wide range in the size of the vines across varieties. Some shorter-vining semi-bush types only trail up to 3 ft (0.9 m), while some of the most vigorous *C. maxima* varieties have vines that trail up to and sometimes beyond 12 ft (3.7 m). The within-row spacing between hills for the semi-bush or short-vined types can be 4 to 6 ft (1.2 to 1.8 m), with 6 to 8 ft (1.8 to 2.4 m) between rows. Larger-vining types have traditionally been planted into hills that are 6 to 10 ft (1.8 to 3 m) apart within the row and 8 to 12 ft (2.4 to 3.7 m) apart between rows. This between-row spacing depends upon the vigor of the vines.

A number of organic squash seed growers are also using an equidistant spacing between hills for "check row" cultivation as is used in a number of cucurbit crops. This is most advantageous with the smaller vining types in squash where a spacing of 5 × 5 ft (1.5 × 1.5 m), 6 × 6 ft (1.8 × 1.8 m), and up to 8 × 8 ft (2.4 × 2.4 m) is used efficiently. Placing the hills equidistant from each other allows for "check row" cultivation, where it is possible to run cultivation equipment through the field in both directions. This method is a very good option for more thorough mechanical weed control right up to the point where the squash vines start to trail out into the open space between the hills.

Seed Harvest

Squash seed is mature when the fruit's color is fully developed and when the stem has completely dried to its mature color. In order to have seed with fully developed endosperm, good vigor, and the highest possible germination rate, it is very important to allow the fruit to achieve these respective colors and not rush the harvest. In drier temperate climates it isn't unusual for squash seed growers to leave the fruit on the plants until light frost occurs. The mature fruit can then be located, harvested, and tossed directly into a mobile seed extractor if one is available for larger commercial seed lots. If the fruit is harvested for hand extraction after frost has killed the vines, it is usually best to get it out of the field in a timely fashion, as repeated frosts and saprophytic organisms can quickly cause excessive rot that may damage some portion of the seed crop. Prolonged exposure to temperatures below 50°F (10°C) generally results in chilling injury. In general, it is never a good idea to allow the fruit to get moldy in the field. In tropical climates, mature squashes are harvested when the rind loses luster and vine tendrils close to the fruit dry and shrivel.

When the fully ripened fruit that is harvested is sound, it is sometimes held for a number of days in a warm, well-aerated place before extracting the seed crop. Unlike many other crops, there is good evidence that for *Cucurbita* an after-ripening period produces an even higher-quality seed than when the seed is extracted immediately after harvest. This post-harvest storage is, of course, only practical when the seed crop is relatively small. If necessary for prolonged storage, winter squash can be kept anywhere from at least 5 weeks (*C. pepo* acorn types) to perhaps 6 months (*C. maxima* Kabocha types), depending on the variety, and on whether there is a suitable rodent-free storage area that is cool

◄ SQUASH SEED HARVEST AND EXTRACTION: (A) Hard-shelled squash are split open in the field before extraction; (B) Squash fruit are thrown into the hopper of the mechanical seed extractor, leading to a series of rotating blades; (C) A screened cylinder separates the seed from the main mass of flesh and rind; (D) Seeds along with small pieces of pulp and rind are diverted to a side trough, with larger pieces of squash ending up in the waste trough at the end of the extractor cylinder; (E) The seed and any adhering pulp are collected; (F) Seed and pulp are poured into containers for the fermentation step.

and dry (temperatures optimally between 50 and 59°F/10 and 15°C, with a relative humidity of 50 to 75%). Rodents can pose a significant problem in squash seed production if they are allowed to consume seed by burrowing into the fruit wall during storage.

Seed Extraction, Fermentation, and Cleaning

Mechanical extraction is used for larger commercial seed lots and is usually done with what is often called a cucurbit seed extractor. At harvest the fruit is put into a hopper that leads to a series of rollers or rotating blades; these macerate the fruit to free the seed as much as possible from the pulp. The pieces of rind, the pulp, and the seed then drop into a large rotating cylinder made of screen material. Much of the small pieces of the rind, pulp, and liquid go through the screen material with the seed into a trough at the bottom of the extractor. Seed that adheres to the pulp is loosened through the spinning motion of the cylinder, passing through the screen. Because the cylinder is at a slight angle, the larger pieces of fruit and pulp slowly work their way to the end of the cylinder and into another trough. Examine the contents of this waste trough before composting them for any appreciable amounts of seed that may have not been captured.

Seed Fermentation: The seed that successfully passes through the screen with smaller pieces of pulp and the peel of the squash fruit is then put into barrels or other suitable containers to ferment in a process similar to that used for tomatoes (see chapter 12, Tomato, "Seed Fermentation" on page 323). The fermentation is activated from native, naturally occurring yeasts that are present on the exterior of the squash. The fermentation in cucurbits starts quickly and in my experience may even be stronger than the fermentation that occurs in tomatoes, but such fermentation is not common for squash seed processing. Certainly, there are very few instances where it is necessary to add baking yeast and sugar to boost the process, so long as the seed mixture is kept within an appropriate temperature range. As with tomato fermentation, there is some debate as to whether adding water to the seed mash hinders the fermentation process. Sometimes it is necessary to add water so that all of the seed is submersed in liquid; however, there is often enough liquid present, especially from overripe squash fruit, to easily ferment the batch. Also, like tomatoes, there is some debate as to whether the addition of water encourages sprouting. While there isn't any published research on this question, there is speculation that naturally occurring sprout inhibitors may be diluted when water is added to the mix.

The temperature range, duration of fermentation time, and separation of the squash seed from the pulp are essentially the same as for tomato.

Seed Cleaning: The final cleaning of squash seed through the use of a sluiceway or with the simpler decanting method is identical to that described for tomato. The main difference may be that when using the sluice to clean squash seed you may need to increase the amount and force of the water to move the heavier squash seed through the sluice (see chapter 12, Tomato, "Washing Seed" on page 323). The drying of the seed needs immediate attention, just as with tomato seed, after this wet seed processing.

GENETIC MAINTENANCE

As with all vining crops where the vines of individual plants tend to intermingle, it is important to consider increasing the spacing between plants in order to clearly identify the phenotype of individual plants if selection to type is necessary. Plants of the largest vining types can quickly become indistinguishable when you're trying to discern the differences between individual plants in the field by the time the crop is flowering.

Several important traits can be checked before the plant begins flowering. As with all crops, seedling vigor and early robust growth are always important traits to select for in all crops under organic production systems. You can evaluate squash plants for the length of the internodes before flowering: Determinate varieties (typically called bush types in squash) have shorter internodes than the indeterminate types. There are also subtle to substantial differences in leaf characteristics such as size, shape, degree of lobing, and color that can often be detected.

Once the plant starts to flower, it is possible to evaluate and select the plants of a given variety for uniformity in the timing of flowering and the production of their first female flowers. There is also variation for the number of male or female flowers per node. In most cases it is easy to distinguish the spine color of the fruit before the ovary has been fertilized and starts to expand. As the fruit develops, the basic shape and the relative length-to-diameter ratio of the fruit are sometimes evident even before the squash have attained a marketable size. The earlier you can eliminate any off-type plants, the less time any off-type individual has to spread its pollen (and hence its genes) around.

As the fruit reach a marketable size it becomes possible to select them for their trueness to type for color and the intensity of color. Shape and size of the fruit should also be rechecked at vegetable maturity. Some squash varieties may have raised ribs, which may be distinct for their width and color and are usually distinct for a particular variety if present. As with all seed crops, routine elimination of the plants that are most susceptible to any endemic diseases is always worthwhile.

ISOLATION DISTANCES

The basic isolation distance that should be used between different squash seed crops of the same species when there are no barriers is 1 mi (1.6 km), especially when these adjoining crops are relatively small (see chapter 13, Isolation Distances for Maintaining Varietal Integrity).

As with all of the cross-pollinated crops it's important to distinguish between the isolation distances necessary when there are two adjacent squash crops of the same species that belong to the same horticultural type and the distance required when growing two different types of squash of the same species. For example, when growing two similar varieties of *C. pepo* jack-o'-lantern pumpkins, the minimum isolation distance between them should be 1 mi (1.6 km) in open terrain. When there are two different types of *C. pepo* squash, if one is a jack-o'-lantern pumpkin and the other is a zucchini, then the minimum isolation distance separating the two crops in open terrain should be at least 2 mi (3.2 km). If the seed crops are being grown in terrain where there are natural barriers on the landscape that hinder the movement of insect pollinators, then you can lessen the isolation distance between two *C. pepo* varieties of the same type to 0.5 mi (0.8 km) and 1 mi (1.6 km) between two different types of varieties.

As mentioned in the descriptions above of *C. pepo* and *C. argyrosperma*, populations of wild or weedy conspecific relatives can grow near crop fields in certain regions and thus can potentially outcross with the crop populations. Such outcrosses can result in contamination due to the transfer of the genes responsible for fruit flesh bitterness into the resultant wild/cultivated hybrid seed. Bitterness is caused by chemical compounds known as cucurbitacins. These are present in fruit and other plant parts of wild *Cucurbita* species but also are present in high amounts in cultivated forms of *C. pepo* ornamental gourds. Thus, make sure that seed production fields of squash are isolated from known populations of wild *Cucurbita* or ornamental *C. pepo* gourds to guard against contamination from such hybridization.

WATERMELON

The center of origin of watermelon (*Citrullus lanatus* [Thunb.] Matsum & Nakai) is the Kalahari Desert region of southern Africa, though its early history is shrouded in mystery. It is one of the oldest cultivated crops of the African continent and has been cultivated in Egypt for at least 4,000 years. The probable ancestor of watermelon, *C. colocynthis,* is a wild perennial with bitter fruit and small seed. It is found in archaeological sites across southern Africa and still grows wild in West Africa. David Livingston, the Scottish medical missionary who worked in Africa in the mid-19th century, found copious wild watermelon populations growing in the Kalahari after heavy rains in 1857. India and China are secondary centers of diversity, as watermelon cultivation spread to both of these regions via trade routes long before the existence of any written accounts of their cultivation. Watermelons have long been coveted by desert cultures as a way of transporting potable water for travel across wide expanses of desert. This is especially important during certain seasons, when the quality and quantity of water at certain points along the route is unknown or questionable.

Watermelon is one of the crops that thrives in hot weather and is best adapted to tropical, subtropical, and many long-season temperate

SEED PRODUCTION PARAMETERS: WATERMELON

Common names: watermelon, citron

Crop species: *Citrulus lanatus* [Thunb.] Matsum & Nakai

Life cycle: annual

Mode of pollination: largely cross-pollinated, with some self-pollination

Pollination mechanism: insect

Favorable temperature range for pollination/seed formation: 80–95°F (27–35°C)

Seasonal reproductive cycle: late spring through late summer or early fall (~4–5 months)

Within-row spacing: regular spacing 18–24 in (46–61 cm)/36–48 in (91–122 cm); hill spacing 4–6 ft (1.2–1.8 m)

Between-row spacing: 8 × 12 ft (2.4 × 3.7 m) for regular and hill spacing

Species that will readily cross with crop: There are feral citrons and feral watermelons, both of which are *C. lanatus* and will readily cross with the cultivated crop.

Isolation distance between seed crops: 0.5–2 mi (0.8–3.2 km), depending on crop type and barriers that may be present on the landscape

zones. Heat and sun are important in developing the sugars that make watermelon so popular across the globe. Watermelon is especially well suited to growing in hot, arid climates, both as a vegetable and as a seed crop. This includes parts of the Middle East, Southern Europe, North Africa, Israel, Indonesia, Taiwan, Korea, China, and Japan. In North America most of the commercial melon seed is currently grown in the Arkansas River Valley of Colorado, in the Central Valley of California, and in west Texas.

▼ Watermelon seeds are mature at "eating maturity" of the fruit, but the ripe fruit are often left in the field for an extra 1 to 2 weeks to ensure good seed quality.

CROP CHARACTERISTICS

Reproductive Biology

Most commercial watermelons are monoecious, though some andromonoecious varieties, with a combination of hermaphrodite flowers and male flowers on the same plant, do exist. Monoecy in watermelon ensures a large degree of cross-pollination in populations of any appreciable size. Flowers are solitary, borne in the leaf axils. They are pale yellow to light green in color, with the pistillate or hermaphroditic flowers formed in every sixth or seventh axil and staminate flowers occupying the other axils.

Each of these staminate flowers has five petals that are fused at their base, like cucumber, and three stamens tightly packed within each staminal collar. The pistillate flowers are easily identified as they have the prominent ovary at the base of the flower. They have three broad stigmatic lobes on top of a short, thick style. The hermaphroditic flowers that occur in some varieties have three stamens tightly positioned around a thick, short style with a three-lobed stigma. The flowers open shortly after sunrise, and the stigma is receptive for most of the day, depending on the environmental conditions. Insect pollinators are attracted to both the pollen and the nectar, which is produced at the base of the corolla. Each flower only opens for one day.

The hermaphrodite or perfect flowers have both stamens (male flower parts) and pistils (female flower parts). The hermaphrodite flowers are borne singularly in leaf axils and are essentially much like the pistillate flowers in monoecious types, except there are also small anthers borne close to the stigma. The anthers of these hermaphroditic flowers do not produce much pollen and therefore only contribute minimally to the pollination of these flowers. The form and structure of the staminate flowers is similar to that of cucumbers (see Cucumber, "Reproductive Biology" on page 213).

Climatic and Geographic Suitability

Successful commercial production of watermelon seed requires hot days and warm nights—temperatures that are consistently warmer than is needed for other cucurbits, even melons, to produce the best-quality seed. Optimum growth and fruit set occur between 80 and 95°F (27 to 35°C) with nighttime temperatures that don't usually dip below 70°F (21°C). Temperatures below 70°F (21°C) during anthesis and early seed growth can hinder the development of vigorous, high-quality seed. As with all crops, temperatures that are too high at the time of the pollination will also retard pollen tube growth and prevent fertilization and subsequent seed formation. In watermelon this can happen at temperatures above 100°F (38°C). For watermelons the best seed production areas are regions with a significant amount of heat during the day, warm temperatures at night, and low humidity. The ancestors of watermelon evolved under very low humidity in the Kalahari Desert; the modern cultivated crop also produces well as desert crop in areas that hold the heat at night. Too high a level of humidity during the growing season should be avoided due to the potential for increased levels of foliar disease.

SEED PRODUCTION PRACTICES

Soil and Fertility Requirements

Watermelons require deep, rich soils with good tilth and superior nutrient- and water-holding capacity to produce a superior crop. Many growers prefer growing watermelons in sandy soils, especially in shorter-season temperate zones where early vigorous growth is essential in maturing a crop. However, heavier soils with good drainage can deliver bountiful fruit and seed crops by supplying the crop with adequate moisture and nutrients, which helps to ensure unchecked growth through the season. As with all cucurbits, watermelons are heavy feeders that thrive when ample amounts of well-decayed organic matter or compost are incorporated into the soil. It has also been reported that ample potassium is important in watermelon for strong rinds that are resistant to cracking at maturity. Watermelons are more tolerant of acidic soils than most other crops and can be grown at a soil pH as low as 5.0. This being said, a soil with a pH of between 6.0 and 6.8 is more desirable. Watermelons grown in arid regions are often grown successfully on slightly alkaline soils.

Watermelons form a large taproot that extends 3 to 4 ft (0.9 to 1.2 m) into the soil. Their secondary root system is well branched and more extensive than that of melons, so they require a wider plant spacing than melons in order to thrive (see "Growing the Seed Crop," below). This extensive root system is concentrated in the top 18 in (46 cm) of the soil. For this reason it is very important to avoid compacted soils and to only cultivate close to the surface of the soil when growing the crop. It goes without saying that the use of any heavy equipment in a watermelon field during the growing season should be avoided to realize the best yields.

Growing the Seed Crop

Watermelons are traditionally direct-seeded in climates that are well suited to seed production. Seed can be planted when warm weather has truly arrived, all danger of frost has passed, and soil temperatures are at least 75°F (24°C); even better is when the soil temperature rises above 80°F (27°C), with 90 to 95°F (32 to 35°C) being the optimum for strong germination. This allows young watermelon plants the unchecked growth that will produce a healthy, vigorous plant if the soil, nutrition, and soil moisture are adequate.

If planted in rows, watermelon seed crops need to be given more space than the 18 to 24 in (46 to 61 cm) between plants often used when growing watermelon crops for fresh market. Watermelon plants for seed should be planted at least 36 to 48 in (91 to 122 cm) apart within the row, depending on the vigor and size of the variety. Plant several seeds at the appropriate interval for the variety, then thin them once you are assured that the selected plants are well established and healthy. The space between rows should be increased to at least 10 to 12 ft (3 to 3.7 m) to allow vines enough room to thrive in production of watermelon seed crops. This wider spacing between rows is very important for watermelon growers who are concerned with diseases—it increases the airflow through the developing crop and slows the progress of any foliar diseases that may develop. This is especially important in growing regions where diseases may be more prevalent.

Alternatively, watermelons can be planted into traditional hills within the rows in the field. The spacing used for watermelon hills can be anywhere between 4 ft (1.2 m) apart within the row by 10 ft (3 m) between the rows to the comparable at 6 × 10 ft (1.8 × 3 m) apart or as much 8 × 12 ft (2.4 × 3.7 m). Seed is often planted into the hills at a rate of six to eight per hill, with seed sown in an area that is at least 6 in (15 cm) in diameter. Plants are then thinned to two or three per hill once it is clear which are the most vigorous and healthiest individuals in each hill; however, make sure that the plants aren't crowded during this pre-thinning period. The idea of a "hill" is also often misunderstood; this area where the seed is sown may not be a noticeable hill, although sometimes there is an actual raised mound when manure is worked into the spot before planting.

Many organic growers transplant watermelons and melons to: (1) get an early start if their spring weather isn't hot; (2) plant well-developed seedlings into fields that are weed-free; and (3) avoid early infestations of cucumber beetles (*Acalymma vittatum* and *Diabrotica* spp.). When transplanting, many growers will plant three or four seeds per cell, thinning to two plants for placement in the hill when transplanting. The direct-seeded row or hill plant spacing can also be used for transplanted crops. In some situations, under heavy cucumber beetle pressure, some growers will cover the seedlings with a floating row cover at planting time to minimize the damage from these pests.

Seed Harvest

For optimum seed quality and ease of harvest, it is best to allow the fruit of all cucurbit seed crops to remain on the plant for at least 1 to 2 weeks after the majority of favorable plants have fully matured. This after-ripening period serves multiple purposes:

1. It allows the ripe fruit to fully mature, as cucurbit fruit often produce a somewhat better seed quality with after-ripening.

2. It allows any laggard high-quality fruit to finish ripening.
3. It softens the placental tissue in the seed cavity to the point where it easily separates from the seed upon cleaning.

Determining when watermelon fruit are mature is accomplished in multiple ways, depending on the variety of watermelon. The most commonly used signs of fruit maturity are to check if the tendril on the shoot bearing the fruit has dried and withered, or to see if the "ground spot" (the spot on the underside of the fruit where the fruit rests on the ground) has turned a warm yellowish cream color. Another important sign of watermelon fruit maturity is when the dominant rind color turns from a bright to a duller, flatter shade of the same color. This duller color of the rind is often described as a "waxiness" that covers the rind as the fruit becomes fully mature. Lastly, many people thump or rap on the fruit to judge ripe fruit. There is a certain resonance of a lower tone that will develop as the fruit ripens, but you must train yourself to hear this with each different variety you grow. None of these signs of maturity is foolproof, especially when considered across different varieties and environments; you must determine which are accurate indicators for your crop (or combination of traits). Don't hesitate to cut open a number of fruit to practice determining full maturity.

Larger watermelon seed crops are almost always extracted using a vine harvester or vine seed extractor directly in the field (see "Seed Extraction, Fermentation, and Cleaning" below). If the seed crop is relatively small and the fully ripened fruit is sound, the fruit can be gathered and held for a number of days in a warm, dry, well-aerated place to complete the after-ripening if the weather is turning unfavorable.

Seed Extraction, Fermentation, and Cleaning

Mechanical extraction is used for larger commercial seed lots in the cucurbit crops and is usually done with what is often called a vine harvester. At harvest, the ripe fruit is put into a hopper that leads to a series of rollers or rotating blades, which macerate the fruit to free the seed as much as possible from the pulp. The pieces of rind, the pulp, and the seed then drop into a large rotating cylinder made

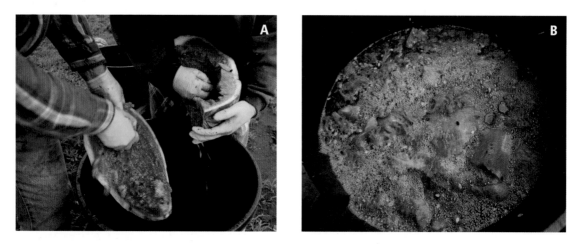

WATERMELON SEED EXTRACTION, FERMENTATION, AND CLEANING: (A) Extracting watermelon seed and pulp by hand; (B) The collected mass of seed and pulp; (C) Tom Stearns of High Mowing Seeds with containers of fermenting watermelon seed; (D) Spray-washing watermelon seed over a wire mesh screen; (E) The washed seed drying on the wire screen.

of screen material. Most of the small pieces of the rind, pulp, and liquid go through the screen material with the seed into a trough at the bottom of the extractor. Seed that adheres to the pulp is loosened through the spinning motion of the cylinder, passing through the screen. Because the cylinder is at a slight angle, the larger pieces of fruit and pulp slowly work their way to the end of the cylinder and into another trough. Examine the contents of this waste trough before composting them for appreciable amounts of seed that may have not been captured.

Watermelon seed is unique among the commonly cultivated cucurbits, as the seed is distributed somewhat evenly in the flesh or endocarp of the fruit. Because of this, hand extraction of the seed is different from scooping of the seed out of a central cavity, as is done with the fruit of other cucurbits. Seed extraction by hand requires that all of the endocarp be macerated and pressed against a screen with a mesh that will allow the seed to pass through, but will catch most of the rind and some of the fiber of the fruit's flesh. Much pulp and all of the juice will pass through as well, which is fine if you choose to use the fermentation method. If you're not using the fermentation method, it is possible to clean out much of the fiber from the flesh by then passing this stew of watermelon juice, flesh, small pieces of rind, and seed across a screen with a fine mesh, one that will not allow the seed to pass through but that will allow much of the other material to pass through when sprayed with a high-pressure stream of water. This process can be repeated a couple of times, with a cleaning of the screen between the spraying events to remove any fiber that gets caught in the screen. At this point the seed can then go to the final seed-cleaning steps using a sluiceway as described in "Seed Cleaning," below.

Seed Fermentation: Watermelon seed does not have a placental sac, so it does not require the fermentation step used with cucumber that frees the seed from this sac. After the extraction described above, most watermelon seed is cleaned under a high-pressure spray of water that removes any fleshy material. While this is often enough to clean the seed, a number of seed growers give melon seed a 1- to 2-day fermentation period, which helps free the seed from all endocarp material. Fermentation of watermelon seed can also help eliminate bacterial fruit blotch of watermelon (*Acidovorax avenae* subsp. *citrulla*), which is the most serious seedborne disease of watermelon and can devastate entire crops if conditions are favorable. This disease grew to epidemic proportions in North America during the 1990s. The result was that most seed companies now require any commercial growers purchasing watermelon seed to sign a waiver stating that they won't sue the seed company if they incur losses from this disease.

Research done in the 1990s demonstrated that a 1- to 2-day fermentation of watermelon seed coupled with 15 minutes of treatment with 1% hydrochloric acid or 1% calcium hydrochlorite, followed immediately with the washing and drying of the seed, is an effective treatment for bacterial fruit blotch of watermelon. There are two questions for organic seed growers and organic seed companies: (1) Are concentrations of 1% hydrochloric acid or 1% calcium hydrochlorite acceptable for organic certification? (2) How effective is the fermentation process by itself?

The problem with fermentation in watermelon, as in all of the cucurbits other than cucumber, is that it can easily damage the seed and lower the germination percentage if it continues for too long. There are always specific varieties that are more easily damaged by fermentation than others in the same crop species. In watermelon, the triploid seedless types are much more susceptible to damage during fermentation than the traditional seeded, diploid types.

If the conditions are right, only about 24 hours of fermentation is necessary for watermelon or melons. It is the same process described for tomato (see chapter 12, Tomato, "Seed Fermentation" on page 323). The main difference is that, once you get the watermelon seed mass fermenting, it only requires a few hours to separate and clean the seed from the placental tissue. Thus you must watch the process closely and separate the seed from the fermentation bath as soon as the desired results are achieved.

Seed Cleaning: The final cleaning of watermelon seed through the use of a sluiceway or with the simpler decanting method is identical to that described for tomato. The main difference may be with the amount and force of the water, which may need to be increased to clean the heavier watermelon seed (see chapter 12, Tomato, "Washing Seed" on page 323). The drying of the seed needs immediate attention, just as with tomato and cucumber seed, after this wet seed processing.

GENETIC MAINTENANCE

An initial selection for seedling vigor and early robust growth is always important to maintain good watermelon stocks, as it will always lead to the establishment of a healthy seedling. This is especially important for varieties selected for organic production methods. It is often easier to identify superior vigor when conditions are less than ideal upon planting or transplanting the crop. The variation in watermelon leaf type is not as pronounced as it is in many of the other cucurbits, but there

may be some variation in the shape and size of the lobes or in their color that could indicate an off-type outcross. Different watermelon varieties may also vary in the length of the internodes; shorter semi-determinate varieties have shorter internodes than the indeterminate types. Variants for all of these traits can usually be detected and rogued out of the field before flowering.

For watermelon, as with all of the cucurbit crops, it is often possible to detect off-type fruit shapes and fruit color at or very near the time of flowering and early fruit set, especially if the off-types are somewhat different from the variety you are growing. As the fruit approaches maturity it becomes easier to discern differences in both shape and color. Deviations from the round, globe, oval, or elongated oval of the larger watermelon types can vary to some degree within a variety, but all seed growers must set a standard range within the fruit shape to which they adhere for varietal integrity. Color variation from dark black-green to very light gray-green can distinguish a variety, and fruit may exhibit anything from solid color to many variations of stripes. The coloration of the fruit should be checked as the crop matures; this is sometimes easier to judge before the waxy coating that accompanies ripeness develops. As with all seed crops where the fruit is the main agricultural crop of interest, the earlier you determine that the fruit is not true to type, the earlier you can rogue out the plant and eliminate it as a source of pollen that might otherwise contaminate the rest of the seed crop.

At fruit maturity it is possible to test a number of fruit quality traits. An important characteristic of watermelon fruit is the thickness and toughness of the rind. There is a standard way to determine this with a handheld punch device that measures the amount of pressure needed to pierce the rind. This is a destructive test, but when harvesting for seed rather than for market sale, it doesn't matter. The simpler, faster, and perhaps easier way to judge the thickness of the watermelon rind is to drop the fruit from approximately knee height and see if it cracks on impact with the ground. This method can be very effective but will vary with the soil type and moisture content of the particular soil where the crop is being grown. Keep in mind that there is probably variation for this rind toughness trait in any open-pollinated watermelon variety; such selection can benefit the commercial value of the variety.

The maintenance of quality traits associated with the flesh of watermelons should also be considered. It is possible to check individual plants for internal traits such as color and intensity of color, as well as sensory characteristics like flavor, sweetness, and texture. Fortunately, selection for these traits at full fruit maturity is a non-destructive test as far as seed harvest is concerned. Fruit evaluated for these internal characteristics can be tossed right into the vine harvester at the time of harvest or into containers for subsequent extraction. Once you find an objectionable fruit, however, you must be sure that its seed, and that from other fruit from the same plant, won't be harvested with the crop.

Watermelons have at least six major classes of flesh color: scarlet red, coral red, orange, canary yellow, golden yellow, and white. The first five colors are all due to the presence of carotenoid pigments, which can vary in concentration, thereby resulting in various shades. Selection for the maintenance of a particular color, and the specific shade and intensity of that color, is important over the long run.

A very important quality trait is the flavor of the watermelon. As with other sweet vegetable crops, there is much more to the flavor of a watermelon variety than just the sweetness. In fact, some of the best breeders do not rely purely on a Brix reading of the sugar content when performing selection for flavor. Watermelon flavor is complex and is a combination of the sweetness and various volatile flavor compounds that constitute the flavor underlying the sugar content of a variety. The taste of these volatiles can vary from what is sometimes called a rich caramel flavor, which some people find syrupy and objectionable, to a range of floral aromatics that are usually desirable but can be overwhelming when present in excess. The only way to distinguish the differences is with the human tongue, experience, and knowledge of what is desirable in the marketplace. A taste test may be necessary, as there may be plant-to-plant variation in flavor or texture in any open-pollinated variety, especially when the variety is not stable or exhibits variation in some other trait.

The texture of the watermelon flesh can also vary quite considerably. The eating texture can range from soft to firm and from crisp to fibrous, even stringy. As with flavor, the best way to judge the texture is with a bite test, especially after you gain experience in sensory evaluation of watermelon. Another trait to be aware of when evaluating ripe watermelons is the tendency of some varieties (or individual plants within varieties) to produce fruit with hollow heart, where the flesh has fractured during the maturing process, leaving a cavity in the flesh at edible maturity. Lastly, the mature color of the seed can vary greatly in color, from white to tan, brown, red, black, speckled, or some combination of these colors. Watermelon seed can also vary considerably in size. Growers have sometimes first detected outcrosses in the previous generation by noticing the unusual variation in seed size at seed harvest.

ISOLATION DISTANCES

For watermelons the basic isolation distance that should be used between different watermelon seed crops without barriers is 1 mi (1.6 km), especially when these adjoining crops are relatively small (see chapter 13, Isolation Distances for Maintaining Varietal Integrity).

As with all of the cross-pollinated crops, it's important to distinguish between the distances that are necessary when the two adjacent watermelon crops are of the same horticultural type versus when there are two different types of watermelons being grown. When you're growing two watermelon varieties of the same type—for instance, two small, round icebox types with red flesh—then the minimum isolation distance between them needs to be 1 mi (1.6 km) in open terrain.

When there are seed crops of two different types of varieties—for instance, if one is a blocky oval 'Crimson Sweet' type with red flesh and the second is an oblong orange-fleshed Texas type—then the minimum isolation distance in open terrain should be 2 mi (3.2 km). Indeed, even if both varieties to be grown were blocky, oval, red-fleshed shipping varieties, but were distinctly different in their days to maturity, size, or disease-resistance profile, they would still require this increased isolation distance if grown in open terrain.

If two watermelon seed crops are being grown in terrain where there are significant natural barriers on the landscape that hinder the movement of insect pollinators, then you can lessen the isolation distance between two varieties of the same type to 0.5 mi (0.8 km); 1 mi (1.6 km) between two varieties of distinctly different types.

Note: It is important to note that there are feral citrons and feral watermelons growing in various areas of the world where watermelons are grown for seed. All watermelon seed growers need to determine if any of these feral forms grow in the regions where they plan on producing watermelon seed.

– 10 –

FABACEAE

The Fabaceae is a very diverse, large family of plants that includes more than 450 genera and over 12,000 species of annuals, biennials, and perennials. The members of this family have traditionally been called legumes. Botanically the term *legume* denotes a simple dry fruit in the form of a pod that develops from a single carpel and usually dehisces, or splits, at maturity into two halves (called valves), with the seed attached to the edge of one of the valves. The French refer to all garden vegetables as *legumes*. Well over 100 leguminous species are cultivated as agricultural crops worldwide, which includes forages, pasture crops, cover crops, pulses, vegetables, and ornamentals. Most of the 40 species used in the human diet are annual **pulses,** which are legumes harvested primarily for their protein-rich dry seed. Indeed, the pulses of the Fabaceae are second only to the cereal grains in their importance as a source of human nutrition worldwide.

The Fabaceae includes several crops used as vegetables. The first class of vegetable legumes includes several crop types where the well-developed immature seed is shelled from the pod and lightly cooked. Examples are horticultural beans, runner beans, garden peas, limas, favas, cowpeas, and edamame soybeans. Another class of vegetable legumes is grown for their succulent, immature, edible pods and underdeveloped seed and includes snap beans, snow peas, yardlong beans, and, to a lesser extent, runner beans and some types of cowpeas. Some version of this diverse group of vegetable legumes, usually more than one, is grown in most cultures around the globe.

Vegetable legumes and sweet corn are collectively called the large-seeded vegetables in the North American seed trade. Due to their large size, planting these crops requires a large amount of seed on a per acre/hectare basis. In North America, Europe, and Japan most of the commercial seed crop of these vegetables is grown in areas with low humidity; clear, warm, sunny days; and cool nights to help ensure disease-free seed. However, in many other parts of the world this is too expensive a proposition for the relatively large amounts of seed required; therefore much of the seed of these crops is still produced regionally.

FAMILY CHARACTERISTICS

Reproductive Biology
Legume flowers have blossoms with an irregular shape that is fancifully described as resembling a butterfly. The typical legume flower is perfect, with five petals consisting of one large, oval banner or standard, two elongate keel petals that are fused together enclosing 10 stamens, and two wing petals. Nine stamen filaments are united into a sheath that surrounds the pistil; one is separate. The petals of the vegetable legumes are often white but may also be many beautiful shades of pastels, ranging from deep violet, purple, pink, and salmon to shades of red. The most

commonly grown runner bean, 'Scarlet Runner', is often grown as an ornamental because it has striking scarlet flowers.

Legume ovules are borne in a simple carpel with a leathery **pericarp** (the wall of the fruit or pod) that becomes dry at maturity and usually dehisces (splits) along two sutures of the valves to release the seed. In some legume seed crops this dehiscence may happen with very little physical prompting; in fact, even a stiff wind rustling through unharvested mature dry pods can cause the pods to pop as they shatter and release their seed to the ground. Among the vegetable legumes this can especially be a problem with many edamame varieties.

Most of the species that make up the Fabaceae vegetables (see the Runner Bean section for a notable exception) are usually self-pollinated, as the anther sacs are borne directly adjacent to the stigma and the pollen is released the day before the flower normally opens. Therefore, under most conditions this self-pollination leads to a full complement of selfed seed that is genetically true to type before the flowers are open and accessible to any insect activity. However, there are several situations that lead to outcrossing between legume plants of the same species. First, if you are growing these crops adjacent to areas with biologically diverse habitats for insects, there is sure to be a higher and more diverse number of insect pollinators that will forage in your fields. These pollinators may be able to get into legume flowers before they fully open, either because the flowers are not completely closed (a genetic trait) or because the insects may be able to pry the flowers open in search of nectar or pollen.

Another common reason for outcrossing involves a combination of excessively hot weather and the fact that, in many of these crop species, certain varieties have pollen that will abort (due to sensitivity to heat) more readily than other varieties of the same crop. So if several varieties of the same crop with genetic variation for this trait are grown in the same field, chances are some will be more sensitive to the heat (and not have viable pollen) than others. If insects are present and actively visiting flowers, then they are apt to move pollen from male fertile varieties to those affected by the heat (and not producing viable pollen) and create a large number of unwanted crosses. It should be noted that heat usually affects pollen viability unfavorably long before the comparable female cells, the ovules, are damaged. (See chapter 2, Reproductive Biology of Crop Plants.)

The threshold for crossing to occur will be at a hotter temperature for some of the vegetable legumes than for the others. For example, cowpeas, yardlong beans, and limas are all originally from tropical or subtropical climates and are able to reproduce successfully under considerably higher temperatures than the other vegetable legumes that are adapted to temperate conditions. Peas and fava beans represent the other end of the spectrum: Both favor cooler temperatures during flowering for ideal seed set.

Climatic Adaptation

There is a broad range in the adaptation to climate among the vegetable legumes. Some species, including favas and peas, can tolerate frost and are often planted 6 to 8 weeks before the last hard frost of the spring. The other vegetable legumes are tender, heat-loving crops; when grown as vegetables, they are usually planted when all chance of frost has passed. But when these species are grown as seed crops, they are often planted early to ensure full maturity before the end of the season. This is especially necessary in temperate climates, or in climates where late-summer or early-fall rains threaten the quality of the crops. This is possible in part because most of the vegetable legumes are more frost-hardy at emergence (although this seedling hardiness doesn't last long) than they are as mature plants at the end of the season.

All of the vegetable legumes are grown as annuals in temperate climates with the exception of fava beans, which are often treated as winter annuals in areas with mild winters. As with all annual seed crops, it is very important to be familiar with the relative maturity of the crop species in the area that you are considering for production. Not only should you be assured that the particular crop type will faithfully mature in your climate, but you should have firsthand knowledge or experience with the specific variety of that crop that you are considering for production. It should be a reliable cropper that will mature satisfactorily in your bioregion, even when faced with a less-than-optimum growing season.

SEED HARVEST AND CLEANING

Proper timing of harvest is important in production of high-quality legume seed that is fully mature, has a high germination percentage, and has maximum storage potential. Gauging the best time to cut, cure in a windrow, and thresh the crop is complicated by the fact that differences between crop species, as well as variety-to-variety differences, abound. The initial signal that the crop is ready to cut is the relative maturity of the pods and their color at or near the time when they are breaking, or when they first turn yellow, mahogany, or buckskin in color, depending on the crop species. The exact desired color is crop-specific and may also be variety-specific within each crop type. For many of the vegetable legumes it is best to cut the crop before the pods have

dried to a papery crisp texture. Maturation to the papery stage may increase the danger of seed shattering during harvest. The seed crop should be cut when approximately 70 to 80% of the pods on the plant are of the desired color and texture.

Harvesting the seed of the vegetable legumes is a multistep process. The first step is to mechanically undercut the stems of the plants just below the soil surface, allowing the plants to wilt in place for a day. The next day the plants are raked into windrows. Depending on weather conditions, the windrows should cure in the field for at least 10 days to 2 weeks, with a possible turning of the rows near the halfway point. Threshing of beans is best accomplished during the heat of the day, when the pods are brittle and easily cracked. Seed should then be further cleaned to separate it from any plant debris that may be moist. If the seed is not sufficiently dry, it may require air- or kiln-drying before being stored for a final cleaning and conditioning.

ISOLATION DISTANCES

While the vegetable legumes are all considered self-pollinated species, crossing will occur in all of them to varying degrees. Much of the seed-growing literature tends to treat these crops as if outcrossing is only a rare occurrence, and the only isolation that is recommended for different crop varieties of the same species is 10 to 15 ft (3–4.6 m).

For the vegetable legumes that are largely self-pollinated—common beans, garden peas, and the edamame class of soybeans—this distance may suffice, especially in conventional cropping systems that are largely planted to one species and have a minimum amount of biological diversity in the form of insect pollinators. Indeed, these crops usually have relatively low incidences of cross-pollination (less than 1%) in most commercial seed production areas of North America. However, the incidence of outcrossing is often higher in self-pollinated vegetable crops with the increase in biological diversity that usually accompanies organic cultural practices or when environmental stresses are present.

For this reason common bean, garden pea, and edamame require a greater isolation distance between varieties of the same species under organic production conditions than what is often stated in the literature. The isolation between different varieties of each of these crops for commercial seed production should be at least 150 ft (46 m) in open terrain. For stockseed or foundation seed this number should be doubled, and a tall barrier crop such as corn or sunflowers planted between crops if no other natural barrier like woods or farm buildings are present.

The other vegetable legumes—cowpeas, favas, limas, and runner beans—are all more promiscuous than the aforementioned crops, with crossing rates that sometimes exceed 5%, especially when grown in biologically diverse settings or under environmental challenges. For this reason special consideration is given for the isolation distances for each of these crops under their respective sections.

COMMON BEAN

The wild ancestors of the modern common bean (*Phaseolus vulgaris* L.) come from Central and South America. These ancestral types are found across a range of environments, from hot, arid climates to humid, lowland tropics, and even into cooler, upland areas of South America. The beans of this species that are grown in North America today are grown in a more limited temperate climatic zone. Nearly all modern agricultural varieties have a determinate bush habit, whether they are dry bean, shell bean, or snap bean types. Older varietals, landraces, and heirlooms are often indeterminate and vining. The snap bean types are grown for a harvest of tender young pods with immature seed that are eaten as a fresh vegetable. While immature pods of any common bean variety were once commonly eaten, there are now literally thousands of varieties that have been selected to have fleshy pods with slow seed maturation for a crop that has become a vegetable mainstay in our culture. Snap beans are alternately called string beans, green beans, and garden beans. The term *snap bean* refers to the way in which fresh garden beans are broken or snapped by hand into short segments before cooking. Garden beans is what they are most often called in the bean-seed-growing regions of the Northwest to distinguish them from the more widespread dry beans or field beans of this same species.

Whether grown as a vegetable or for seed, beans are produced as an annual crop that matures in one growing season. Bean seed crops

SEED PRODUCTION PARAMETERS: COMMON BEAN

Common name: common bean, snap bean, green bean, shell bean
Crop species: *Phaseolus vulgaris* L.
Life cycle: annual
Mating system: primarily selfing
Mode of pollination: closed perfect flower; no need to stimulate pollen shed
Favorable temperature range for pollination/seed formation: 60–74°F (16–23°C)
Seasonal reproductive cycle: late spring through late summer or fall (4–5 months)
Within-row spacing: 1–1.5 in (2.5–4 cm)
Between-row spacing: 22–30 in (56–76 cm)
Species that will readily cross with crop: All true *P. vulgaris* types, including those grown for dry beans
Isolation distance between seed crops: 20–100 ft (6–30 m), depending on crop type and barriers that may be present on the landscape

may require a long season to mature (on average 90 to 120 days) and must be planted early enough to mature and dry prior to fall frosts or rains. Commercial bean seed production is focused in the western region of the United States, as the long dry summers and arid climate create good growing conditions with lower disease pressure and a long season for maturation.

CROP CHARACTERISTICS

Reproductive Biology

Nearly all modern agricultural varieties of common bean are daylength-neutral in their flowering response. In other words the length of day does not determine flower initiation, and plants flower as soon as they are physiologically ready. There are exceptions to this in some bean varieties from the tropics and some temperate dry bean varieties. Many tropically adapted common beans are short-day plants and fail to flower until the days shorten in late summer. Because there is varietal variability in flowering response, it is recommended that you grow a small plot of any new seed crop to see if it matures optimally in your particular region.

Common bean flowers have the shape and structure of a typical legume flower. Their petals are usually whitish but may be tinged deep violet, purple, or red. There are five petals consisting of one large, oval banner or standard, two elongate keel petals that are fused together enclosing the stamens, and two wings. Nine stamen filaments unite into a sheath that surrounds the pistil, with one stamen filament separate from the fused nine. Common beans are highly self-pollinated, as the anther sacs are borne directly adjacent to the stigma (the receptive part of the female flower) and the pollen is normally released the day before the flower opens. Under most conditions the flowers are self-pollinated before

Common bean flower (*Phaseolus vulgaris*). ▶

they open and become accessible to insect activity. However, there are some environmentally influenced situations that can cause an increased rate of cross-pollinations (see "Isolation Distances," page 261).

Climatic and Geographic Suitability

Common bean is a tender warm-season crop that requires warm, well-drained soils for germination. Temperatures of 70 to 80°F (21 to 27°C) are preferred for optimum crop growth. Temperatures below 50°F (10°C) or above 90°F (32°C) during flowering may adversely affect pod set and seed yields. Most snap bean cultivars germinate best when soil temperatures are at or above 65°F (18°C), but germination may be inhibited at temperatures above 95°F (35°C). There are instances when seed growers must plant with soil temperatures that are below optimum in order to fully mature a seed crop by the end of the season. Considerable differences exist between cultivars in their ability to germinate in cool, moist soils and to resist common root rot organisms that can damage or destroy seedlings. It is commonly believed that the French filet types and the modern white-seeded snap bean types are the most susceptible to these maladies.

SEED PRODUCTION PRACTICES

Soil and Fertility Requirements

Soil conditions and types may vary and still produce healthy plants and superior seed, but adequate drainage is essential, as beans are sensitive to both moisture stress and waterlogging. Early plantings for long-season varieties

are best done on light sandy soils. Soils that easily crust should be avoided, and irrigation and tillage managed to avoid crusting; seedling emergence may otherwise be impaired. All crop residues should be thoroughly incorporated into the soil prior to planting. Residues on the surface can interfere with seed placement during drilling and result in inadequate coverage of weeds with soil during an early cultivation. Incorporation of residue is additionally crucial in the availability of nitrogen (and other nutrients) to the young bean plants before their development of symbiosis with the *Rhizobium* bacteria.

Growing the Seed Crop

Common beans are frost-sensitive, so the usual recommendation is to plant after all danger of frost is past. However, it is often necessary to plant a seed crop 7 to 10 days before the last anticipated frost date, either to allow time for full maturity and harvest prior to inclement weather in the fall, or in some cases to avoid the heat of summer during flowering. Soil temperatures do need to be warm, approximately 65°F (18°C), so the seed will germinate quickly and not succumb to root rots in cool soils. However, young plants with fleshy cotyledons are more frost-tolerant than mature bean plants and can tolerate a light frost soon after emerging.

Bean seed should be planted at a depth of 1 to 1.5 in (2.5 to 4 cm) below the surface of the soil after the press wheel has packed the soil. Common beans for seed are routinely grown on 22 in (56 cm) row centers with an in-row density of six to eight bean plants every 12 in (30 cm). Seed may be inoculated with *Rhizobium* bacteria prior to planting to enhance nitrogen availability. This may be particularly beneficial when planting on soils where beans have not been produced in recent years. Prior to planting beans, soils are

commonly pre-irrigated, allowed to dry to moderate soil moisture, planted, and then not irrigated again until seedling emergence. This avoids soil surface crusting prior to seedling emergence. Alternatively, soil crusting is mitigated by a very shallow dragging of soil with a harrow prior to emergence.

Weed control is particularly critical in organic bean seed production to achieve optimum growth and prevent weed seed contamination in harvested seed. One very successful method used for weed control under organic production is as follows: Plant single rows in hills that are 6 to 8 in (15 to 20 cm) high. After 4 to 5 days, when seed has sprouted but the epicotyl is still 1 in (2.5 cm) belowground, cultivate the hills for the first time using a drag harrow. The drag harrow will scrape off the top surface of each hill, exposing the first set of tiny weeds before they can become established in the bean rows. Even if done properly, an occasional bean plant will be lost, but it is well worth the effort to eliminate the first within-row weeds.

The next cultivation is done after the bean plant has emerged and has one set of true leaves. Set up two tool bars on a cultivating tractor. The first tool bar should have shanks with chisel points that are set as close to the bean rows as possible. These chisel points will disturb the soil (killing weeds) close to the plant as well as throwing some soil up onto the base of the plant to smother other weed seedlings. Follow these shanks with a second tool bar with "duck feet" sweeps, which will disturb the soil, kill weeds between the rows, and throw soil back onto the hills to rebuild them. Subsequent cultivations are done with knives that cut soil away from the bean plants, and sweeps between the rows to rebuild the hills. In addition to disturbing as much of the soil surface as possible to uproot or bury weed seedlings, it is important to create hills high enough to facilitate undercutting the bean plants at harvest. Cultivation should be done with attention to avoiding major root disturbances as the plants get larger.

Seed Harvest

Proper timing of harvest is important in order to produce high-quality bean seed that is fully mature, has a high germination percentage, and has maximum storage potential. Each variety has its own specific harvest timing, and while this makes overall recommendations for gauging cutting, curing, and threshing difficult, there are basic signs that indicate maturity. The initial signal that the crop is ready to cut is the relative maturity of the pods and their color at or near the time when they are breaking, or when they first turn yellow, mahogany, or buckskin in color but haven't gone to the tan, papery shade. Maturation to the brown, papery stage increases the danger of seed shattering during harvest. Pods should generally be yellow at harvest in order to mature properly in the windrow, but the exact desired color may be variety-specific. The crop should be cut when approximately 70 to 80% of the pods on the crop are of the desired color and point of breaking. The stems of the crop are undercut mechanically just below the soil surface and left in place for a day. The next day the plants are raked into windrows. Depending on weather conditions, the windrows should cure in the field for at least 10 to 14 days, with a possible turning of the rows near the halfway point. Threshing of beans is best accomplished during the heat of the day when the pods are brittle and easily cracked. Seed should then be further cleaned to separate it from any plant debris that may be moist. If the seed is not sufficiently dry it may require air- or kiln-drying before being stored for a final cleaning and conditioning. Physical mixing of seed varieties can be a source of variety contamination. Attention should also be given to thoroughly cleaning out harvesting and cleaning equipment between harvests when you're working with different varieties.

GENETIC MAINTENANCE

There are very few stages in snap bean growth in which genetic differences within a variety are apparent. If you are able to establish a uniform stand, then it is possible to perform selection for speed of seedling emergence, seedling health, and overall vigor—all essential traits for organic production. If you routinely rogue out late-emerging, low-vigor seedlings, you'll see improvement over cycles of selection for these traits. Many bean growers can easily identify differences between pod characteristics, flower color, and the presence of runners, but other plant characteristics are only apparent to experienced farmers who specialize in the crop. For some traits, such as leaf shape, color, and plant stature, it is possible to evaluate the plant and rogue it to type before flowering has begun. Flower placement, timing, and duration of flowering can all be selected during flowering. Pod characteristics such as color, shape, and length are also notable to the experienced grower and can be selected. Additionally, it is critical to rogue for disease in bean crops in order to produce quality seed, minimize risk of seedborne diseases, and select or maintain for a cultivar's disease resistance.

ISOLATION DISTANCES

Common beans are considered to be highly self-pollinating by most commercial seed producers. In the seed production areas of southern Idaho the isolation distance between different varieties of common bean is usually

261

no more than 10 to 15 ft (3 to 4.6 m), with a few rows of a taller crop such as corn planted in between to act as a barrier to deter pollinator movement. This minimal distance and physical barrier between common bean varieties is even used in stockseed production. And indeed, in the Treasure and Magic valleys of Idaho, this seems to be adequate isolation, as only a very low rate of crossing is ever detected in the seed lots produced there. This is probably due in large part to the almost ideal environmental conditions that exist in this region during the growing season for the reproductive ability of this crop (not excessively hot, cool nighttime temperatures, low relative humidity, and a low incidence of wild pollinators interested in this species).

However, there are several situations that can lead to outcrossing between bean plants, including these:

1. When beans are grown adjacent to areas with biologically diverse habitats for insects, there is usually a higher and more diverse number of insect pollinators present. These pollinators may be able to get into flowers before they fully open, either because the flowers are not completely closed (a genetic trait) or because the insects may be able to pry the flowers open in search of nectar or pollen.
2. Variety differences of floral structures exist, and some varieties (particularly pole beans) have longer pistils that extend farther out of the flower, making cross-pollination more likely.
3. An increase in rate of crossing can occur due to a combination of excessively hot weather and the fact that some bean varieties are more sensitive to heat than others. The pollen of many bean varieties can be damaged and become non-viable if temperatures are too high at the time of flowering, whereas the stigma will usually remain unaffected. There is genetic variation for this condition among bean varieties, so the pollen of one variety may lose viability while another does not under similar conditions. If multiple varieties are being produced during a hot period, it is possible that pollinating insects may move pollen from a fertile variety to another variety lacking fertile pollen and create unwanted crosses.

The common recommendation for isolation distance between common bean varieties is usually to allow "enough room to avoid mechanical mixing." However, as with the other largely self-pollinating species of the Fabaceae, it is appropriate to observe the minimum 150 ft (46 m) isolation distance if you are growing more than one variety in open terrain with no natural barriers to ensure a high level of genetic purity. If producing multiple common bean crops in areas with natural barriers (see chapter 13, Isolation Distances for Maintaining Varietal Integrity), then the distance can be dropped to 50 ft (15 m) or so to minimize crossing.

It is often stated by experienced seed growers that the vining or pole varieties of common bean are more prone to outcrossing than bush varieties. Definitive information on this is not readily available, although the strict standards stated above for common beans in general should minimize the extent of crossing in all forms of this crop species.

EDAMAME

Edamame is a type of soybean (*Glycine max* [L.] Merr.) that is harvested at an immature stage and used as a vegetable. This crop, which is sometimes known as green vegetable soybean in English, is called *mao duo* in China and *poot kong* in Korea but is best known internationally by its Japanese name *edamame*. The pods of these vegetable types are harvested when the seed has attained 80 to 90% of its full size within the pod. Edamame seed is larger and usually has a greener color than the more common oilseed soybean, both at this stage and when mature. Historically, the soybean is one of the oldest crops known to Chinese agriculture, predating written records. The vegetable types of soy are presumed to be a later-derived crop from the staple soybean. The earliest recorded use of *mao dou* in China dates to 200 BCE; the earliest Japanese reference to "aomame" is in a guide to agricultural commodities in 927 CE. Edamame was first introduced in the United States in 1902 by the famous plant explorer of the USDA, David Fairchild, who brought seed back from Japan. However, the real interest in this crop outside of Asia came only with an explosion of interest in international foods at the turn of the 21st century.

CROP CHARACTERISTICS

Reproductive Biology

Edamame soybeans have flowers that in form and structure are typical of the pulse crops of the Fabaceae, though they are smaller and

SEED PRODUCTION PARAMETERS: EDAMAME

Common names: edamame, vegetable soybean
Crop species: *Glycine max* L.
Life cycle: annual
Mating system: mixed self-pollinated
Mode of pollination: closed perfect flower; no need to stimulate pollen shed
Favorable temperature range for pollination/seed formation: 60–74°F (16–23°C)
Seasonal reproductive cycle: late spring through late summer or fall (4–5 months)
Within-row spacing: 1–1.5 in (2.5–4 cm)
Between-row spacing: 22–30 in (56–76 cm)
Species that will readily cross with crop: All soybean types will cross with edamame.
Isolation distance between seed crops: 10–50 ft (3–15 m), depending on the crop type and barriers that may be present on the landscape

less conspicuous than the flowers of the other vegetable crops in this family. Flower color varies from white to lilac and other shades of purple. Soy flowers are highly self-fertile and are not regularly visited by insect pollinators in most situations. The anthers dehisce and shed pollen before the flower opens, with pollination and fertilization occurring before insects have access to flower parts. However, crosses between varieties do occur when large numbers of pollinators are present.

The flowering habit of most edamame varieties is strongly indeterminate, as this type of soybean has been selected for extended harvest as a vegetable crop. The pods are produced in clusters. Each pod usually contains from two to three seeds. Pods that retain a green pod color as the seed matures are desirable for the fresh market; many modern varieties retain their green color through to full maturity. Edamame seed is often globe-shaped and may be slightly flattened at maturity.

Climatic and Geographic Requirements

Optimum growing conditions for edamame seed production include mean temperatures between 68 and 86°F (20 to 30°C) and seasonal dry conditions during seed maturation and harvest. Temperatures below 68°F (20°C) will slow vegetative growth and the development of flowers and subsequent seed. As with most heat-loving crops, temperatures between 50 and 54°F (10 to 12°C) and below may inhibit pollen development, fertilization, and seed set.

While many soybean types are short-day plants, most edamame varieties are daylength-neutral and can be produced across a wide range of latitudes. It is often hard to find information on whether a particular edamame variety is daylength-sensitive. Ultimately, the best way to determine if there is a problem for your latitude is to grow a small amount of a given variety, making sure it flowers in a timely fashion and matures a seed crop during the normal harvest period for your climate.

SEED PRODUCTION PRACTICES

Soil and Fertility Requirements

Soil conditions and fertility requirements are the same as those recommended for garden peas (see Garden Pea, "Soil and Fertility Requirements" on page 276). If you are growing soybeans in a particular field for the first time, then inoculating the seed with *Rhizobium japonicum* before planting is recommended.

Growing the Seed Crop

Edamame-type soybeans are a frost-sensitive crop that have a bush stature and form a canopy much like most bush bean varieties; therefore, it is recommended that you plant and cultivate the crop per the recommendations for bush-type common beans (see Common Bean, "Growing the Seed Crop" on page 260). Similarly, for many growers in temperate regions, the ability to mature a seed crop of this indeterminate-flowering type of soybean in a timely fashion (before fall rains and possible frost) may require planting the crop 7 to 10 days before the last anticipated frost date for your region. This type of early planting should only be attempted when the weather is warm and clear during the day and soil temperatures are at or approaching 65°F (18°C), which helps ensure that the seed germinates quickly and will not succumb to root rot organisms in cool soils.

Seed Harvest

As with all Fabaceae crops, proper timing of harvest is important in producing high-quality seed. In fact, the timing of the edamame seed harvest may be more sensitive than other vegetable legume seed crops. Edamame varieties have been selected for easy pod separation, as the fresh crop is traditionally eaten from the cooked green pod by squeezing open the pod and popping the beans into your mouth. This characteristic also influences the behavior of the pod on the growing plant. As edamame pods approach full seed maturity and become dry and brittle, they can easily split apart along the suture that separates the pod halves. This seems to happen almost spontaneously, and farmers standing in the field on a hot, dry day are always amazed at the audible popping sound of pods shattering around them when they hear it for the first time. Whether it is the breeze or some other slight movement that provokes this response, it is a flaw that can cause appreciable amounts of seed to shatter in the field before harvest unless you monitor the crop carefully.

In order to harvest edamame seed successfully you must watch closely for the first plants that will invariably mature their pods before other plants in the population. The pods will have the unmistakable leathery, tough-skin look as they turn to a tannish brown color that many growers in the western United States refer to as buckskin. The transition of a few plants to a majority of plants expressing this change is quick and dependent on the weather. Under warm, sunny conditions the crop will need to be cut when a majority of the pods (70 to 80%) have this buckskin appearance but haven't yet dried to the point of shattering.

Much of the high-value conventional edamame seed harvest in Asia is done with the crop standing at or near this stage. This is possible largely through the use of synthetic desiccants that are sprayed on the crop to dry the foliage several days before the crop is combined directly. This enables the pods to be easily threshed without green foliage interfering, as

Edamame seedpods at the dry harvest stage.

the combine is able to break the dry foliage into small pieces that are easily winnowed out during the threshing process. At the time of this writing the best edamame harvest method for organic production is currently used by

several pioneering organic seed growers in southern Idaho and is closer to the methodology used with common bean (with a couple of notable twists).

In organic edamame seed production the initial step in harvesting requires undercutting the plants as described for common bean. Because losing seed due to shattering is the main concern during harvest, always complete this step in the morning to take advantage of the moisture on the pods from the morning dew. Next, instead of mechanically raking the cut plants into windrows, it is preferable to manually lift or fork the plants onto a porous agricultural fabric like Reemay or landscape cloth for curing. This cloth, which is rolled out in strips at regular intervals in the field, will catch seed from any pods that shatter when moved and during the curing process. Use of a porous material is recommended in case of rain during the 10 to 14 days that are required for curing the crop, as any moisture will easily drain away from the crop. The expense of the material is justified to offset what can be heavy losses in seed yield during the curing of the edamame seed.

◄ An edamame field trial.

Threshing the seed is best accomplished by gently forking the cured piles into a stationary thresher in dry weather to minimize losses. Edamame seed is more easily damaged during threshing and cleaning than most of the other legume seed crops. Therefore, threshing should be watched closely and adjusted early in the process so that the rotors do not cause splitting of the seed as it is threshed.

GENETIC MAINTENANCE

Phenotypic differences for foliar and flower traits between edamame varieties are usually subtle and not easily distinguished until a grower has experience with a variety. Differences in the timing of pod ripening and the ability of a pod to stay green during ripening do vary from one variety to another. Selection for uniform pod ripening and roguing of any off-colored pods during the ripening process is important, especially when finding yellowing pods in varieties that have pods that usually hold their green color through their peak edible stage. As always, it is key to select for early vigor and overall health, roguing any diseased plants and becoming intimately familiar with what seems to be an ever-expanding profile of viral pathogens that affect soybeans.

ISOLATION DISTANCES

Soybeans are highly self-pollinated with a crossing rate of less than 1% in most production settings. However, as with the other vegetable legumes, a number of environmental factors and genetic differences between varieties may contribute to increased cross-pollination. Therefore the recommended isolation distance between varieties to maintain a high level of genetic purity is 150 ft (46 m) in open terrain or 50 ft (15 m) when significant physical barriers separate varieties.

FAVA BEAN

The fava bean (*Vicia faba* L.) has been an integral part of the agriculture of the Mediterranean basin and the Near East for at least 8,000 years. These beans are now widely grown in temperate climates, most notably Western Europe and China. They are probably the most cold-hardy of the common vegetable legumes: Some varieties can withstand temperatures of 14°F (−10°C) during early vegetative stages of their life cycle. In warm temperate and subtropical areas they are grown as winter annuals, planted in autumn, putting on rapid growth and flowering in early spring, and harvested before the heat of summer. In cool temperate zones favas are planted early in the growing season, several weeks before the last frost, and grown as a summer annual, much like other vegetable crops of the Fabaceae.

Fava beans fall into two categories based on their use. The large-seeded type, which is often called broad bean, faba, or fava, is grown for its large, flattened, disc-shaped seed, which is 0.75 to 1.25 in (2 to 3 cm) long and is eaten fresh when slightly immature as a shell bean or is harvested at full maturity as a dry bean. The color of the large-seeded types varies from a buff or solid tan to a medium brown with darker spiral markings. The small-seeded form, which is grown as a cover crop, green manure, or fodder, is known by various names in English, including horse bean, tick bean, field bean, and bell bean. Its seed is 0.5 to 0.75 in (1.3 to 2 cm) in diameter, has

SEED PRODUCTION PARAMETERS: FAVA BEAN

Common names: fava, faba, bell bean
Crop species: *Vicia faba* L.
Life cycle: annual
Mating system: mixed selfing and crossing
Mode of pollination: closed perfect flower; no need to stimulate pollen shed
Favorable temperature range for pollination/seed formation: 60–74°F (16–23°C)
Seasonal reproductive cycle: late spring through late summer or fall (4–5 months)
Within-row spacing: 1–1.5 in (2.5–4 cm)
Between-row spacing: 22–30 in (56–76 cm)
Species that will readily cross with crop: Bell beans, which are usually used as cover crop seed, are *Vicia faba* and are fully sexually compatible with all fava types.
Isolation distance between seed crops: 20–100 ft (6–30 m), depending on crop type and barriers that may be present on the landscape

◄ The 'Chak'rusga' fava bean is a tall, prolific, multibranched, hardy, and drought-tolerant variety that comes from northern Bolivia.

Fava beans are linked to a medical condition known as favism, an enzyme deficiency in humans that is an inherited trait among a small percentage of individuals of Mediterranean descent. People who eat favas that are deficient in a particular enzyme may suffer from a hemolytic anemia, which can cause death in some cases. It is therefore advisable for anyone of Mediterranean descent to always start with only a small bite the first time that you eat favas.

CROP CHARACTERISTICS

Reproductive Biology

Fava plants stand upright and range from 2 to 6 ft (0.6 to 1.8 m) tall at full maturity. Stout, square stems bear pinnate leaves with flowers borne in clusters of two to four in the axils of the leaves. Flowers are 0.4 to 1 in (1 to 2.5 cm) long and have white petals with deep purple or black markings with all of the characteristic legume flower parts. Pods are inflated, are 3 to 8 in (7.6 to 20 cm) long, and usually produce three to four seeds. A large number of the flowers or young pods usually abort, and there is some speculation that presence of pollinators will increase the average number of pods that set on favas. While this species is considered to be primarily self-pollinated, there can be quite a bit of crossing with pollinator pressure. There is evidence that insects can be responsible for up to 30% cross-pollination in some locations.

an irregular round shape, and is mahogany brown to black in color. The widespread use of the small-seeded type as a green manure or cover crop is based on its ability to produce luxuriant spring growth and fix large amounts of nitrogen when overwintered in relatively mild temperate zones.

◀ Fava bean flowers (*Vicia faba*).

Climatic and Geographic Suitability

Favas require cool-season conditions for best development. In hotter climates they are planted in the fall as a winter annual so that they can take full advantage of the moderate growing conditions of spring, flowering and developing seed before the heat of summer. In cooler climates, where daily high temperatures don't regularly exceed 84°F (29°C), it is possible to grow a fava seed crop that matures in summer from a spring planting. As with other vegetable legume seed crops, it is desirable to have a seasonal dry period and lower humidity to ensure a lower incidence of seedborne pathogens.

SEED PRODUCTION PRACTICES

Soil and Fertility Requirements

Favas can be grown on nearly all soil types, though they seem to grow best on agricultural loams. They are more tolerant of acidic soils than other vegetable legumes. Available soil calcium is important, as with all Fabaceae crops, for maximum seed yields. Favas are one of the best legumes for their ability to fix nitrogen through their relationship with *Rhizobium* bacteria. Seed should be treated with the appropriate species of *Rhizobium* for this genus before planting, especially if you are planting on ground that hasn't been used for fava production previously.

Growing the Seed Crop

The fava seed crop can be planted in either of two seasons depending on your climate. If sown in fall to be overwintered, you must plant favas early enough to become well established before low temperatures and short daylengths curb growth. When spring-sown, it is important to plant as early as possible to ensure that flowering and seed development take place before hot weather arrives. Fall planting is desirable to help ensure an early seed harvest in regions where winter temperatures don't usually drop below 14°F (−10°C).

Large-seeded vegetable fava types should be planted at a depth of 2 to 4 in (5 to 10 cm), while smaller-seeded bell bean types can be planted at half this depth. Favas are usually drilled with row spacings that are 22 to

Immature fava bean seedpod.

271

28 in (56 to 71 cm), and 4 to 6 in (10 to 15 cm) between plants within the row. In Europe seed crops are sometimes planted in double rows that are 10 in (25 cm) apart and are separated from the next pair of rows by 24 to 28 in (61 to 71 cm). Cultivation can be practiced early in the season in a similar fashion to the method described for common bean (see Common Bean, "Growing the Seed Crop" on page 260) but can only be done early in the growth of the crop, as the stems of the fava plant are easily damaged, resulting in lodging.

Seed Harvest

Maturation of the fava seed crop is first accompanied by a darkening of the pods, which slowly blacken in color. This is followed by a drying and loss of sponginess in the characteristically thick fava pods. Watch the lower-setting pods closely; they will mature first and can shatter if you wait for the later-setting, upper pods to fully mature. You must cut the crop when the upper pods are fully formed but less than fully ripe. Swathing the crop should be done during cloudy, cool weather or early in the morning when shattering can be minimized. Mechanically raking the crop into windrows should follow the same protocol. Smaller growers in Northern Europe traditionally place the crop into shocks for the final drying before threshing. Threshing is easily done with a stationary field thresher or a bean combine that has been adjusted for the large seed size of the vegetable favas.

GENETIC MAINTENANCE

Selection and maintenance of key agronomic traits for overall health and vigor is always desirable. Maintaining uniformity of plant stature (bush versus vine) and general morphological traits is important; variation of any of these traits could be an indication of an unwanted outcross between different varieties. Because of the propensity of this species to cross-pollinate, growers should be especially vigilant in monitoring for plants that are the result of these crosses. Variations in seed coat color and seed size are probably the two most easily recognized traits that indicate probable outcrossing between fava bean varieties in a previous generation.

ISOLATION DISTANCES

In the Fabaceae, fava beans are second only to runner beans for their degree of promiscuity, and therefore require a fairly large isolation distance. Informed organic seed producers prefer to treat them conservatively, as they know how many insect species will visit their flowers. For this reason they will separate varieties of this species by a minimum of 0.5 mi (0.8 km) in open terrain, with a 0.25 mi (0.4 km) minimum distance if physical barriers exist between production fields. This is the sensible recommendation for isolation of a crop that is so influenced by insect activity in its pollination.

GARDEN PEA

The garden pea (*Pisum sativum* L.) is one of the oldest cultivated pulses. N. I. Vavilov, a Russian ethnobotanist, recognized four centers of origin for this crop spanning an area from Central Asia through the Mediterranean basin to Abyssinia, indicating that the dissemination of pea germplasm must have occurred early in the spread of agriculture. Among the pulses the pea is one of the most widely grown species on Earth. Various forms of the crop are used as a vegetable, fodder, and green manure; its dry mature seed is used for soups and stews. The vegetable forms include: (1) shelling peas, which have well-developed, immature seed that is "shelled" from the pod when still soft and succulent; (2) snow peas, which have edible, fiberless flat pods that are harvested when the seed begins to swell; (3) snap peas, which have thick-walled, edible, fiberless pods much like snap beans and which are harvested when the peas are swollen but still succulent; and (4) pea vines, which includes the young shoots, with tendrils and sometimes flowers that are used in Asian-style stir-fry dishes and as a garnish.

While these vegetable types are grown seasonally in many diverse climates around the world, peas are best adapted to environments with a relatively cool growing season, especially for seed production.

SEED PRODUCTION PARAMETERS: GARDEN PEA

Common name: garden pea
Crop species: *Pisum sativum* L.
Life cycle: annual
Mating system: primarily selfing
Mode of pollination: closed perfect flower; no need to stimulate pollen shed
Favorable temperature range for pollination/seed formation: 60–74°F (16–23°C)
Seasonal reproductive cycle: late spring through late summer or fall (4–5 months)
Within-row spacing: 1–1.5 in (2.5–4 cm)
Between-row spacing: 22–30 in (56–76 cm)
Species that will readily cross with crop: All true pea types, including Austrian field pea types and those grown for dry peas and split peas
Isolation distance between seed crops: 20–100 ft (6–30 m), depending on crop type and barriers that may be present on the landscape

Snap peas are a new version of an ancient crop with seeds that can be more fragile during harvesting and threshing than some of the older forms.

CROP CHARACTERISTICS

Reproductive Biology

Peas are self-pollinated annuals with a short, intermediate, or long indeterminate vine growth habit. Short vine or dwarf varieties may be 2 ft (0.6 m) tall or less with a concentrated fruit set, while the taller indeterminate types can easily reach 6 ft (1.8 m) and bear over a longer season. Pea flowers are characteristic of most species of the Fabaceae as described under the "Reproductive Biology" section of the Fabaceae family introduction. The pistil of each pea flower contains a single carpel and a single ovary, which may bear up to 13 ovules attached to two adjacent placentae that occur along its length. A sticky, bearded stigma is receptive to pollen for several days before the maturation and release of pollen: However, the petals of the pea flower are usually tightly closed during this period and only open 24 hours after fertilization. While the petals of the pea flower are usually tightly closed during this period, this is when peas are most apt to cross if pried open by an insect species with this ability. There are several species of insects that can cross-pollinate leguminous flowers by cutting through the petals to access pollen and nectar before self-pollination has occurred. While the incidence of cross-pollination is usually well below 1% in most seed production areas, there are areas with a wide diversity of pollinator species present that may have much higher outcrossing rates in this crop. This may be the case in many organic production fields, especially on farms with diverse cropping and close proximity to native ecosystems.

Climatic and Geographic Suitability

Production of pea seed is best accomplished in regions with cool, moist spring weather

A purple-flowered pea (*Pisum sativum*). Pea flowers can either be purple or white. ▶

that gradually turns warm in spring. The availability of ample moisture early in the growing season enables the crop to establish a vigorous, fully grown plant by the onset of flowering, which supports a bountiful pod set and seed yield. Peas are adapted to a moderate climate for both their vegetative growth and the development and early growth of the seed crop. In early spring the pea plant is sensitive to frost once blossoms and flowers are present. Exposure to temperatures above 84°F (29°C), especially during anthesis, fertilization, and early endosperm development may interfere with normal seed development and lower seed yields. Hotter summer weather is acceptable after the seed is fully formed—during seed maturation and harvest.

Pea seed produced in areas with a seasonal dry period during seed maturation and harvest has a much better chance of being free of seedborne pathogens and blemishes. Seed production areas that produce the highest-quality seed have a lower incidence of destructive pathogens and insects. Several serious insect pests of peas—including pea aphids, pea borers, pea leafminers, pea weevils, and thrips—are usually less persistent in these western valleys, although the incidence of damage from the pea weevil has greatly increased since North American pea seed production first moved to the West after World War II. Seed production areas in North America include portions of the Snake River Valley of southern Idaho, the Columbia Basin of eastern Washington, and interior valleys of southern British Columbia. The seasonal dry weather and low humidity in these locations at the time of harvest greatly enhance the quality of the crop.

As a major international crop, appreciable quantities of pea seed are grown in a large number of areas around the globe. In Europe commercial pea seed is produced in parts of the United Kingdom, the Netherlands, Denmark, Germany, France, and Spain. Commercial seed is also grown in Poland, Hungary, Russia, and Greece. Peas are grown widely in the Indian subcontinent, with large-scale commercial seed production in India and Pakistan. Much of the enormous quantities of pea seed used across China and Southeast Asia is grown locally, with commercial production in Thailand, Taiwan, and China.

SEED PRODUCTION PRACTICES

Soil and Fertility Requirements

Peas are grown on a wide range of soil types, from lighter sandy and silt loams to heavier clayey soils, but any soil must be well drained to minimize the chances of root rot organisms, which can cause crop losses. As with all legumes, peas are best grown on soils with a good, balanced fertility that do not have an excess of nitrogen. This is especially true in producing the seed crop, where excess foliar growth will only be a detriment in the early and even maturation of a superior seed crop. Compost or well-composted manure should supply plenty of fertility, especially if the ground has a good long-term program for adequate phosphorus, which is a must for good legume seed yields.

Inoculation with the proper nitrogen-fixing *Rhizobium* species should be done the first time you grow a pea crop in a particular field. After that, however, further inoculation may not be necessary as long as you grow another pea crop in the crop rotation within the next several years.

Growing the Seed Crop

The pea seed crop should be planted in spring after the threat of severe freeze has passed. While young pea plants are quite cold-hardy,

a heavy freeze can damage the apical growing points of the young shoots. Early plantings also run the risk of suffering losses from damping-off organisms under prolonged cold, wet conditions. Therefore many experienced growers will not plant the crop as early as possible, which is often recommended. However, planting must still be done early enough for the crop to complete flowering and early pod formation before hot temperatures occur. In most pea seed production areas of the Intermountain West (eastern Oregon and the Palouse region of eastern Washington and northern Idaho) the crop is planted in mid- to late April, but it can planted in late March to early April in the warmer areas of the Columbia Basin of Washington and in both the Treasure Valley and Magic Valley of southern Idaho.

Pea seed can be planted at a depth of 1 in (2.5 cm) or less if steady, even moisture is the norm in spring. A shallow planting depth will help promote quick seedling emergence and a healthy stand if cold, wet conditions persist. Deeper plantings of up to 2 in (5 cm) can be used if the crop is grown on lighter soils and if precipitation is erratic in spring before irrigation is in place.

Pea seed crops have been grown in the western United States using either a row-cropping method similar to common bean or by planting the seed with a grain drill at 7- to 8-in (17- to 20-cm) centers where they eventually produce a solid closed canopy that is harvested standing, without swathing or windrowing. The latter method is not suited to organic production, as it relies heavily on pre-emergence herbicide treatment. Also, for organic culture it is essential to cultivate at least two to three times before the pea plants sprawl and flower. The close spacing used with the solid canopy method hinders many cultivation possibilities.

◀ Fully mature pea pod ready to be threshed.

The planting specifics for organic pea seed production are very similar to the methodology described for common bean, using the same 22 in (56 cm) row spacing but with a higher within-row density of 8 to 12 seeds per 1 ft (30 cm), or one seed every 1 to 1.5 in (2.5 to 4 cm). Seeding rates may vary according to type, although in pea seed production the recommended spacings are quite similar for the dwarf and vining types, as vining varieties are grown without trellising. Dwarf types can be grown at the higher end of this density of 1 seed/in (1 seed/2.5 cm) and vining types at 1 seed/1.5 in (1 seed/4 cm). While cultivation can be done in much the same way as you would for common beans, the earliest "blind cultivation" that is done after planting and before the crop emerges is often not as effective as it is in beans. This is because peas are planted earlier in the growing season than beans, and the earliest flush of weeds frequently only emerges with the warm spring days that also bring the pea seedlings out of the ground.

Seed Harvest

The pea seed crop is ready to harvest when a majority of pods develop a lighter color that is either yellow or light brown, and when they develop a rough, leathery appearance. The crop is then swathed and the plants are cured in windrows. In the western United States pea seed growers use pea lifters that are attached to the swather and hold the crop off the ground as it is cut. This minimizes damage to the crop and is best accomplished in the morning when the dew is still on the ground. Curing usually takes 7 to 10 days, or until the plants are completely dry and easily threshed. The windrows are then threshed with a grain combine using a pickup attachment, or the crop can be manually fed with pitchforks into a stationary combine. The spacing of the cylinder and speed of the rotor

should be closely monitored to minimize damage to the seed.

GENETIC MAINTENANCE

Selection for seedling vigor, speed of emergence, and overall health of seedlings is always desirable, especially in a crop like peas, which is often planted during cold, wet periods that can cause the crop to languish both before

'Oregon Giant' is a popular disease-resistant variety of snow pea that produces prolifically over an extended season.

and after emergence. Routine selection for the strongest seedlings will definitely improve these traits over generations of selection.

Peas have several key traits that should be observed when monitoring the crop for varietal integrity. Roguing should be done at an early vegetative stage, at flowering, and at full pod development. Leaf shape, size, and color can vary considerably, and variants should be eliminated before flowering. The size, color, and position of the flower on the plant can vary, as can the size, shape, color, and curvature of the pods, and off-types for these traits or other noticeable variants should be rogued. Another common pea off-type that may appear has the colorful moniker *rabbit-eared rogue*. The rabbit-ear rogue is best described by its leaves, which are smaller and narrower than the leaves of almost all commercial pea varieties of the past 100 years. It also is usually a more rangy plant that is taller, spindlier, and later flowering than most named pea varieties. These can usually be detected before flowering and possible crossing can occur, which is always desirable.

ISOLATION DISTANCES

As with all self-pollinated vegetable crops in the Fabaceae, there are a number of environmental factors and varietal variations that can lead to cross-pollination in peas. While peas are usually highly self-pollinated, crossing less than 1% of the time in climates that are conducive to seed production, they can also cross to the extent that the recommended isolation distance of 150 ft (46 m) in open terrain is appropriate to maintain a high level of varietal purity. When barriers, natural or artificial, are present (see chapter 13, Isolation Distances for Maintaining Varietal Integrity), it is still advised to maintain at least 50 ft (15 m) of isolation to minimize crossing. In some areas where crossing has been noted to be significantly less than 1% between adjacent pea varieties grown for seed, growers have planted several rows of a taller unrelated crop to keep them isolated from potential insect pollination or mixing at harvest.

Importantly, the threat of mechanical mixing in a round-seeded crop like peas should not be dismissed. Mechanical mixing between two different varieties can always present a problem during harvest when self-pollinated crops have been planted in adjacent plots with no isolation breaks or natural barriers. Mechanical mixing can also happen when harvesting and seed-cleaning equipment is not thoroughly cleaned between any step in harvesting or processing different varieties of the same crop (see chapter 17, Stockseed Basics). However, with round-seeded vegetable crops like peas, edamame, or the small, round seed of the brassicas, it is possible for mixing to occur in seed-cleaning facilities, sheds, or any other space where seed is cleaned, conditioned, or bagged. The fact that round seed can easily roll across a table, floor, or any work surface means that seed from one lot can end up on the other side of the room near another lot by rolling from one place to another. Many experienced seed workers will acknowledge the fact that this has certainly been the cause of countless genetic mixes through the years. Therefore, anyone coming into seed growing should heed the well-worn phrase of old-timers in the seed business: "Seed on the ground, leave it down!"

LIMA BEAN

Lima beans (*Phaseolus lunatus* L.) are native to the northern reaches of South America, where they have been cultivated for at least 7,000 years. They were extensively cultivated in the coastal valleys and western central area of what is modern-day Peru when the European explorer Francisco Pizarro first collected them in the area that would become known as Lima. The limas originally from Peru are the larger, thick-seeded variant of the modern crop and historically known as the Potato Lima type in North America. This type quickly spread across the globe, although its use never rivaled that of the hardier common bean (*P. vulgaris*). This type of lima is one of the oldest cultivated crops of the New World.

The small-seeded or Sieva type of limas were domesticated from wild ancestors of the crop only 2,500 to 3,000 years ago in what is now southern Mexico and Guatemala. These small-seeded types produce the popular butterbeans that are ubiquitous in the southern United States. Unfortunately, both these and the larger-seeded types have gotten an unfavorable reputation in most of North America in modern times, as they have been grown as a dry bean, harvested on a large scale and eaten as a rather pasty mush by most people. The real treat that makes this crop such an outstanding vegetable among the initiated is the fresh-shelled, bright green, sweet and nutty beans that are picked and eaten long before the pods turn yellow and the seed gets starchy.

SEED PRODUCTION PARAMETERS: LIMA BEAN

Common names: lima bean, butter bean
Crop species: *Phaseolus lunatus* L.
Life cycle: annual
Mating system: primarily selfing, though higher crossing rate than common bean
Mode of pollination: closed perfect flower; no need to stimulate pollen shed
Favorable temperature range for pollination/seed formation: 70–80°F (21–27°C)
Seasonal reproductive cycle: late spring through late summer or fall (4–5 months)
Within-row spacing: 1–1.5 in (2.5–4 cm)
Between-row spacing: 22–30 in (56–76 cm)
Species that will readily cross with crop: All true *P. lunatus* types, including those called butter beans in the southeastern United States and limas grown as dry beans
Isolation distance between seed crops: 150–300 ft (46–91 m), depending on crop type and barriers that may be present on the landscape

CROP CHARACTERISTICS

Reproductive Biology

Limas are self-pollinated annuals with a determinate or indeterminate growth habit. Determinate or dwarf varieties may be 2 to 3 ft (0.6 to 0.9 m) tall with a concentrated fruit set, while the taller indeterminate types can easily reach 6 to 10 ft (1.8 to 3 m) tall and bear over a longer season. Lima flowers are characteristic of most species of the Fabaceae as described under the "Reproductive Biology" section of the Fabaceae family introduction. The flowers are borne on racemes that are 2 to 4 in (5 to 10 cm) long, and even under the best environmental conditions only a minority of the flowers set and produce pods. Anthesis has been reported to occur early in the morning. Limas are notoriously fickle in setting seed, as they are real heat lovers and probably suffer from inadequate fertilization of the ovaries when environmental conditions are not just right in the morning hours after anthesis.

As with many other legumes, limas have a sticky, bearded stigma that is receptive to pollen for some period of time before its maturation and release. While the anthers surround the style and stigma, there are times when the stigma will project out past the end of the keel petal, especially when insects land on the wing petals. This usually happens after pollination has taken place, but if it occurs before the pollen has been shed, then there is an opportunity for cross-pollination from an insect visitor. Limas have white to cream-colored flowers and have nectaries that secrete a high-quality nectar, both of which are known to attract a variety of pollinating insects, including honeybees, bumblebees, assorted wild bees, and many types of butterflies. Several studies have shown that insect visitation during pod set increases the number of pods, beans per pod, and total seed yield. This appears to be just another case where insects can increase the overall amount of seed set in self-pollinated crops by stimulating the effectiveness of pollen shed through physically moving the flowers.

Climatic and Geographic Suitability

Lima bean is a tender warm-season crop that requires warm, well-drained soils for germination. The optimum soil temperature for vigorous germination is 85°F (29°C), and soil temperatures below 65°F (18°C) can encourage root rot. Ambient temperatures of 70 to 80°F (21 to 27°C) are preferred for optimum growth of the lima crop. Temperatures below 55°F (13°C) or above 90°F (32°C) during flowering may adversely affect pod set and seed yields. Lima beans require a longer season and warmer climate than common beans to mature their seed crop. Limas routinely require a minimum of anywhere from 110 to 140 frost-free days to produce a satisfactory commercial seed crop, with pole types generally requiring a longer season. For this reason much of the lima bean seed produced in North America has long been produced in the warmer interior valleys of California, with some of the earlier-maturing varieties grown in the seed production valleys of southwestern Idaho. However, growers must be careful in choosing the warmer-season climates in places like the Central Valley of California, as higher temperatures during flowering can reduce pod set and subsequent seed yield.

SEED PRODUCTION PRACTICES

Soil and Fertility Requirements

Soil conditions and fertility requirements are the same as those recommended for common bean (see Common Bean, "Soil and Fertility Requirements" on page 258). The main difference is that lima seed needs to be planted into soils at least

5 to 10°F (3 to 6°C) warmer than is needed for planting common bean (see "Climatic and Geographic Suitability," page 281). This means that in many regions lima seed is planted a full 10 to 14 days later than common beans, ensuring that soils are sufficiently warm for rapid germination. Also, planting into moist soil is preferred to irrigating the crop after planting, both to minimize soil crusting and to minimize the chances of rotting the seed before emergence. If you're growing lima beans in a particular field for the first time, then inoculating the seed with the appropriate *Rhizobium* species is recommended.

Growing the Seed Crop

As most commercial lima bean varieties are bush varieties and have a stature and form a canopy much like most common bean bush varieties, it is suggested that you plant and cultivate the crop per the recommendations for bush-type common beans (see Common Bean, "Growing the Seed Crop" on page 260). However, you can plant bush limas at a wider spacing between rows in climates where there is some concern of getting enough heat to mature the crop. Also, a wider spacing can accommodate some of the larger bush lima types, which can produce quite a canopy under ideal conditions. Between-row spacing for limas can be anywhere from 22 to 36 in (56 to 91 cm). The within-row planting density should be no more than six to eight plants every 12 in (30 cm).

Pole lima varieties require a much wider spacing, with rows at least 36 to 60 in (91 to 152 cm) apart and with a 4 to 8 in (10 to 20 cm) within-row spacing between plants. Large-scale seed production of many pole or vining type of limas is generally done without staking or trellising the plants.

Seed Harvest

As with all Fabaceae crops, proper timing of harvest is important in order to produce high-quality seed that is fully mature, has a high germination percentage, and has maximum storage potential. Each lima variety and type has its own specific harvest timing, though there are basic signs that indicate maturity. The initial sign for all beans that indicates a crop is approaching maturity is the color of the pods as they are breaking, or after they have turned yellow and are turning mahogany or buckskin in color, but before they dry completely to a light tan or brown and the pods become papery thin. Maturation to the brown, papery stage increases the danger of seed shattering during harvest. Pods should generally be yellow to buckskin at harvest in order to mature properly in the windrow, but the exact desired color may be variety-specific.

The crop should be cut when approximately 70 to 80% of the pods on the crop are of the desired color and point of breaking. The stems of the crop are undercut mechanically just below the soil surface and left in place for a day. The next day the plants are raked into windrows. Depending on weather conditions, the windrows should cure in the field for at least 10 to 14 days, with a possible turning of the rows near the halfway point. Threshing of limas is best accomplished during the heat of the day, as is done with other beans, when the pods are brittle and easily cracked. Seed should then be further cleaned to separate it from any plant debris that may be moist. If the seed is not sufficiently dry it may require air- or kiln-drying before being stored for a final cleaning and conditioning. Physical mixing of seed varieties can be a source of variety contamination. Thoroughly clean out harvesting and cleaning equipment between working with different varieties.

GENETIC MAINTENANCE

There are very few stages in the growth of most legume crops in which genetic differences within a variety are readily apparent. If you are able to establish a uniform stand, then it is possible to perform selection for speed of seedling emergence, seedling health, and overall vigor: all essential traits for organic production. If you routinely rogue out late-emerging, low-vigor seedlings, you'll see improvement over cycles of selection for these traits and improvement in the overall seedling vigor of the variety. Many growers of legume crops can identify obvious differences that might appear in pod characteristics, flower color, and presence of runners, but other, more subtle plant characteristics may only be apparent to experienced farmers who specialize in the crop and know a specific variety. For some traits, like leaf shape, color, and plant stature, it is possible to evaluate the plant and rogue it to type before flowering has begun. Flower color, placement, and the timing and duration of flowering can all be selected during flowering. Pod characteristics such as color, shape, and length are also notable to the experienced grower and can be selected. Additionally, it is critical to rogue for disease in all bean crops in order to produce quality seed, minimize the risk of seedborne diseases, and select or maintain a variety's disease resistance.

ISOLATION DISTANCES

While lima beans are considered highly self-pollinating by most commercial seed producers, they have been reported to outcross at a higher rate in some environments than the common bean (*P. vulgaris*). This may be due to the unique morphological characteristics of the stigma being exposed and the higher nectar content than is found in some of the other *Phaseolus* species (see "Reproductive Biology" in the Fabaceae introduction on page 253). For this reason, many researchers recommend increasing the minimum isolation distance between two lima crops over what is recommended for common beans.

Because my isolation recommendations uphold a very high standard for all of the highly self-pollinating crops of the Fabaceae (including common bean, garden pea, and edamame), the minimum isolation distance recommendations for lima beans of the same horticultural type do not need to be higher than what I am recommending for these crops.

As with the other largely self-pollinating species of the Fabaceae, it is appropriate to observe the minimum 150 ft (46 m) isolation distance if you're growing more than one lima variety of the same type in open terrain with no natural barriers to ensure a relatively high level of genetic purity. If you are producing multiple lima crops of different horticultural types (say, a small-seeded Sieva type versus a larger Potato Lima type or a bush type versus a vining type), then it is recommended that you produce these a minimum of 300 ft (91 m) apart in open terrain.

If you're producing multiple lima bean crops in areas with natural barriers between crops (see chapter 13, Isolation Distances for Maintaining Varietal Integrity), then the minimum isolation distance needed between two limas of the same type can be dropped to 75 ft (23 m). For two different lima types with a substantial physical barrier between them, then the isolation distance should be increased to 150 ft (46 m).

It should be noted that these recommendations are based in large part on information from seed production areas in North America. If lima seed is produced in more tropical regions, especially in areas with both the kind and breadth of insect biodiversity that is present in the ancestral homeland of the various forms of this crop (southern Mexico, Guatemala, or northern South America), then it may be necessary to reassess these minimum isolation recommendations and increase them to maintain varietal purity.

RUNNER BEAN

The runner bean (*Phaseolus coccineus* L.) is a warm-season legume native to high-altitude regions of Guatemala and southern Mexico. It is a vining herbaceous perennial with a large tuberous root that is usually grown as an annual. In temperate climates with only light frosts, runner beans may regrow for a number of seasons from these perennial roots. It is prized for its large showy flowers, its flavorful immature pods, and the shell beans that are harvested when the beans are well filled but still immature and succulent. The indigenous people of Central America, who have cultivated runner beans for at least 2,000 years, utilize both the large, fleshy roots and the dried beans in their diet. There are reports that the roots may contain toxins and require special preparation if they are to be used as food.

Runner beans have become very popular as an ornamental in the United States and Europe because of their attractive flowers, which range in color from bright shades of scarlet or crimson to a cream white. Robust vines easily reach from 12 ft (3.7 m) to as much as 16 ft (5 m) in length and may have hundreds of blooms at their peak. Europeans also favor runner beans for the unique flavor of the immature pods, which are eaten like snap beans. Dwarf or true determinate varieties have been developed for the culinary markets of Europe.

SEED PRODUCTION PARAMETERS: RUNNER BEAN

Common name: runner bean, scarlet runner
Crop species: *Phaseolus coccineus* L.
Life cycle: annual (perennial in the subtropics)
Mating system: mixed selfing and crossing
Mode of pollination: perfect flowers that shed pollen before opening, but tripping is required to transfer pollen to stigma
Favorable temperature range for pollination/seed formation: 65–86°F (18–30°C)
Seasonal reproductive cycle: late spring through late summer or fall (4–5 months)
Within-row spacing: bush type 8–12 in (20–30 cm); vining type 6–8 in (15–20 cm)
Between-row spacing: bush type 3–4 ft (0.9–1.2 m); vining type 5–6 ft (1.5–1.8 m)
Species that will readily cross with crop: In North America there are a number of runner beans that include the name *lima* due to the size and shape of their seed.
Isolation distance between seed crops: 0.5–2 mi (0.8–3.2 km), depending on crop type and barriers that may be present on the landscape

CROP CHARACTERISTICS

Reproductive Biology

While runner beans have perfect flowers that are much like standard Fabaceae blossoms in form, they are probably the most promiscuous of the vegetable species of this normally self-pollinating family. This species certainly has blossoms that are very attractive to a number of pollinators, with individual flowers measuring almost 1 in (2.5 cm) across. These flowers are borne in clusters with upward of 20 blossoms on a single raceme. This attracts a diversity of pollinating species, from hummingbirds to honeybees, bumblebees, assorted wild bees, and many types of butterflies.

Indeed, this species has co-evolved with pollinators. Unlike most other legumes many runner beans are unable to self-pollinate unless a pollinator lands on the wing petal and "trips" the flower. This causes the stigma and anthers to extrude through the keel, allowing pollen to reach the stigma and self-pollinate the flower. A number of breeders have noted a marked reduction in yield in runner beans when pollinators are artificially excluded from the plants during flowering. For optimum conditions for seed production it is advised that growers ensure that a large number of honeybees or bumblebees are present to adequately visit the plants during flowering.

Another important issue in growing runner beans is the fact that pollen is often deposited on the abdomen of pollinators during the tripping of the flowers, thereby increasing natural cross-pollination. Jim Myers, a bean breeder at Oregon State University, has found that the rate of cross-pollination among runner bean varieties is "highly variable" and reports that rates of crossing up to 50% per generation are not uncommon in temperate climates, when different varieties are grown in close proximity. Researchers in the tropics have witnessed crossing rates as high as 87% with native pollinators. Many seed growers who have grown more than one type of runner bean in close proximity have also reported crossing between varieties far above acceptable levels for commercial production. Because insects are integral to the pollination of this crop, the isolation distances must be increased significantly over the standard distances set for most self-pollinated legumes (see the "Isolation Distance" section, page 287).

Climatic and Geographic Requirements

Runner beans are a warm-season legume and need a minimum of 110 to 120 frost-free days to produce a satisfactory commercial seed crop. Runner bean varieties grown in higher latitudes are day neutral in their flowering response and are most prolific in setting fruit during hot weather with summer daytime temperatures in excess of 86°F (30°C). The types grown in the tropics are short day plants in their reproductive response. A seasonal dry period for seed maturation and harvest is desirable for all types to minimize any rot or seedborne pathogens.

SEED PRODUCTION PRACTICES

Soil and Fertility Requirements

Soil conditions and fertility requirements are the same as those recommended for garden peas (see Garden Pea, "Soil and Fertility Requirements" on page 276). If you're growing runner beans in a particular field for the first time, then inoculating the seed with the appropriate *Rhizobium* species is recommended.

Growing the Seed Crop

As with other bean crops, seed of runner beans is tender and should only be planted when soil temperatures are above 65°F (18°C). Most runner bean varieties have a vining habit, and many small-scale specialty producers in North America trellis their crop with much hand labor. Trellised crops are routinely planted at 5 ft (1.5 m) row centers with a 6 to 8 in (15 to 20 cm) within-row spacing. Larger scale seed productions of vining types in Europe are grown without staking. Bush runner bean varieties grown for seed are planted at 3 ft (0.9 m) row centers with 8 to 12 in (20 to 30 cm) spacing within the row due to their large bush frame.

Seed Harvest

Seed of bush-type runner bean varieties is harvested in a similar fashion to bush snap bean harvest (see Common Bean, "Seed Harvest" on page 261). For seed production of vining types grown without trellising, the use of pea lifters attached to the swather is necessary to prevent excessive cutting and shattering of mature pods when swathing the crop. In addition to lifters, one also needs vertical shears on each end of the cutter bar since vines intertwine and form a continuous sheet at harvest. The crop is then placed into windrows and cured for 7 to 10 days (weather depending) before threshing and cleaning. Cutting and threshing is best done in the morning, when the moisture from the dew helps prevent excessive shattering.

A number of small-scale US seed producers of the vining types do trellis the crop. The advantage over allowing them to sprawl on the ground is that there is usually very little rot associated with the drying pods; however, stripping the vines from the trellis is both time- and labor-intensive. Vines should still be cured in windrows before threshing. Take care to set the speed and spacing of rotors of the threshers, as with all large-seeded legumes, in order to not damage seed.

GENETIC MAINTENANCE

Selection and maintenance of key agronomic traits for overall health and vigor is always desirable. Maintaining uniformity of plant stature (bush versus vine) and general morphological traits is important; variation of any of these traits could be an indication of an unwanted outcross between different varieties. Because of the propensity of this species to cross-pollinate, be especially vigilant in monitoring for plants that are the result of these crosses. Variations in flower color and in seed coat color are probably the two most easily recognized traits that indicate probable outcrossing between different runner bean varieties in a previous generation.

The most common flower color of runner beans is scarlet, although there are variants in the shade of this color in the petals of different scarlet runner varieties. The second most prevalent flower color of this species is white. If crosses occur between a scarlet-flowered type and a white-flowered type there are two different outcomes in detecting the outcrosses, depending on which of the two populations you monitor in the subsequent generations. (*Note:* Flower color in runner beans is a qualitative, simply inherited trait similar to the qualitative traits Mendel used in his classic work with peas that laid the foundation of modern genetics. Hence, a review of Mendel's work with monohybrid crosses from any number of high school or college biology texts may be of value in understanding this information on roguing based on flower color.)

If you grow the seed of the scarlet-flowered variety after an outcross with a white variety there will be no indication in the next generation that the white flower trait is in the scarlet population. This is because the white petal color trait is genetically recessive to the scarlet petal trait (or any other pigmented petal shades that occur in runner beans). The white trait will only be expressed again in the following (F_2) generation when 25% of all progeny resulting from any initial crosses between the two types will have two copies of the white color gene and have white flowers. While these plants with white flowers are easily rogued from a scarlet-flowered population, there will be twice as many plants where the white petal gene is in the heterozygous state and is not expressed. Through normal segregation of the genetic material there will always be a percentage of the plants in this scarlet population with the white flower gene in the heterozygous recessive condition that you will be unable to detect (and unable to rogue) through a normal visual inspection. In this case the best solution is to return to a stockseed lot of the scarlet variety or perform a controlled progeny test (see chapter 17, Stockseed Basics).

Alternatively, when you grow the white-seeded runner bean variety after it has outcrossed with a scarlet-flowered variety you will be able to easily identify the plants that have resulted from cross-pollinations with the scarlet type. Because the gene for scarlet flower color is dominant, it will be expressed in all progeny of these outcrosses in the first generation after the cross. Therefore, by eliminating all of the scarlet-flowered plants you will completely rid the white variety of all plants that resulted from the outcrosses with the scarlet variety. With a dominant gene for a simply inherited trait crossing into a new population, "what you see is what you get!"

The seed coat color of the scarlet-flowered types is a mosaic of purple and black; the white-flowered runner beans have a pure white seed coat. The inheritance in crosses between these two types seems to follow exactly the same pattern as the corresponding flower color trait.

ISOLATION DISTANCES

Of all of the vegetable crops of the Fabaceae, runner beans may be the most promiscuous, and therefore they require the largest isolation distance. Indeed, a number of seed producers prefer to treat them as if they are cross-pollinated crops and separate varieties of this species by a minimum of 1 mi (1.6 km) in open terrain, with a 0.5 mi (0.8 km) minimum distance if physical barriers exist between production fields. This is the sensible recommendation for isolation of a crop that is so influenced by insect activity in its pollination.

– 11 –

POACEAE (SWEET CORN)

Maize, or corn as it is known in North America, is the most valuable crop plant worldwide to come from the New World. Corn was derived from a closely related species, teosinte (*Zea mays* var. *mexicana*). Modern sweet corn most likely descends from the Northern Flint corns of the northeastern United States, but we know that the sweet corn gene (*sugary1*, *su1*) was selected by Native Americans in at least four additional locations; the highlands of Peru, the central plateau of Mexico, northwestern Mexico/southwestern United States, and the northern Great Plains.

Corn has been cultivated for at least 5,000 years, with the probable center of diversity in the lowlands of western Mexico. By the time of European conquest of the Americas corn had spread as far south as Argentina in South America and as far north as Saskatchewan in Canada. Columbus found large fields of corn being grown by the local farmers on the island of Hispaniola in 1492. The diversity of types

SEED PRODUCTION PARAMETERS: SWEET CORN

Common name: sweet corn
Crop species: *Zea mays* L.
Life cycle: annual
Mating system: largely cross-pollinated, some percent self-pollination
Mode of pollination: wind
Favorable temperature range for pollination/seed formation: 65–86°F (18–30°C)
Seasonal cycle: late spring through late summer or fall (4–6 months)
Within-row spacing: 8–12 in (20–30 cm)
Between-row spacing: 3–4 ft (0.9–1.2 m)
Species that will readily cross with crop: Other types of corn (*Zea mays* L.)
Isolation distance between seed crops: The standard isolation of 660 ft (201 m) between similar types of corn or 1,320 ft (402 m) between different types requires that you avoid harvesting at least four to six buffer rows from the periphery of the seed crop.

After allowing corn seed to mature and dry as much as possible on the plant, it is important to harvest the ears and put them on racks like this multicolored corn. This crop is drying naturally under low-humidity conditions in New Mexico, but sweet corn often needs supplemental heat to finish this process.

across the hemisphere is amazing in both form and utility. Many tropical corns are 12 to 15 ft (3.7 to 4.6 m) tall, while some of the Northern Flints can be as short as 3 ft (0.9 m). There have long been many kernel sizes and starch types within the kernels, with very specific culinary uses for these different types. One of the pre-eminent maize geneticists of the 20th century, Paul Mangelsdorf, once proclaimed that 90% of the breeding work in corn had already been completed by the time the Europeans arrived.

There are essentially five classes of corn that are widely cultivated today. These types differ in the makeup of the endosperm portion of the kernel. They are: (1) flour corn, (2) flint corn, (3) dent corn, (4) popcorn, and (5) sweet corn. All are sexually compatible and will cross, though popcorn has a relatively strong gametophytic self-compatibility that prevents most crosses with other types of corn. Traditional sweet corn is the result of a genetic mutation that has resulted in the *sugary1* gene (*su1*) that prevents the normal conversion of sugar into starch during endosperm development, and the kernels accumulate phytoglycogen, which give classical sweet corn its creamy texture. In the second half of the 20th century corn breeders in the United States also introduced two other important sugar-producing genetic variants that have become important in commercial sweet corn varieties, the *shrunken2* (*sh2*) and *sugary enhancer1* (*se1*) types. Commercial sweet corns in the early 21st century are almost exclusively F_1 hybrid varieties. However, collaborative breeding work by the Organic Seed Alliance and the University of Wisconsin is producing promising open-pollinated stocks at the time of this writing.

Sweet corn seed production requires a warm, moderate climate with lots of sun and a good, steady source of moisture. Almost all commercial sweet corn seed is grown in the Treasure and Magic valleys of the Snake River in southwestern Idaho in the United States. Recent years have seen an increasing amount of seed grown in the arid valleys of the western slope of Colorado and the Columbia Basin of Washington. All of these areas promise dry weather for seed maturation and harvest as well as lower disease and insect pressure on the crop.

CROP CHARACTERISTICS

Reproductive Biology

Sweet corn is a wind-pollinated, monoecious, heat- and water-loving annual that requires good, vigorous growth at the time of planting to produce a full-sized flowering plant by midsummer in order to bear a fully developed seed crop before frost.

In this monoecious plant with separate male and female flowers, the male or staminate

Selecting fully filled ears with straight rows of kernels is always a worthy pursuit in sweet corn.

flowers are borne on a terminal cluster (tassel) with lateral branches, which is capable of producing more than 20,000 pollen grains for every female ovule produced. The female or pistillate flowers are borne on spikes or cobs in the axils of lateral leaves of the middle nodes of the plant. Each female flower produces a single kernel, which is an individual fruit known as a **caryopsis.** Paired pistillate spikelets are borne in paired rows on the **cob.** The number of rows of spikelets can range from 8 to 20. The collection of these fruits on a single cob is known as an ear of corn. There are often

Corn tassels bear the male (staminate) flowers, which produce a large amount of pollen.

Each corn silk is from a single female (pistillate) flower, each of which becomes a kernel. The silks emerge from the husks of the ear to receive the pollen, which is dispersed on the wind.

multiple ears on a single plant. Sweet corn varieties will frequently produce two ears; the second and smaller ear, called a **nubbin,** usually does not produce a full-sized ear and should not be relied on for seed production.

From each pistillate flower a silk develops. The silk is essentially a style with the end of the stigma that extends from the husk surrounding the ear. The silks usually emerge from the husk within 3 days after the pollen starts to dehisce from the tassel. Temperatures of at least 65°F (18°C) facilitate good pollination and fertilization of the ovules. Temperatures above 86°F (30°C) can shorten the viability of pollen grains, especially under arid conditions.

Corn is a wind-pollinated species, and while this mode of reproduction favors a high degree of cross-pollination, the corn plant is self-fertile and will receive its own pollen, successfully producing both self- and sibling pollinations. In fact, when producing seed of an open-pollinated corn population it is important to place the crop where there is good wind flow to encourage cross-pollination, in order to ensure good genetic mixing for a more resilient commercial corn crop.

Climatic and Geographic Suitability

The optimum conditions necessary to produce a bountiful sweet corn seed crop include reasonably warm days and nighttime temperatures that average between 68 and 72°F (20 to

22°C) through the growing season. The crop does exceptionally well when daytime highs regularly exceed 78°F (26°C). Sweet corn also needs an abundant and steady supply of water and fertile soils with good nutrient- and water-holding capacity. Corn is a tender annual and is easily damaged by frost or excessive heat.

While there are many modern sweet corn varieties that will produce a vegetable crop in 80 days or less, you must realize that producing seed of this crop requires a climate that can supply at least 120 frost-free days with reasonable amounts of warmth to mature a high-quality seed crop. Late-spring weather should be warm enough to safely plant seed and expect vigorous growth, with minimum soil temperatures above 60°F (16°C) at the time of planting.

During pollination and anthesis, kernel formation can be affected by a combination of temperatures above 95°F (35°C), low humidity, and drought stress. When these environmental stresses come together, the stigmatic surface on the silk and the pollen can desiccate to the point where pollen germination will not occur, leading to unfertilized ovules and poor kernel set.

The interior valleys of the northwestern United States have been particularly good for sweet corn seed production with their relatively hot summers, ample irrigation water, and seasonally dry late-summer weather, which helps to mature high-quality seed crops with a low incidence of foliar diseases.

SEED PRODUCTION PRACTICES

Soil and Fertility Requirements

Sweet corn will grow on many different soil types, but heavier fertile loams with good moisture-holding capacity are conducive to high yields of high-quality seed. Some growers in cooler, shorter-season districts may want to plant onto lighter sandy soils in order to get onto warmer soils earlier than would otherwise be possible. In this case you must be sure to incorporate plenty of compost or well-decomposed manure to deliver steady nutrients and hold moisture between irrigation or precipitation events. Sweet corn is moderately tolerant of acidic soils, though the best results are attained when the pH is kept in the range of 6.0 to 6.8.

Growing the Seed Crop

Corn should be planted within the row at the rate of two seeds every 8 to 12 in (20 to 30 cm) based on the size and robustness of the variety at full size. This affords you the opportunity to thin every other plant for vigor and health. The rows should be at least 30 in (76 cm) apart; some seed growers will space the crop rows at 4 ft (1.2 m) in order to better inspect the crop for selection during the season. It's always very important to place any corn planting in a block pattern as wide as it is long across the field. This ensures not only more thorough pollination of the plants in the population but also better genetic mixing among plants when growing an open-pollinated population.

Seed Harvest and Seed Cleaning

As the seed matures, it increases in weight as it accumulates dry matter. It is always best to allow the seed to reach its full maturity on the plant if possible. If you have the ability to test the moisture of the kernels, it should ideally be 35 to 40% for regular sugary (*su1*) types, 40 to 45% for sugary enhancer (*se1*) types, and 50% for the supersweet (*sh2*) types, according to the Crookham Seed Company of Idaho.

If you do not have the capability to measure the moisture content, then it is important to closely watch the yellowing or fading color of the ear, as the lack of chlorophyll in the ear's husk leaves is a good indication of the amount of carbohydrate that is still going into the endosperm. When the husk leaves lose all chlorophyll, you can assume that there is no longer any carbohydrate being moved into the kernel. In many climates the ears may have to come out of the field before the seed has fully matured, to avoid the wet, cold weather of fall. The ears should then be husked immediately for proper drying. Sweet corn seed is always dried at first on the ears in mesh cages or on mesh racks, always with good airflow and often with added heat. Drying the seed is often done at 95 to 100°F (35 to 38°C), though when the seed comes out of the field early it is best to lower the drying temperature until the seed gets harder. When the kernels are hard and brittle and the surface of the seed doesn't give when you try to stick your thumbnail into it, it is time to shell the seed from the cob. This is done with a machine called a corn or maize sheller, which can be adjusted for the diameter of the ear. Experience using and adjusting the sheller to the proper setting is very important in order to avoid breaking or cracking the seed. Seed is then screened manually or with a Clipper-type fanning mill for the final cleaning and sorting of the seed by size.

GENETIC MAINTENANCE

Sweet corn has a large number of traits that are quite important to evaluate through regular selection if you hope to maintain a high degree of varietal integrity. The horticultural traits associated with the eating quality of the crop are ultimately of great importance: If the variety isn't sweet and tender, then very few people will be interested in growing it. But all organic seed growers need to be fully aware of the agronomic traits associated with the plant and its ability to perform and deliver a robust crop under various environmental challenges.

There are several traits important in an organic system that can be selected for before the plant produces tassels and starts the reproductive phase of its life cycle. Sweet corn seed is very susceptible to the damping-off complex of soilborne diseases before it emerges as a seedling. Selecting for good emergence and seedling vigor as the crop emerges from the soil is always important in corn varieties that will be used for organic production. As the plant develops it must be strong and resist lodging (falling over) during strong wind or heavy rain events. Plants that easily fall over or bend while others in the population stand strong should be eliminated. Corn varieties are also susceptible to a host of foliar diseases in every climate in which they are grown. Most of the resistance to these diseases is durable horizontal resistance that has been developed by the slow, methodical selection of plants with some level of partial resistance over time. In North America this has resulted in decent levels of resistance to the fungal diseases: common rust, smut, and northern corn leaf blight. There are also very good levels of resistance to the bacterial disease Stewart's wilt in many sweet corn seedstocks. But it is important for all serious seed growers to become familiar with the signs of these diseases and to eliminate any plants that exhibit more than minor symptoms.

When selecting for eating quality, familiarize yourself with the range of possibilities in good texture (silky vs. gritty), tenderness, flavor, and sweetness. These traits are best learned from discerning growers who have been in the business of growing and marketing high-quality sweet corn for a number of years. Other ear characteristics that are important to consider are the shape, diameter, and length of the ear, as well as how straight the rows are and the number of rows. The ability of a plant to bear an ear that adequately fills the tip of the cob is in part a heritable trait and needs constant selection to maintain. Also, the ability of the husk to fully cover the ear and tightly protect the tip, thereby guarding it from the elements and from corn earworms, is crucial. Even the color of the husk at the fresh market stage is important—a good dark green color connotes freshness.

ISOLATION DISTANCES

Determining adequate minimum isolation distances for corn seed production continues to be a lively topic, with few people involved in organic seed production wanting to pin their reputations on specific recommended distances.

The published minimum isolation distance for corn seed crops that has been used for many years by the conventional seed industry is 660 ft (201 m), which seems to be very close when considered by seed growers who have worked with other wind-pollinated crops. However, the fact is that corn pollen both is rather heavy and dries out fairly quickly in the air after being shed by the anthers. Thus a number of studies have shown that corn pollen doesn't routinely pollinate neighboring crops even at roughly half these distances. Combine this with the fact that corn seed growers will routinely not harvest seed from at least four to six rows on the outer edges of the seed crop and you should find little, if any, crossing taking place.

If you're producing more than one type of corn, of either a different kernel type, starch type, or color, then this standard minimum isolation should be doubled to 1,320 ft (402 m) between crops. Of course, if you're producing sweet corn seed in an area with lots of field corn growing on all sides, then achieving this type of isolation is difficult. The real problem then also becomes isolation from any fields with genetically modified (GM) corn. Even when a standard field corn crop is not intended to be a GM crop, there often is a low level of GM contamination in the standard crop. This situation should be viewed with great concern, as the risk of GM contamination in an organic seed crop is a very serious breach of the standards that have been set for certified organic seed. Under these conditions it is recommended that organic sweet corn seed crops need to be produced in non-corn-growing regions.

SOLANACEAE

The Solanaceae or nightshade family is a family of 90 genera and at least 2,600 species. This includes five of the most important agricultural crops in the world, tomato (*Solanum lycopersicum*), potato (*S. tuberosum*), sweet and chile peppers (*Capsicum* spp.), eggplant (*S. melogena*), and tobacco (*Nicotiana tabacum*), as well as a number of minor crops that are important as ornamentals or as sources of medicine. A number of members of the Solanaceae family have psychoactive alkaloids that have sedative properties in mammalian systems. The etymology of *Solanum* is thought to be the Latin word *solari*, which means "to soothe." Many of the alkaloids in members of this family are also toxic to varying degrees, and some species like jimsonweed (*Datura stramonium*) or deadly nightshade (*S. nigrum*) can be lethal. The common name for the family, the nightshades, may be derived from the fact that many of its species are night blooming, with blossoms that only begin to open at sunset.

The incredibly quick ascent of at least four of the five most important crop plants of this family to prominence worldwide is a testament to how quickly humans and their agricultural systems can adapt to newly introduced crops. Tomato, potato, chile pepper, and tobacco are all New World crops, and in less than 400 years they all became commercially vital crops across most of the agricultural societies on Earth. All four of these crops were first carried to the Old World within the first 20 years or so of European conquest near the beginning of the 16th century. While there is scant documentation of the paths these crops traveled once they were introduced to Western Europe, they must have passed quickly from hand to hand, region to region, and rapidly spread via seafaring vessels once they arrived in the Old World.

Commercial seed production of the Solanaceae crops is done in climates where hot daytime temperatures and warm nights contribute to a high rate of fertilization of ovules, rapid endosperm development, and high levels of fruit maturation. All of these factors result in strong yields of seed with high germination rates. Seed crops for peppers, tomatoes, and eggplants are best grown in regions with dry summer weather to minimize the exposure to many diseases of the fruit and foliage that can impact production. Drier climates can also help you avoid potentially devastating seedborne diseases like bacterial spot and bacterial speck. Hybrid seed production of these crops is frequently done in Thailand, Taiwan, Indonesia, and China because they all require hand pollination. However, many agricultural enclaves across the globe still rely on regionally grown seed of well-adapted, non-hybrid varieties that are often culturally significant.

FAMILY CHARACTERISTICS

Reproductive Biology

The Solanaceae family has flowers that are either conical or funnelform with five petals that are fused or partially fused. The flowers can be quite large, with fused petals, and are quite showy in genera such as *Datura, Nicotiana,* and *Petunia.* In the crop plants of the *Solanum* and *Capsicum* genera, the flowers are much smaller and the corolla is bell- or wheel-shaped. The petals can be predominantly white (most chiles) to yellow (tomatoes) to purple (eggplants). The flowers are perfect, with stamens that are nestled into the throat of the corolla. The anthers form a cone in the crops of this family with pollen that is shed from slits or pores situated at different points at or near the end of the anthers. The style can be of varying lengths in each of these species, and in some cases it's long enough to extend beyond the end of the anther cone and be more apt to cross-pollinate, especially in the presence of pollinating insects. (See the "Reproductive Biology" section for each of the crops in the Solanaceae.) The flowers are usually borne in or near the leaf axils, except with eggplant, which bears its flowers opposite from the leaf axils. The fruit is a true berry with pulpy flesh and placental tissue that can be dry, as with peppers, or wet, as in the locules of tomatoes.

Life Cycle

Tomatoes, chiles, and eggplants are all tender, warm-season crops. They are all sensitive to frost and require a reasonable amount of heat, depending on the crop type and variety for each species, to mature their fruit. To take advantage of the warm summer growing season, all of these crops are generally started in the greenhouse and transplanted after all danger

▼ 'Serrano' chile pepper flower (*Capsicum annuum*).

of spring frost has passed. These crops are usually started under glass 6 to 8 weeks before transplanting to the field. Plants are routinely hardened off to both the cool of the night and to full sun for several days before transplanting.

Climatic Adaptation

These three crops all produce their best seed crops with warm nighttime temperatures that will support pollen tube growth, to ensure a high percentage of fertilization of the embryos in each fruit that is set. Adequate heat units during the day are also important to stimulate rapid fruit growth and maturity. In summary, the best regions to produce seed of these crops need to have: (1) a relatively long frost-free growing season; (2) enough hot weather for good fruit production; and (3) warm night-time temperatures for maximum seed set (see individual crop sections for specific temperature ranges for each crop).

Seed Harvest

Perhaps the most important aspect of seed harvest for these crops is making sure that the fruit that is harvested for seed is fully ripe, even overripe, but also making sure that it isn't rotting and moldy to the point where the seed is damaged. In fact, allowing the fruit of all of these crops to go a bit beyond full ripeness may ensure that the seed is fully mature. In most wet-seeded crops the overripe fruit also usually "release" their seed from the flesh much more easily than the slightly underripe or even perfectly ripe fruit.

Fruit of the Solanaceae crops is usually harvested by hand for seed production in order to have human eyes make a final assessment as to whether: (1) the fruit is fully ripe; (2) there are marked levels of disease or rot on a particular fruit or plant; and (3) the fruit from a particular plant is true to type. Harvesters can then reject individual fruit if they're rotting or not ripe, and they can make a final act of selection if the plant has excessive disease or off-type fruit. There are also some seed-growing operations where the fruit is machine-harvested for seed extraction. However, this one-time-only harvesting method has obvious disadvantages that make it acceptable only to seed companies selling lower-quality seed.

Historically, tomato seed was sometimes harvested in joint tomato juice/tomato seed operations in North America, at a time when most commercial seed was still non-hybrid. This practice is not done with hybrid tomatoes, as they are almost always hand-pollinated and marked, with many instances where non-hybrid fruit is set that is not a result of a cross-pollination. I recommend that all organic Solanaceae seed crops be hand-harvested to ensure their seed quality and genetic integrity.

Seed Extraction and Cleaning

The three main crop types of this family that are grown from seed are all considered wet-seeded crops. Wet-seed extraction means that water is used in either the extraction or cleaning steps, or both. Tomato seed extraction doesn't usually use water, but water is always used during the cleaning process. The seed extraction and cleaning of both chiles and eggplants often use water for both steps, especially on a larger commercial scale of seed production. However, many seed growers around the world also use a dry-seed extraction technique for eggplants and many kinds of chiles that is described for both crops in their respective sections.

ISOLATION DISTANCES

There has been a lively debate as to what the appropriate minimum isolation distances are for the commonly cultivated seed crops of the

Solanaceae. Tomatoes, chiles, and eggplants are all considered self-pollinated species. But as is commonly known, all species that are designated as selfers do outcross to varying degrees. Many seed savers have discovered the hard way that growing more than one heirloom tomato or hot chile variety in close proximity to another can easily lead to plants in next year's progeny that are the result of an obvious cross-pollination event. Also, the rate of crossing that occurs can change significantly under a number of circumstances that are discussed in chapter 13, Isolation Distances for Maintaining Varietal Integrity.

While a number of factors come into play to increase the rate of crossing in all selfers, there are thought to be two major reasons why many growers of the Solanaceae seed crops have noted an increase in the rate of crossing over what has been reported over the past 50 to 80 years in the literature:

1. Many seed growers have noted that pollinating insects visit the flowers of these crops more than is commonly recognized. This is especially true in the biologically rich environments of many of the diversified organic farms, where there is often much more habitat for pollinators and much less use of insecticides.

2. A number of crop types within these three species groups share a morphological trait that makes the plant much more conducive to outcrossing. Hot chiles, heirloom and potato-leaved tomatoes, and most eggplants have styles that grow beyond the end of the anther cone of the flower. This exposes the stigma to the open air, and to the possibility of cross-pollination when insects visit the flowers. Peppers seem to be the most notorious for outcrossing. Under favorable conditions the crossing rate of some hot peppers has been measured to be anywhere from 8% to over 60% by a number of researchers, who were trying to get an idea of just how extensive crossing can be when more than one pepper variety is planted directly adjacent to another.

For this reason the recommendations for a minimum isolation distance for these crops for organic production systems will be increased from the isolation distances that have routinely been published and used in recent years. They will be based, in part, on some of the excellent investigative work done by pollination biologist Jeff McCormack. Jeff has compiled a set of isolation distances in a series of publications that takes into account the environmental reality of organic farms and the fact that the exerted stigma in some crop types of the Solanaceae contributes to higher-than-expected crossing rates. The recommendations that I am making for commercial production of these crops under organic conditions are somewhat stricter than those recommended by Jeff. I firmly believe that, with the exception of modern tomato and sweet pepper varieties, all of the other most commonly grown seed crops of the Solanaceae should have an absolute minimum of 150 ft (46 m) between crops with a barrier present to minimize the chances of a cross when producing commercial seed. This distance should be increased to 300 ft (91 m) when these promiscuous crops are grown in open terrain. (See specific information under "Isolation Distances" for each crop type.)

EGGPLANT

Eggplant (*Solanum melongena*) is a heat-loving crop that is grown in many tropical, subtropical, and warm temperate regions across the globe. It is known as *aubergine* in most of Europe and *brinjal* in India, where it is one of the most popular cultivated vegetables. It is a native of the South Asian subcontinent, with its center of origin believed by Nikolay Vavilov to be in the Indo-Burma center, which includes modern-day Myanmar and Bangladesh. A number of closely related wild species of *Solanum* still grow across this area, though all have bitter alkaloids that largely prevent their use. Early efforts in the domestication of eggplant undoubtedly concentrated on eliminating the bitter glycoalkaloids that are shared by most of the relatives of this important vegetable crop. Eggplant traveled to China very early in its history, and its diversity flourished under the selection of many hands across a number of climates and geographic regions. The diversity that developed there is so profound that southern China is recognized as a secondary center of variation for eggplant.

The color of eggplant fruit can vary from white to yellow, orange, green, and purple and can sometimes have streaked pigmentation. The dark purple types can appear almost black when the anthocyanin pigmentation is most intense. The shape of the fruit can range from small oval or round, to long with a club shape, to the large oval or globe-shaped types that

SEED PRODUCTION PARAMETERS: EGGPLANT

Common names: eggplant, aubergine
Crop species: *Solanum melogena* L.
Life cycle: annual (perennial in the subtropics)
Mating system: largely self-pollinated, with increased crossing based on type and climate
Mode of pollination: closed perfect flower that requires stimulation for pollen shed
Favorable temperature range for pollination/seed formation: 60–74°F (16–23°C)
Seasonal reproductive cycle: late spring through late summer or fall (4–5 months)
Within-row spacing: 1–1.5 in (2.5–4 cm)
Between-row spacing: 22–30 in (56–76 cm)
Species that will readily cross with crop: Several forms of wild eggplant (*Solanum* spp.)—which can cross with the cultivated crop—grow across the subtropics of Asia, from South Asia to southern China.
Isolation distance between seed crops: 10–50 ft (3–15 m), depending on the crop type and barriers that may be present on the landscape

been used in Europe and North America for decades. Commercial growers across Asia who are producing eggplant as a commercial vegetable for markets outside of their local communities are increasingly using hybrid seed. Hybrid seed is produced in several of the areas where hybrid tomato seed is also produced, including China and India. At the time of this writing there is increasing pressure to produce and use GM hybrid eggplant seed in India, with considerable resistance from many Indians, who fear the potential consequences of this fundamental change in their agricultural sovereignty. South Asia still has an incredible diversity of open-pollinated eggplant varieties that are grown regionally, and there are also wild related species that can potentially cross with the cultivated types. Eggplant can experience considerable cross-pollination, and the danger is that the flow of the transgenes from these GM hybrids could easily spread through both the traditional varieties and the wild populations, which would further spread these transgenes across the landscape.

CROP CHARACTERISTICS

Reproductive Biology

Eggplant has perfect flowers that are often solitary, or occur in sets of two to five, and are borne opposite the leaves on the main stem of the plant. The six violet petals are fused; the number of anthers that form the anther cone can vary from 6 to 20, depending on the genetic background. The pores that facilitate the release of the pollen are at the end of each anther. The pistil, which has a style surrounded by the anther cone, is usually exerted out beyond the end of the cone in most eggplant types. The pollen released from the anther pores adjacent to the stigma will usually more than saturate the stigma and lead to a high

These eggplant flowers clearly exhibit the exserted stigmas that result in higher rates of cross-pollination.

are the norm in North America. The common name of eggplant most probably comes from the phenotype of varieties that are white and bear the smaller oval or round types, which are much the same size as a chicken or duck egg.

Eggplant is such an important staple vegetable crop throughout most of tropical and subtropical Asia that locally produced seed of many of the typical, open-pollinated regional varieties is grown locally by small, diversified farmers in many parts of Asia, as well as in other parts of the world that still depend on decentralized agriculture. Hybrid eggplant seed is becoming more ubiquitous, having

An eggplant flower with an inserted stigma that will not be as prone to cross-pollination.

percentage of self-pollination. However, as the style and the stigma are exerted into the open, out of the confines of the cone, there is a much greater chance of cross-pollination when an insect visits and brushes up against the stigmatic surface. This is much like what occurs with heirloom and potato-leaved tomatoes, and it contributes to a higher percentage of outcrossing than what is normally expected in a self-pollinated species.

Climatic and Geographic Suitability

Eggplant is solidly in the class of heat-loving crops. Optimum growth and fruit set occur between 72 and 86°F (22 to 30°C), with nighttime temperatures that average at or above 65°F (18°C). Nighttime temperatures below this can cause erratic pollen tube growth, resulting in lower fertilization rates and ultimately lower seed yields. As with all crops, temperatures that are too high at the time of the pollination will also retard pollen tube growth and prevent fertilization and subsequent seed formation. In eggplant this can happen at temperatures above 95°F (35°C). The most favorable climates to grow eggplant seed are regions with a reasonable amount of heat during the day and enough moisture in the air to hold some of the heat through the night. Regions with too high a level of humidity during the growing season should be avoided due to the potential for increased levels of foliar disease.

SEED PRODUCTION PRACTICES

Soil and Fertility Requirements

Eggplant is able to grow in a wide variety of soils, but the soil must be well drained to avoid root rot problems. This crop has an extensive root system and can develop a deep, strong taproot if grown in a rich loam soil with good tilth. Eggplants are heavy feeders and thrive when ample amounts of well-decayed organic matter or compost are incorporated into the soil. A soil pH of between 5.5 and 6.5 is desirable. In the era before fungicides there were eggplant growers in North America who attested to the practice of raising eggplants on soils with a pH no higher than 6.0 in order to control verticillium wilt (*Verticillium dahliae*), claiming that this fungus could not thrive under acidic soil conditions. Long crop rotations are advised if verticillium wilt is present.

Growing the Seed Crop

Eggplant seedlings are usually started in the greenhouse or in cold frames. It is important to seed them about 8 weeks before transplanting. Most growers seed them into flats and prick them out into individual cells when they form their first true leaves. Growers need to be careful and not transplant the plants to the field until the warm weather has settled and summer has truly arrived. Eggplant is particularly sensitive to arrested growth; it needs steady, unchecked growth to develop a full frame that will support a high number of fruit and produce good seed yields.

As with other seed-producing members of the Solanaceae, eggplant seed crops are usually grown at the same spacing as the eggplant vegetable crop. Plants can be spaced at 14 to 18 in (36 to 46 cm) between plants, with 30 to 36 in (76 to 91 cm) between rows. If you prefer growing the crop on beds, the space between plants should be increased to 18 to 24 in (46 to 61 cm) and then planted in an offset pattern in rows 24 to 28 in (61 to 71 cm) apart on wide beds. In some cases growers will stake the plants, which can be effective in slowing fruit rots that can occur over the long season that is necessary to mature a seed crop.

Seed Harvest

As with all of the Solanaceae seed crops, the first step in seed harvest is to harvest the truly ripe fruit by hand. This allows you to harvest the ripest fruit possible multiple times if the season allows. Eggplant fruit should be allowed to ripen to full seed maturity. At maturity all eggplant types have fruit that changes colors. The dark purple types first change to a duller shade of purple and then to a dingy brown color. The other common pigmentation types include the lightly streaked Italian purple types, as well as the green, yellow, and white types, all of which turn to some shade of yellow or gold as they reach seed maturity. With all eggplant varieties an abscission layer forms at some stage in the maturation process, allowing the harvest crew to easily pull the fruit from the plants as opposed to cutting them with clippers. To avoid damaging the plant, all ripe fruit that isn't easily pulled off the plant should be cut off cleanly with clippers.

Eggplant fruit is routinely allowed several days of after-maturing off the plant in a warm, dry place. This enhances the maturity of the seed, usually improving the germination percentage. In dry, long-season climates eggplant fruit can be left on the plant till it has shriveled, which is often done across many dry areas of Asia when harvesting limited amounts of seed for on-farm or local use.

Seed Extraction and Cleaning

When the fruit has after-matured, the seed can be separated from the fruit by two different methods: wet-seed and dry-seed extraction.

Wet-Seed Extraction: Wet-seed extraction is the method most favored for the commercial extraction of relatively large amounts of seed. The overripe fruit is crushed or macerated mechanically in order to free the seed from the pulp in which it is embedded. This eggplant pulp is then run across screens with a stream of water that helps to loosen the seed from the flesh and force it through the screens; most of the pulp stays on top. This can be done with the help of human hands, rubbing the material across the screens. Larger operations will sometimes employ a large cylindrical screen unit that when spinning will force the seed through the screen under centrifugal force. Both this method and the more manual method require water to loosen the seed from the dry eggplant flesh. The seed that passes through the screen can then be separated from the small pieces of flesh and skin using the same decanting process with water that is used in the final stages of tomato cleaning. Fermentation is not necessary in cleaning eggplant seed and in fact may be harmful to the viability of the seed.

Dry-Seed Extraction: The second method is used when the fruit is allowed to dry. In long-season climates where fruit can be fully dried in a combination of on- and off-the-plant maturation, growers have the option of harvesting the seed from completely dried fruit. The fruit are then beaten to split them open, and seed can be picked from them with much hand labor. This method is only appropriate for extraction of relatively small amounts of seed, though there are probably tricks I am not aware of that are still practiced by farmers in South Asia, the homeland of eggplant.

GENETIC MAINTENANCE

Several important traits can be checked before the plant begins flowering. As with all crops, seedling vigor and early robust growth are always important traits to select for in organic production systems. Eggplant prospers from increased vigor, as the seed is often planted into less-than-ideal greenhouse conditions. I have seen many cases where farmers have selected for seedling vigor over a number of generations, and the resultant variety appeared to pass more quickly through a stage of susceptibility to the damping-off complex of seedling diseases. Selecting for vigor may be accomplished by eliminating any slower, poorly formed seedlings, after all of the seedlings have germinated and emerged.

The next traits to observe for off-types are stem color, leaf color, leaf shape, and plant stature. Sometimes there will be an obvious difference for one of these characteristics that will warrant the removal of a plant from the population. While genetic variability of this type isn't necessarily bad in a genetically elastic crop population, it can be a good indicator of an off-type, outcross, or seedling mix. By paying attention to the basic plant characters it is possible to rogue out a potential off-type plant, instead of not recognizing a plant that can potentially spread pollen to other plants after the seed crop is flowering. Once the crop has started to produce fruit, it is possible to eliminate plants with misshapen or off-type fruit. After the fruit has reached vegetable maturity, you can evaluate plants and perform a final selection for fruit shape or size. Fruit color at vegetable maturity should be at its truest shiny luster. This is the best time to eliminate plants with fruit that varies too much from the norm for color, color intensity, or degree of striping for the particular variety you are growing.

ISOLATION DISTANCES

Eggplant is traditionally thought of as a self-pollinated species, but the morphology of the flower in most eggplant varieties allows for an appreciable amount of outcrossing, especially under organic production practices (see "Isolation Distances" in the Solanaceae chapter introduction on page 299). Elegant research conducted in the 1920s by Kakizaki in Japan demonstrates that crossing rates averaging over 6% and reaching as high as nearly 47% occurred in some instances when the progeny of white-fleshed plants that had been interplanted with purple plants was grown the following year. Kakizaki also reported that no crossing occurred at 164 ft (50 m) between these two different-colored types of eggplant.

From reports of crossing that has occurred in organic production of eggplant seed in recent years, I believe it is important to be even stricter than Kakizaki's research indicates. I recommend that when growing two eggplant varieties it is wise to isolate them by a minimum of 300 ft (91 m) when growing the seed crops in open terrain. This minimum isolation distance can be decreased to 150 ft (46 m) when natural barriers are present. This also fits with Jeff McCormack's recommendations, which are based on his thorough investigations into the isolation distances of the Solanaceae crops.

PEPPER

Peppers or chiles (*Capsicum* spp.) are a clan of at least 25 different domesticated, semi-domesticated, and wild species fruiting plants that have become among the most important culinary crops in almost all agricultural societies worldwide. Members of the *Capsicum* genus are native to the tropical and semi-tropical regions of South and Central America. There are probably still new species of this diverse genus that are awaiting discovery as ethnobotanists venture farther into the wild, rugged, and diverse landscapes of this part of the world. With such a wealth of species occupying different niches in these radically varied ecological zones, it is no wonder that early farmers domesticated at least five different species, selecting them for many of the same forms resulting in shapes, colors, and culinary qualities that are surprisingly similar across the species.

The five domesticated species of the *Capsicum* genus that account for the vast majority of the agricultural pepper production across the globe are:

1. *C. annuum,* which is by far the most widely cultivated pepper species and includes sweet peppers, cayenne, paprika, and most of the varied hot chiles grown worldwide.
2. *C. frutescens* is from the lowlands of the Amazon basin and thrives in hot climates.

SEED PRODUCTION PARAMETERS: PEPPER

Common names: pepper, chile
Crop species: *Capsicum* spp.
Life cycle: annual (perennial in the subtropics)
Mating system: largely self-pollinated, with increased crossing, especially in hot types due to flower structure
Mode of pollination: closed perfect flower that requires stimulation for pollen shed
Favorable temperature range for pollination/seed formation: 60–74°F (16–23°C)
Seasonal reproductive cycle: late spring through late summer or fall (4–5 months)
Within-row spacing: 1–1.5 in (2.5–4 cm)
Between-row spacing: 22–30 in (56–76 cm)
Species that will readily cross with crop: All of the *Capsicum* spp. have the potential to cross, though some do so more readily than others.
Isolation distance between seed crops: 75–300 ft (23–91 m), depending on the crop type and barriers that may be present on the landscape

4. **C. baccatum** is from low to mid-elevations in Bolivia, and its aji chiles are grown widely in South America, where they are prized for their subtle bouquet and fruit-like flavors that complement their pungency.

5. **C. pubescens,** which is named for the hairy pubescence of the leaf veins, is native to the mid-elevations of the Andes, making it adapted to cooler weather than other chile species. It is also grown in mountainous areas of Mexico and Central America. Its plants are striking for their beautiful purple flowers and black seed.

The domestication of peppers was already quite advanced and had spread across the tropics of the Western Hemisphere when Columbus landed in the Caribbean on his first voyage in 1492. Columbus famously returned to Spain with seed of what he called *pimiento,* the Spanish word for "pepper," and declared his trip to find a faster route to the Far East for spices a success. This story is often told as if Columbus were trying to deceive his benefactor, Queen Isabella, by overstating the value of the chiles as being the equivalent of black pepper (*Piper nigrum*), which was

It includes the well-known Tabasco type and the so-called squash peppers.

3. **C. chinense,** also from Amazonia and closely related to *C. frutescens*, includes habanero and Scotch bonnet, varieties famous for very high concentrations of capsaicin, the compound responsible for the pungency in peppers.

literally worth its weight in gold at the time.

In retrospect, the seed of chile peppers was indeed a treasure, as this crop is widely adapted to any climate with hot summers and can be grown in many regions where the woody perennial black pepper would never survive. In Europe chiles were first grown in monastery gardens, but unlike tomatoes their cultivation spread quickly across the tropical and subtropical regions of the Eastern Hemisphere by land and by sea. The adoption and selection of this crop was so extensive across East Asia that the region became a secondary center of diversity for peppers. In fact, when early taxonomists classified the peppers growing in China, they named one of the types *S. chinense*, assuming that the extensive diversity they found there meant that this type must be native to this part of Asia.

Bell peppers (*C. annuum*) and other modern sweet peppers are less likely to cross-pollinate than the hot chile types, but they will still cross at higher rates than is normally thought.

CROP CHARACTERISTICS

Reproductive Biology

Peppers are perennials that are usually grown as annuals, because they are easily killed by frost. Peppers have perfect flowers that are often borne singly in the axils of the leaves, though some of the less common species can have multiple flowers at each axil. Most commonly the flowers have white petals, though they can exhibit a purplish blush. The

The 'Peruvian Purple' chile (*C. frutescens*) has fruit that changes in color from purple to red when ripe and is mildly hot.

Unlike most *Capsicum annuum* chiles, 'Serrano' types are closer to the wild with their dark purple anthers and grayish pubescence on their calyx. ▶

normal state is to have five stamens with long blue-tinged anthers that are fused at their base but do not form a unified anther cone like tomatoes, eggplants, or many other wild members of the Solanaceae. The stamens split along their length to release pollen. The pistil has an ovary with two to four carpels, and the style and stigma collectively are a little longer than the stamens. The length of the style and the absence of an anther cone exposes the pepper flowers to a higher incidence of cross-pollination when insects are present.

Subtle differences in these floral traits are used to distinguish the five major species of cultivated peppers; however, they are anatomically very similar.

There is some debate on the ability of all five of these species to freely cross and produce viable offspring. In general, while there may be some crossing barriers between specific species it should be assumed that all of these species are potentially sexually compatible whenever you grow a seed crop of any of them. Therefore, always observe the minimum isolation distances recommended in the "Isolation Distances" section, page 315.

Climatic and Geographic Suitability

Peppers are generally considered among the "moderate heat-loving" seed crops, requiring a reasonable amount of heat during fruit set and maturation, but definitely a rung below eggplants or melons in their requirement to mature a good seed crop. Importantly, with the large amount of genetic breadth across all of the cultivated peppers, there are undoubtedly differences in the ideal temperatures for seed production among the various types of peppers.

Optimum growth and fruit set occur at temperatures between 68 and 80°F (20 to 27°C), depending on the type and variety of pepper, with nighttime temperatures that are at or above 60°F (16°C) but don't regularly exceed 72°F (22°C). As with all heat-loving seed crops, temperatures at or near the accepted low temperatures for seed production will often cause erratic pollen tube growth, resulting in lower fertilization rates and ultimately lower seed yields. Temperatures that are too high at the time of the pollination can cause low pollen production and also retard pollen tube growth, preventing fertilization and subsequent seed formation. High temperatures can also cause flower abscission. These heat-related problems can manifest at temperatures that persist above 90°F (32°C). Sweet peppers

are generally more sensitive to these heat-related problems than the hot chile types.

Peppers are such a ubiquitous crop throughout most of tropical and subtropical areas of the world that locally produced seed of many of the typical, open-pollinated regional varieties is still grown locally in many decentralized agricultural systems. While many of these regions may not be ideal for pepper seed production, it is possible to produce plenty of seed for regional cropping needs if growers are careful not to allow the fruit to become moldy while ripening or before seed extraction.

Much of the large-scale commercial seed production for pepper seed is done in the moderately dry climates of northern Chile, northern Thailand, the northern Philippines, southern Taiwan, and more recently parts of mainland China and India. Much of this seed is high-value hybrid seed that is produced through emasculation and pollination by hand.

SEED PRODUCTION PRACTICES

Soil and Fertility Requirements

Peppers thrive on lighter soils such as sandy or silt loams. These soils warm up faster than heavier clay soils, allowing vigorous early growth before flowering that can subsequently support a sizable fruit crop. Peppers need good, balanced fertility and even moisture throughout the season; however, the soil must be well drained to avoid root rot problems.

Peppers will develop an extensive root system in a rich loam soil with good tilth. The root mass will reach to at least 3 ft (0.9 m) in all directions when healthy. Therefore, cultivation needs to be relatively shallow to avoid excessive root damage. While peppers are heavy feeders, excessive levels of nitrogen

should be avoided; this can contribute to plants that lodge (topple over) in extreme weather. According to Jeff McCormack, the incidence of lodging under windy conditions in southeastern United States increases when the organic matter content of the soil exceeds 6%. A soil pH of between 5.5 and 6.8 is desirable for peppers.

Growing the Seed Crop

Pepper seedlings are usually started in the greenhouse or in cold frames, but in the most favorable long-season climates they can be direct-seeded to the field. If you're transplanting, it is important to seed them under cover up to 10 weeks before transplanting; they often need a couple of weeks more than tomatoes or eggplants to produce a stout transplant. Most growers seed them into flats and prick them out into individual cells when they form their first true leaves. It is important to not transplant the plants to the field until the warm weather has settled and summer has truly arrived.

Pepper seed crops are usually grown at the same spacing used when growing the crop as a vegetable. Planting densities can vary, as there is a significant range in the size, height, and shape of plants among the different types and varieties of peppers. Plants can be spaced at 16 to 24 in (41 to 61 cm) between plants depending on the size of the variety, with 30 to 36 in (76 to 91 cm) between rows. If you prefer growing the crop on beds, increase the space between plants to 18 to 24 in (46 to 61 cm) and then plant in an offset pattern in rows 24 to 28 in (61 to 71 cm) apart on wide beds. In some cases growers stake the plants, which can be effective in slowing fruit rots that can occur over the long season necessary to mature a seed crop. This may be especially important when growing a high-value sweet pepper, as sweet types are often more prone to fruit rots than chiles.

Seed Harvest

As with all of the Solanaceae seed crops, the first step in seed harvest is to harvest the truly ripe fruit by hand. This allows you to harvest the ripest fruit multiple times if the season allows. Pepper fruit need to fully ripen to harvest mature, high-germinating seed. At maturity all pepper types have fruit that change color to red, yellow, gold, orange, red, purple, or chocolate brown. The best indicator that they have reached full seed maturity is when they've become uniformly pigmented with one of these colors.

At fruit maturity many growers allow the fruit to remain on the plants to overripen past their edible peak, as this ensures that the seed will be fully mature when picked. This practice is usually successful in seasonally dry climates, where there is less chance of fruit rots developing; however, as you gain experience growing pepper seed in any particular climate, you are advised to monitor the crop closely for fruit rot at or near the time of maturity. Any rot that forms can quickly enter the interior cavity of the fruit and damage the seed. Fruit is routinely hand-harvested, and at the time of harvest the crew should eliminate any fruit with noticeable signs of rot. If the seed is not going to be extracted from the fruit within 24 hours, the fruit should be harvested in such a way as to minimize any damage. This is essential, as fruit rots can easily spread through harvested fruit if held for several days and care is not taken.

Seed Extraction and Cleaning

After harvest, pepper seed can be separated from the fruit by two different methods: wet-seed and dry-seed extraction.

Wet-Seed Extraction: The wet-seed extraction method is most favored for the commercial extraction of relatively large amounts of pepper seed. This process begins in a similar fashion to the method used for both tomato and eggplant, where the ripe fruit is either crushed or macerated. The resultant pulp is then run across screens with a stream of water that helps to loosen the seed from the flesh and force it through the screens while most of the pulp stays on top. This can be done in a very low-tech fashion with the help of human hands, rubbing the pulp across the screens. The workers must be very careful in this situation, though, if they're handling hot peppers. Gloves, respirators, and eye protection should be used in all cases when processing any peppers that have any detectable levels of heat in any hands-on methods of seed extraction. In fact, everyone from the harvest crews to the seed extractors needs to protect all mucosal tissue against particulates from all peppers that contain capsaicin.

Larger pepper seed operations will sometimes employ a large cylindrical screen unit that when spinning will force the seed through the screen under centrifugal force. Both this method and the more manual method require water to loosen the seed from the flesh. The seed that passes through the screen can then be separated from the small pieces of flesh and skin using the same decanting process with water that is used in the final stages of tomato cleaning. The seed can also be run through a sluice like tomatoes to separate it from fragments of the pulp. Fermentation has been used to a limited extent by some growers; most pepper seed producers agree that it not only is unnecessary but may also be harmful to the viability of the seed.

When using the wet-seed extraction method it is important to dry the seed quickly and efficiently as soon it's as clean as possible from the decanting process. The seed should be spread thinly on screens and placed in a warm spot with good air circulation. Sun-drying is acceptable. Supplemental heat may be necessary under cool conditions at the end of the season, being careful to not heat the seed much above 95°F (35°C). Seed should always be stirred at least twice a day when drying.

Dry-Seed Extraction: The dry-seed method is only used for thinner-fleshed hot chile types, as sweet pepper types usually have a thicker fresh that is much harder to dry without mold forming. Of course, there are also some hot types with thicker flesh that will not dry easily.

In long-season, arid climates the thin-fleshed hot types can often be left on the plant to partially dry before harvest and then be dried fully in the sun. But in less-than-optimal climates it is best to pick the fruit at or soon after full maturity, spread them thinly across a clean surface with plenty of good air circulation, and dry them in the sun. The fruit must be protected from precipitation; you may need to supply airflow via fans if you're drying them in a greenhouse, high tunnel, or dryer.

When completely dried and shriveled the dried fruit are then threshed or flailed to free the seed from the pods. As with all of the steps when handling hot chiles, it is very important for any workers to cover all mucous membranes with protective clothing or equipment to guard against any painful inflammation from the capsaicin found in the dried flesh. During the steps of threshing the dried pods and the subsequent cleaning of the seed it is especially important to avoid the copious hot pepper dust that will become airborne using this dry-seed extraction method.

Seed Cleaning: After the seed is dried using either of the above methods, you may need to winnow it or pass it across the appropriate-sized screen to eliminate any pieces of dried pulp that might remain.

GENETIC MAINTENANCE

Several important traits can be checked before the plant begins flowering. As with all crops, seedling vigor and early robust growth are always important traits to select for in organic production systems. Selecting for vigor may be accomplished by eliminating any slower, poorly formed seedlings after all of the seedlings have germinated and emerged.

The traits to observe for possible roguing are stem color, leaf color, leaf size, plant stature, and plant height. Sometimes there will be an obvious difference for one of these characteristics that will warrant the removal of a plant from the population. Leaf size is often correlated with the pungency of the variety, as hot varieties usually have significantly smaller leaves than the sweet types. Therefore, it may be possible to eliminate a small-leaved plant that might carry the pungency trait from a large-leaved sweet pepper crop before flowering and prevent this trait from being crossed into multiple sweet plants in the field.

Once the crop has started to produce fruit, it is sometimes possible to eliminate plants with

'Corno di Toro' is an example of a class of thick-walled peppers that are both sweet and mildly hot. These types often exhibit variation for shape and color that can be selected as novel types by the seed grower.

obviously misshapen or off-type fruit. When fruit reach their vegetable maturity you can rogue for fruit shape, size, and color as peppers turn to their true mature color. The color at full fruit maturity should be at its true shade and luster. This is the best time to eliminate plants with fruit that varies too much from the norm for color and color intensity for the particular variety that you are growing.

ISOLATION DISTANCES

Peppers may be the most obvious example of a presumed self-pollinated species that should nonetheless always be treated as if there were a high likelihood that they will cross at an appreciable rate. This is especially true under organic production practices (see the "Isolation Distances" section in the Solanaceae family introduction on page 299). As explained in the "Reproductive Biology" section earlier, the morphology of the pepper flower lends itself to cross-pollination in the presence of insects. This is especially true of hot chile types as well as the semi-domesticated and wild types of peppers. While the sweet pepper types may not outcross as readily as the pungent types, they are frequently involved in unwanted cross-pollination events in the field.

It should also be repeated that all five of the commonly cultivated species of peppers may be able to cross under certain circumstances and therefore should all be isolated appropriately when producing seed on any of them. Also, anyone growing a seed crop in or near any of the ancestral centers of diversity of peppers, including western South America or much of Central America, should be conscious of any populations of wild *Solanum* species that may be in the area. There is also potential crossing that can occur with the wild chiltepin pepper (*C. annuum* var. *aviculare*),

which is traditionally wild harvested and grows in many areas of northern Mexico and the southwestern United States. According to ethnobotanist Gary Nabhan, the range and proliferation of the chiltepin may have increased since the 1980s as farmers in the Sonoran Basin initiated plantings to increase the commercial harvest of this prized chile.

It is surprising that so many seed growers have treated peppers as if they were faithful self-pollinators, because a number of research teams going back to the 1940s have found cross-pollination common when two distinct varieties were planted in close proximity. In a study in New Mexico where plants of a test variety were interplanted with a commercial chile at the normal within-row spacing, Steve Tanksley found an average crossing rate more than 40% of the time, and some plants had up to 90% of seed resulting from cross-pollinations! While this is an extreme example, where plants of two different varieties are literally touching in the field, it reveals how much crossing can actually happen in a self-pollinated species when both the morphological and environmental factors are conducive to it.

I recommend that when growing two hot pepper varieties (or a combination of one hot and one sweet type), it is wise to isolate them by a minimum of 300 ft (91 m) if growing the seed crops in open terrain. This minimum isolation distance can be decreased to 150 ft (46 m) when natural barriers are present for any combination that includes a hot pepper type. If you are growing seed of two sweet pepper varieties in the same season, then it is possible to reduce this to a 150 ft (46 m) minimum isolation distance when growing them in open terrain and 75 ft (23 m) apart when natural barriers are present. This fits with Jeff McCormack's recommendations, which are based on his thorough investigations into the isolation distances of pepper crops.

315

TOMATO

The cultivated tomato (*Solanum lycopersicum*) is one of the most widely grown vegetables on Earth. It is among the five most important vegetables of commerce in many agricultural societies. This is remarkable when we consider that this humble fruit was only grown in limited areas of what is now southern Mexico and Guatemala in the early 1500s when invading Spanish conquistadores found it growing in Aztec villages. This fruit was called *tomatl*, or "swelling fruit," in the Nahuatl language of the Aztecs, which became *tomate* for the Spanish. The origin of this cultivated form is mired in mystery, as all of its wild relatives are native to mountainous regions of western South America, from Ecuador and Peru to northern Chile, including two species that are endemic to the Galapagos Islands. For many years the weedy cherry tomato of Mexico, *Solanum lycopersicum* var. *cerasiforme,* was thought to be the progenitor of the cultivated type, but recent genetic profiling at Cornell University indicates that this weedy form is in fact a feral mixture between wild and cultivated types.

Tomatoes were brought back to Spain by Hernán Cortés and were grown as a botanical curiosity. Within very few years they made their way to Italy. In 1544 Pietro Andrea Matthioli, a Tuscan physician who studied the medicinal value of plants, described them in an herbal text he was writing and suggested that tomatoes might be edible. In the second edition of this

SEED PRODUCTION PARAMETERS: TOMATO

Common name: tomato
Crop species: *Solanum lycopersicum* L.
Life cycle: annual (perennial in the subtropics)
Mating system: largely self-pollinated, with increased crossing in some types due to flower structure
Mode of pollination: closed perfect flower that requires stimulation for pollen shed
Favorable temperature range for pollination/seed formation: 60–74°F (16–23°C)
Seasonal reproductive cycle: late spring through late summer or fall (4–5 months)
Within-row spacing: 1–1.5 in (2.5–4 cm)
Between-row spacing: 22–30 in (56–76 cm)
Species that will readily cross with crop: In subtropical areas there is a weedy cherry tomato (*S. lycopersicum* var. *cerasiforme*) that is fully sexually compatible.
Isolation distance between seed crops: 10–200 ft (3–61 m), depending on the crop type and barriers that may be present on the landscape

text 10 years later Matthioli first used the term *pomo d'oro,* golden apple. While many have speculated since that time that the first Italian tomatoes must have been yellow-fleshed, it seems that *pomo d'oro* was a generic phrase used for all soft tree fruit at the time and wasn't specific to the color of the fruit. In Italy, the tomato was embraced as food in the poorer southern regions of the peninsula as well as in Sicily before becoming widely grown. There is also some evidence that tomatoes were grown as a vegetable in parts of Spain relatively soon after their introduction, but many parts of Europe only grew the crop as an ornamental for many years, fearing that the fruit was poisonous like many of the wild native members of the Solanaceae. Northern Europeans were late to the party as far as accepting tomatoes as a food plant, with the British and their colonies (and former colonies like the United States) not eating them widely until early in the 19th century.

Tomato seed has historically been produced wherever tomatoes are produced. As tomatoes spread across the globe people easily saved seed from the fruit and adapted the crop to their regional and climatic needs. The specialization of growing tomato seed in ideal environments didn't start until well into the 20th century. As several seedborne diseases, including fusarium and verticillium wilts, along with bacterial spot and speck, became more prevalent in commercial tomato acreage, it became obvious that growing tomato seed in drier climates and controlling the irrigation after fruit set was definitely desirable for controlling the spread of these diseases via the seed.

CROP CHARACTERISTICS

Reproductive Biology

The cultivated tomato is a short-lived perennial that is grown as an annual crop under most cultivation systems. The plant has a range of sizes, from extremely small dwarf determinates that can be 10 in (25 cm) tall to some very vigorous indeterminate types that easily reach 72 in (183 cm). Most modern commercial tomato varieties are determinate or vigorous determinates with an inflorescence borne between each leaf and range in size between 18 to 36 in (46 to 91 cm) in height. Determinate types also have terminal flower clusters at the end of each shoot. Indeterminate varieties usually bear an inflorescence every three to four leaves along the length of their shoots, and apical growth continues throughout the season until frost kills the plant.

Each tomato inflorescence usually has between 4 and 12 flowers that are formed and mature sequentially on a raceme. Individual flowers are perfect, with six bright yellow petals that curve outward, away from the flower as the flower matures. The ovary can have anywhere from 2 (especially in cherry types) to 15 or more locules, which contain the ovules. The six stamens have compact fused anthers that form a yellow cone, 0.5 to 0.75 in (1.3 to 2 cm) long, that surrounds the pistil, with its style and stigma that usually terminates within the cone but can occasionally extend slightly beyond the tip of the cone, which has a small opening. The anthers have slit openings on the interior of the cone, and when pollen dehisces it will shower out of these pores with any kind of motion of the flowers, whether from wind or insect visitation.

As the anther cone of the flower usually points downward, the pollen will thoroughly cover the bulbous stigma, it is well within the anther cone as it is with most modern tomatoes, or the cone is exerted out of the tip of the cone as it often is with many heirlooms. The pollen, which is shed over a 2-day period, will usually pollinate its own stigma within the anther cone, supplying the pistil

with plenty of pollen to fertilize a full complement of ovules.

◀ The flower of a modern tomato variety (*Solanum lycopersicum*), with an inserted stigma that is well within the anther cone, resulting in lower rates of cross-pollination.

▼ This flower of a 'Pruden's Purple' heirloom tomato shows the exserted stigma that makes cross-pollination of such varieties by insects much more likely.

However, the stigma is often receptive a day before pollen shed and remains receptive 2 or 3 days after the pollen from its flower has shed. This means that there are opportunities for crossing to occur, especially with the exerted stigma of the older varieties. When the style pushes the stigma out of the end of the anther cone, it is exposed to possible insect

activity. While tomato flowers are not visited by a wide number of insect species, they are often visited by several types of bumblebees (*Bombus* spp.). Bumblebees have a unique way of clinging to the flowers upside down while vibrating their wings rapidly and shaking the pollen out of the cone onto their abdomen. If the stigma is exerted then it is possible that pollen on their abdomen from a previous flower can be transferred to the flower they are currently visiting, producing a cross-pollination. This is obviously much less likely to occur with more modern tomato varieties, which have stigmas that are well encased in the anther cone; other insect pollinators, however, will sometimes pry the flowers open and cause a cross to occur.

Climatic and Geographic Suitability

Tomatoes can have problems setting seed at temperatures that are too high or too low. At temperatures above 90°F (32°C) and below 60°F (16°C) the pollen of many varieties will be affected and fertilization of ovules will be impeded, both resulting in poor seed set. In extensive experiments with tomato pollination in the 1930s, Ora Smith of Cornell University found that the optimum temperature for pollen to germinate on the stigmatic surface is 85°F (29°C); at 100°F (38°C) or 50°F (10°C) pollen germination was virtually stopped. Smith found that even at favorable temperatures pollen tube growth is slow, taking 2 to 3 days to reach the ovules following pollination. This means that, even if temperatures are favorable at the time of pollination, any temperature swings below 60°F (16°C) or above 90°F (32°C) may severely slow or stop the growth of the pollen tube on its journey to the ovules. Therefore, even when the temperature for pollen tube growth is at or near the optimum during the day, if

the temperature drops to lows at or near 50°F (10°C) during the night, any of the pollen tubes that started their journey within the last day or two can stop growing. Alternatively, in hot climates the pollen can germinate and start growing during the cooler temperatures of the morning or evening and then be stifled when hot temperatures approach or exceed 100°F (38°C) in the middle of the day. Once the pollen tube stops it usually will not resume growth. If this happens repeatedly over the course of the several days that the flower is receptive then there is a good chance that most of the embryos won't be fertilized; hence the fruit won't "set" and will abort (see chapter 2, Reproductive Biology of Crop Plants).

A potentially worse situation may occur when a tomato seed crop sets a full complement of fruit, but because of less-than-optimum environmental conditions the fruit can have very little seed at harvest. This can happen when the temperatures at or during the pollination and subsequent fertilization of the embryos are either too hot or too cold, as in the previous example, resulting in only a minority of pollen tubes reaching their destination. Some tomato varieties will successfully set fruit with only a fraction of the available embryos becoming fertilized, resulting in a fruit with potentially very few seeds. This often occurs when seed growers attempt to grow the so-called parthenocarpic tomatoes that can be seedless due to their unique ability to set fruit even when temperatures are too cool for most tomato varieties to successfully do so. A number of farmers have been sadly disappointed when growing a parthenocarpic tomato for a seed crop and having a good fruit set, but then finding the crop is barren of seed (or yielding very little) when they crush the ripe fruit to extract it. (See the "Genetic Maintenance" section on page 325 for possible solutions to this problem.)

SEED PRODUCTION PRACTICES

Soil and Fertility Requirements

Tomatoes can be grown on all types of soils, but they seem to benefit greatly from growing in a good agricultural loam, ranging from sandy loams to heavier clay loams that are well drained. These soils, if well drained to avoid root rots, can supply continuous moisture and fertility, which promotes good seed yields in tomato. A soil pH of between 6.0 and 7.0 is desirable, and excess fertility, especially nitrogen, should be avoided; it can promote luxuriant foliar growth to the detriment of fruit growth. Phosphorus should be readily available, which is sometimes problematic under organic fertility regimes, especially under cooler-than-optimum temperatures. In most cases high-quality compost with a good humus fraction is adequate to meet the needs of a tomato seed crop.

Growing the Seed Crop

Planting the Crop: The first step in growing a successful seed crop is to produce healthy, disease-free tomato starts. Seed can be sown into greenhouse hot beds or seedling flats; in warmer climes it can be planted into cold frames. Plants are then frequently pricked out from these initial thick plantings and transplanted into flats or individual pots within a couple of weeks at 6 to 10 in (15 to 25 cm) apart on the greenhouse bench. Seed can be sown as early as 10 to 12 weeks before the projected transplant date. However, many growers prefer a transplant that is stocky and no more than 8 to 10 in (20 to 25 cm) tall when setting them out into the field. Producing a stocky transplant that resists getting leggy can be accomplished by subjecting the plants to good, strong air currents on a regular basis over the duration of their time growing under cover. Seedlings should be grown at a moderately rapid rate and hardened off without exposing them to extremes of cold, below 55°F (13°C), or extremes of direct sun when first being moved outdoors. Lastly, starting with seed that is free from seedborne diseases is very important, and appropriate steps to avoid contaminated seed should be strictly adhered to by all tomato seed growers (see chapter 16, Seedborne Diseases).

Crop Spacing: When transplanting to the field tomato plants can be spaced at much the same spacing as when growing the crop as a vegetable. Spacing depends in large part on the harvest methodology and the degree of vigor in the tomato variety being produced. Determinate tomato varieties vary greatly in their size and the extent of the canopy they produce. Some vigorous determinate types are in fact intermediate in their stature between the average determinate types and the indeterminate types. Indeterminate tomato varieties can also vary considerably in their size and stature and will require a range of plant spacings in order to grow the best seed crop.

Upon transplanting into the field plants of determinate varieties are routinely planted anywhere from 14 to 24 in (36 to 61 cm) apart within the row if staked, and often at increased spacing if no support system is used and plants are allowed to sprawl on the ground. The between-row spacing is anywhere from 36 to 72 in (91 to 183 cm) for determinate varieties, depending on how vigorous the variety is that is being grown. Some growers will also plant two rows on a bed with support to increase the population. Compact determinate varieties that don't require any staking can be grown at even tighter spacing based on their size and stature.

Indeterminate tomato varieties usually require greater spacing and are usually staked for seed production. Staking tomatoes for seed production is always desirable when possible as it minimizes the soilborne diseases that may possibly infect the crop. The spacing between plants within the row for determinate plants is from 24 to 36 in (61 to 91 cm), and the spacing between rows is similar to the spacing for determinate types, though usually starting at 48 in (122 cm) and going up to 72 in (183 cm), depending on the vigor of the crop and the spacing required for your harvest methods.

Seed Harvest

Tomato harvest for seed is done in essentially the same way that it is done for the fresh market. The major difference is that fruit harvested for seed is often picked at full dead-ripe maturity, where fruit harvested for the fresh market is often harvested at various stages of immaturity, even when sold locally. Damage to the fruit at harvest should be minimized to allow for roguing out diseased or rotting fruit at the time of processing for seed (this is also very important if you want to extract an additional product such as sauce or juice from the fruit pulp). While the tomato seed may be at its peak maturity when the fruit is dead-ripe or even a little overripe, it is important to not leave the fruit on the vine to the point where it is rotting, as any excessive rotting can either damage or discolor the seed.

Once the tomato fruit is harvested it is also important to extract the seed in as short a period of time as possible, as the inevitable rotting of the fruit from saprophytes (bacterial or fungal organisms that feed on damaged or dead tissue) will grow quickly, especially on any damaged fruit. To harvest the seed the fruit needs to be crushed and mashed until the seed is largely freed from the locules, which are the cavities in which the seed is borne

inside the tomato. In many of the machines used for wet-seeded crops, the seed is then forced through a circular wire mesh screen by centrifugal force. This allows the seed, the juice, and small pieces of pulp to pass through and go into a container but catches larger pieces of pulp and most of the tomato skin. This fruit debris then comes out the end of the cylinder, and if the operation is done with properly maintained food-grade equipment, the pulp can be used for salsa or sauce. Historically there were companies that harvested tomato juice in connection with seed production using specialized equipment. If you are processing small batches of fruit it is possible to simply press the tomatoes through a wire mesh screen that allows the seed to easily pass through, while catching much of the pulp.

Seed Extraction Methods: Tomato seed is enclosed in a gelatinous sac that clings tightly to each seed. The traditional method used to break down and eliminate this sac is to ferment the seed, juice, and pulp with the endemic yeast that occurs naturally on the skin of the fruit. The fermentation of tomato seed is an effective way to separate it from the gel and one that is acceptable under organic farming practices and organic certification. In contrast, most of the conventionally grown tomato seed since the 1970s has been extracted using an acid separation technique. It is often used in conjunction with partial fermentation and has become popular, as it saves time and produces a very clean-looking, light tan–colored seed. In this method a controlled amount of hydrochloric acid is added to the seed pulp, mixed thoroughly, and only allowed to interact with the seed for up to half an hour. However, it is a potentially dangerous process, and policymakers agree that it doesn't conform to most organic standards worldwide. At the time of this writing it is still the industry norm, but there is debate over whether it should be acceptable for certified organically grown tomato seed.

Seed Fermentation: The fermentation process is a fairly straightforward process using tubs or tanks to hold the seed that is still suspended in the tomato juice and pulp that has been run through a screen. Some authors encourage adding water to this mash, but this dilution is seen as a hindrance to achieving the full potential of the fermentation process. Some tomato seed growers also believe the presence of water increases the likelihood of germination occurring during this process. This may be based on the fact that the juice of the tomato, which is largely from the locules of the fruit, has sprout inhibitors that are diluted when water is added to the mix.

An important factor in encouraging fermentation is to ensure that temperatures are held at between 72 and 80°F (22 to 27°C) during most of the time that the seed is fermenting. The time it takes to achieve full seed extraction using fermentation is in large part dependent on the temperature that the seed mash is exposed to. The fermentation period will only take 2 to 3 days if temperatures are maintained at the upper end of this scale. Most tomato seed producers agree that it is desirable to have this process take no more than 4 days. If the temperature of the fermenting mash remains much below 70°F (21°C) for any appreciable period of time the gel may not fully separate from the seed, and undesirable fungal or bacterial growth can affect seed quality and lower the germination percentage. If the fermentation temperature goes much above 82°F (28°C) for any appreciable length of time during the process, then the viability of the seed can also be lowered. Therefore, fermenting seed in very hot climates will need to be done within a cooled environment.

Conversely, in cooler climates it may be necessary to ferment the seed in a greenhouse or heated building.

From the beginning of the fermentation process, the fibrous pulp and the enclosed seed float to the surface of the fermentation vats. As the seed separates from the gelatinous sacs it will sink to the bottom if it is sound, while the pulp, remnant sacs, and any unviable seed will float to the surface. To encourage this separation regular stirring of this mash should occur at least twice a day and sometimes more often in rapidly fermenting batches. This stirring is also very important to discourage the formation of mold on the pulp at the surface of the mash, which can discolor or damage good seed in this floating material.

Fermentation and Seedborne Diseases: Fermentation of tomato seed is often purported to kill a number of seedborne diseases that may affect tomatoes. Unfortunately, the only pathogen that seems to be affected is the bacterium that causes bacterial canker (*Corynebacterium michiganense*), and the effectiveness of fermentation on it is dependent on the temperature that is maintained and the duration of time that fermentation occurs (see chapter 16, Seedborne Diseases).

Washing Seed: When the seed has fully separated and collected on the bottom of the fermentation vessel, it is time to wash out all of the pieces of pulp that remain. There are at least two methods commonly used for commercial quantities of seed. For both methods the first step in washing the seed is to remove as much of the floating mass of pulp as possible. After a final thorough stirring to release good seed that will sink from the floating mass, it is important to either scoop and discard as much of the pulp as possible, or to slowly and gently add water to the vessel till the top few inches of pulp spills over the lip of the vessel. This latter option must be done quite slowly and gingerly with a low-volume stream of water so as to not disturb the good seed that is on the bottom of the vessel.

The first seed-washing method uses a sluiceway or flume, which is a long narrow trough, built at a slight decline of approximately 1 in 50. This is the same technology used by gold miners during the California Gold Rush of 1849. These seed-cleaning sluices are usually made of either stainless steel or wood, with a trough 12 to 18 in (30 to 46 cm) wide and anywhere from 10 to 25 ft long (3 to 7.6 m). On the bottom of the trough are a series of riffles or crosspieces that are 2 to 3 in (5 to 8 cm) high and run diagonally every 12 to 24 in (30 to 61 cm) along the length of the flume.

As the contents of the fermentation vessel are mixed with water and gently poured into the top of the trough, the heavier, viable seed is caught behind the riffles and the lighter pulp is easily carried down the length of the trough until it is eliminated over the spillway at the end of the sluice. After the riffles have accumulated a good amount of seed, only water is run until all of the pulp and debris has cleared. Then the riffles are taken out and the seed is washed down onto a clean fine-mesh screen at the spillway. Controlling the flow and speed of the water to properly clean tomato seed with a sluice requires practice and experience, and the trough should always be equipped with a fine-mesh screen at the spillway during the cleaning to capture any good seed that may be washed through the system due to error.

The second seed-washing method is useful for smaller batches that are frequently done in varied-sized pails, from 5-gallon (19-liter) buckets to 55-gallon (208-liter) drums. Be sure to be conscious of the previous contents of these vessels. For organic seed producers it is important to remember that any recycled receptacles used for seed cleaning have to meet organic certification standards.

At the end of the fermentation process the floating pulp that has accumulated at the top of these smaller vessels is easily eliminated by tipping and scooping off as much of the pulp as possible before the actual washing of the seed begins. Always make sure you have stirred this pulp one final time, and give the good seed that has been freed a minute to settle to the bottom of the vessel.

After scooping off the bulk of the pulp and debris, it is time to rinse and decant the seed mass in repeated cycles with cool, clear water. This is done by first filling the vessel with water, stirring, allowing the good seed to settle, and then gently decanting off the pulp that is suspended in the water without disturbing the good seed sitting on the bottom of the vessel. At first the liquid will be quite cloudy with debris, but upon repeated cycles the water will get clearer each time. The idea is to stir up the debris every time you add water, then wait long enough to allow the seed to settle, and then pour off the liquid while the pieces of pulp, placenta, and non-viable seed are suspended in the liquid. This process usually needs to be repeated at least 8 to 10 times to eliminate most of the debris before the water runs clear and the seed is clean. The debris at the end of the process is the hardest to decant off as it is the heaviest and will sink almost as fast as the good seed; it takes a deft hand to get these pieces out of the vessel without losing any good seed.

Drying Seed: Wet-seed extraction requires that the seed be dried as soon as possible after cleaning. When processing large quantities of tomato seed it is actually desirable to wring or squeeze the excess moisture out of large seed masses, which can be done after the seed is placed in strong cloth sacks. In fact, tomato seed in these cloth sacks can be run through unheated spin-dryers, then placed on drying screens. Seed racks need to be elevated to allow airflow both below and above the seed, which should be spread out evenly and fairly thinly on the racks. Tomato seed can be dried in direct sunlight as long as the heat at the surface doesn't exceed 90°F (32°C); higher temperatures can damage the seed. The important factor in drying all seed crops is to always have good airflow. Never hesitate to use fans when good airflow is lacking, even sometimes if you're drying seed outdoors. As the seed dries on racks, stir it at least twice a day.

In humid climates or during cool, wet weather, tomato seed is often dried with the help of controlled, supplemental heat. In large seed-processing facilities there are often cabinets where seed racks are placed on the top of long wooden or metal open-topped cabinets in rows, then warm, dry air is forced up through the racks. This warmed air must be blown through the cabinet with enough force to reach all of the seed racks along the length of the structure.

GENETIC MAINTENANCE

Tomatoes are largely a self-pollinated species, so they do not usually exhibit as much genetic variation as many cross-pollinated crops. Because they are a fruiting crop, most of the evaluation selection of the characteristics that distinguish each particular variety is done after the crop has matured fruit. However, a number of traits can be checked before the plant sets fruit.

It is usually possible to determine differences in leaf shape or leaf color when the plants are still in the pots in the greenhouse with only the first few sets of true leaves. When the plants are forming their first flower clusters it is usually quite easy to see the proportion of flowers to leaf nodes. Determinate types usually have one flower cluster per node, while indeterminate types often bear only one flower cluster every third node. Some tomato breeders evaluate tomato fruit shape shortly after flowering has started, by examining the very small, newly formed fruit shortly after fertilization has occurred. The shape of these small immature fruit is essentially the same as what they will become when they mature. Seed growers have used this method to their advantage by growing tomato plants to the point of flowering while still in pots, then selecting for shape before transplanting to the field.

As the fruit attains full maturity it is possible to judge if each plant's fruit is true to type. Important traits to consider when roguing to type include fruit shape, color, and relative size. Quality traits such as flavor, texture, and juiciness can also be evaluated when the fruit ripens. The fruit of any plant should always be judged collectively and as an average of all of the fruit on that plant, as any one fruit may differ from the others due to the environmental conditions during its period of initial growth.

ISOLATION DISTANCES

Determining adequate minimum isolation distances for tomato seed production has become a lively topic, with few people involved in organic seed production wanting to pin their reputations on specific recommendations. The published isolation distances that have been used for years by the conventional seed industry have been discredited with the advent of unacceptable levels of crossing between adjacent tomato crops. There are many factors that contribute to increased cross-pollination in all crops (see chapter 13, Isolation Distances for Maintaining Varietal

Integrity). The accepted theory to explain this is that two fundamental differences in organic production methodology increase the likelihood of cross-pollination in this largely self-pollinated species.

The first factor that contributes to higher rates of cross-pollination is that there is often more biological diversity among plant species within their farms. Many organic farmers strive to have more crop diversity in their fields with both a more diverse number of crop species on their acreage at any given time and more active rotations across years. Both of these factors will usually lead to a richer, more diverse cohort of insect species. This is almost always true when a number of these crops are flowering crops that attract insect pollinators. Many organic growers also try to have some percentage of their diverse crop mix flowering at all times to attract beneficial insects.

The second factor is that organic farmers also often use less insecticide than their conventional counterparts. Many insecticides are broad-spectrum compounds that have the effect of killing a wide range of insect species, including the beneficial pollinators. Also, when organic farmers choose to use an insecticide that is certified by organic standards, it is likely that these products will degrade quickly and have less of an overall impact on the populations of beneficial insects on their farms. The potential increase in beneficial pollinators on organic farms from these two factors must certainly increase the number of cross-pollination events in any given season.

Tomato flowers are of two somewhat distinct types:

1. Modern tomato varieties, usually those bred and released after about 1920, have styles that are shorter than their wild ancestors did and that are usually well encased inside the anther cone. This

This sliced 'Cherokee Purple' tomato shows nicely the gel-like liquid in the carpels, which serves as the medium and food for the yeast during the fermentation process. ▶

greatly reduces the chances of the stigma coming in contact with any foreign pollen, even if an insect visits the flower.

2. Heirloom types or varieties with wild ancestry often have styles that are long enough to extend past the end of the anther cone, with their stigma clearly exposed to contact with a visiting insect. Many heirloom tomatoes, especially the potato-leaved and older beefsteak types, have this phenotype that can often be crossed, especially by diligent bumblebees that grab the cone from the bottom and push it right up against their abdomens. The tomatoes with wild ancestry—both the currant tomato (*S. pimpinellifolium*) and most cherry tomatoes, which are derived from the feral cherry of Mexico and Guatemala (*S. lycopersicum* var. *cerasiforme*)—also usually have the exerted styles.

The recommended minimum isolation distances between two different tomato varieties for the modern types should be 75 ft (23 m) if the crops are planted out in the open with no barriers and can be reduced to 50 ft (15 m) between them, when the two crops are separated by a significant barrier of the landscape (see chapter 13, Isolation Distances for Maintaining Varietal Integrity, "Physical Barriers" on page 337). While these distances are much increased over the isolation distances published in many older seed guides, they are currently being used by a number of the larger conventional seed production companies in Europe.

The isolation distance will need to be doubled for the heirloom class of varieties, which includes many of the potato-leaved types and

◄ The flowers of currant tomatoes (*S. pimpinellifolium*) often have stigmas that are slightly exserted or that are flush with the apex of the anther cone, as seen in this currant tomato. In either case, much cross-pollination can occur, as bumblebees seem to be particularly fond of the currant types.

beefsteak types. These varieties have long been known to cross-pollinate at higher rates than the modern types that are more commonly grown for commercial production. In a thorough study of cross-pollination in tomatoes from the 1920s that was based on differences in style length, J. W. Lesley found that varieties with exserted styles could cross-pollinate up to 5% per generation. For this reason it is important to separate any heirloom varieties from other heirlooms or any other tomato varieties by at least 150 ft (46 m) in open terrain, or by 75 ft (23 m) when natural barriers are present.

Jeff McCormack, a pollination biologist who has thoroughly examined the research into cross-pollination in the Solanaceae, believes that the minimum isolation distance used for the currant types and most cherry tomatoes should be at least as great as the minimum distance used between the heirloom types.

'Chadwick's Cherry' produces an abundant crop of large, red "English-style" cherry tomatoes that are prized by many home and small-scale growers for their exceptional flavor.

PART III

Seed Grower Fundamentals

- 13 -

ISOLATION DISTANCES FOR MAINTAINING VARIETAL INTEGRITY

One of the main requisites of growing high-quality seed is to maintain the varietal integrity of the seed crop you are producing. To do this you must make sure that the genetic makeup of the crop isn't compromised with any genetic mixing via pollen from other plants of the same species. This requires producing a seed crop at an adequate spatial isolation from any cultivated, wild, or weedy plants of the same species that are capable of intermating with your crop via their pollen. You must consider a number of biological and environmental factors that can affect pollen movement by insects and wind when determining the minimum isolation distance for each crop and cropping system.

THE SPECIES BOUNDARY

First, it's important to know which crops are capable of sexually intermating or "crossing" with the seed crop that you are planning to produce. The word species means "kind" in Latin. In general, each species represents a distinct kind of organism that is able to intermate and produce fertile offspring only with other individuals of its own kind or species. Of course, there are exceptions. For example, a number of species of deciduous trees will sometimes cross with related tree species and produce interspecific hybrids. Thankfully, though, there are very few exceptions among vegetable species.

In the vegetable crops the species designation almost always defines a boundary to potential crossing when determining placement of seed crops for isolation. The two notable exceptions among the vegetables are the several species of peppers (*Capsicum* spp.) that are sexually compatible and easily cross and at least two wild species of tomato that will readily cross with the cultivated tomato (*Solanum lycopersicum*). (see chapter 12, Tomato, "Isolation Distances" on page 325.) Another exception is crossing among squash species in the *Cucurbita* genus, though it does not occur nearly as readily as with tomatoes and peppers (see Chapter 9, Cucurbitaceae, "Isolation Distances," page 211). Knowing the species designation of each crop you work with also allows you to learn a great deal about the reproductive habits of the crop. When questions arise as to identifying the species of a given crop, it is recommended that you contact researchers or seed professionals who are well versed in the crop before planting.

INCREASED ISOLATION CONSIDERATIONS FOR THE ORGANIC SEED GROWER

The importance of maintaining adequate spatial isolation has increased significance in the modern era for the organic seed grower.

Pollen contamination from genetically modified (GM) crops will disqualify the seed crop from consideration as a certified organically grown product. Many of the opponents to GM crops see the resultant genetic mixes as fraught with potentially damaging effects on the ecological balance in organic farming systems and having potentially harmful effects on human health. Certainly, everyone can agree that the introgression of modified genetic traits into seedstocks intended for organic agriculture may be very hard to get out, resulting in long-lasting effects on the integrity of the germplasm. Most professionals involved in the production of organic seed advocate for at least a two- or threefold increase in the minimum isolation distance when GM crops of the same species are in the vicinity over the normal minimum isolation distances stated for each particular crop species in this text.

THE MYTH OF STANDARDIZED ISOLATION DISTANCES

The first thing that most farmers learn when growing a seed crop is that their crop must be isolated from any other crops of the same kind in order to produce seed that is genetically pure and hasn't crossed with a neighbor's crop. For most farmers, the best information that they can get on how much isolation they'll need for their seed crop is often vague and not very specific to their environment or the crop species that they are growing. The recommendations that are most often repeated is that seed crops require either 0.5 mi (0.8 km) or 1 mi (1.6 km) between cross-pollinated crops of the same species, and that self-pollinated crops need little or no isolation (although a barrier crop is sometimes recommended) from other crops of the same species. This

information is often just hearsay from farmers or local agricultural officials, or it is found in much of the published literature from older seed saving publications and on the Internet. It often leads to poor planning of seed crops, which in turn can lead to pollen contamination and genetic mixing and outcrossing.

If you're located in a region where there is a concentration of production for the particular seed crop of interest, isolation guidelines have often been worked out between regional growers' cooperatives and local governmental organizations, like the state Extension Service of the Land Grant Universities in the United States. Unfortunately, much of the specialized information generated by these groups is difficult to access, is unknown outside of these regions, and is not easily found in the literature.

THE MYTH OF PURE SEED

The other common misunderstanding among seed growers has to do with the idea of varietal or genetic purity in crop varieties. In most of the seed-growing literature there seems to be the implication that if growers follow the recommendations for isolating their seed crop, then they have eliminated any chance of an outcross and can expect to harvest a seed crop that is genetically pure. In reality there is no such thing as perfect isolation, unless perhaps truly extraordinary steps are taken. Even when the strictest isolation distances are observed by the seed companies for stockseed and foundation seed, cases of genetic crossing have occurred that can only be explained by pollen mixing. While pollen of any given crop has an average range of movement based on the types of insects or direction and strength of wind that carries it, there are certainly the unusual events where the pollen is carried several times farther than its usual trajectory due to an extraordinary insect pollinator or wind gust that may come along. Many seed workers know that pollen is easily carried on clothes, on the fur of animals, and in water. Ultimately, having a pollen-proof space to produce truly pure seed would require high-tech equipment comparable to some of the gadgetry of NASA and would be prohibitively expensive. All experienced seed growers learn how to deal with the occasional crosses that occur in their stocks; in fact, they expect it to happen and select or "rogue" against it in every generation. Indeed, sometimes these chance crosses result in new combinations that the savvy grower can isolate and develop into a new variety. Remember, nature is always creating new genetic combinations to be tested in the great experiment of life on Earth.

WHAT IS THE INTENDED USE OF YOUR SEED CROP?

The first question that seed growers must ask themselves when determining the placement of a crop is, "What is the intended use of the seed that I am producing?" The relative genetic purity of seedstocks necessary for a seed company's commercial sales is quite different from the increased level of purity required for its foundation seed or stockseed. The precautions that are often taken when growing a seed crop for genetic preservation purposes will invariably meet a stricter standard than the isolation requirement for farmers saving seed for their own production needs. The relative nature of this determination must always be thought through and decided by the individuals who are upholding the quality standards for the farm, seed company, or public institution that they represent. The isolation distances may frequently vary based on the regional fluctuations of the biological and environmental factors discussed in this chapter. Those experienced in seed growing in any given region

will be the most qualified to take the isolation guidelines proposed at the end of each crop section and amend the distances based on their practical knowledge of the regional factors that exist. This requires experiential knowledge to be collected among seed growers and seed-company fieldworkers who aren't afraid to get their boots dirty, who have good observational skills, and who are real students of nature. Many of the minimum isolation distances given for the different crops in this book are greater than many experienced seed growers believe are necessary. The minimum distances presented here are only guidelines, to ensure relatively small levels of crossing between moderately large production fields, and should be treated thusly.

The fact that there is no such thing as perfect isolation can be intimidating to anyone seeking genetic purity in seed. But it can also be liberating once everyone involved in seed production realizes there is no such thing as absolute purity and that some genetic mixing is inevitable whenever seed is produced. It then becomes your responsibility to be much more involved in the process of determining the isolation distance based on the biology of the crop and the environment and topography of the location where you're growing it. You must also get more familiar with each seed crop, realizing that your selection of off-types, outcrosses, and seed mixes is the only way in which the integrity of the crop will be maintained as it passes through your hands to farmers who will grow the crop now and in the future.

Knowing that genetic mixing and variation is inherent to the process puts us back in touch with our true role in the process. This is the same role that our ancestors who first domesticated these plants had. This is the way that we integrate these crops into our lives. The genetic integrity of the crop then becomes a reflection of our commitment and involvement in the process. The crops that we grow and use thus become woven into our communities and into the "culture" of our agriculture.

FACTORS THAT DETERMINE ISOLATION DISTANCES

First and foremost we must remember that what we are working with is a biological system that is dynamic and highly interactive with other biological entities in the ecosystem. The next thing to keep in mind is something all good farmers know: Crops interact with the surrounding environment. Certainly climatic factors of temperature, precipitation, wind, daylength, and relative humidity play a central role in the overall success of a crop, but they also can determine what amount of isolation distance is necessary. Along with these physical manifestations of climate are the physical elements of the surrounding landscape. The terrain—whether it's hilly or flat, covered with lush vegetation or open—can play a very important role in determining the isolation distance that's needed between crops capable of crossing. Lastly, the physical size of the crop, its configuration on the land, and the type of pollen, whether airborne or insect-transmitted, are all crucial in any determination of isolation distances. All these factors will be considered for the two reproductive classes of seed crops, selfers and crossers.

Is the Crop a Selfer or a Crosser?

The crops that are grown from seed are usually categorized as either self-pollinated or cross-pollinated when referring to their mode of reproduction. The assumption is that selfers always self and crossers always cross, with only rare exceptions. In fact the scope of reproductive behavior for either of these reproductive types is a continuum, with some selfers crossing upward of 50% of the time and some crossers that routinely self at a significant rate under certain environmental conditions. (Indeed, all selfers will cross to some degree in almost any situation, and all crossers, even those with strong physiological mechanisms to enforce outcrossing, will self at some small percentage during reproduction.) As with many mechanisms in biology, discrete groups do not exist, and exceptions to any general category abound. As a seed grower you must learn the reproductive peculiarities of each seed crop for your environment through your own observation and experience, as well as the shared knowledge of others.

Self-Pollinated Crops

Common Misinformation: The literature concerning isolation distances for self-pollinated species usually has little more than cursory mention of the various ways that selfers naturally cross-pollinate (if this is mentioned at all). It is presumed that these crops faithfully self-pollinate when grown for seed, with rarely a mention of the possibility of crossing. When a distance is given it is usually specified as the distance necessary to prevent mechanical mixing between two varieties at harvest. In a number of instances in North American literature the minimum separation is given as 10 ft (3 m), 12 ft (3.7 m), or 25 (7.6 m) ft between varieties of selfers with this intent. However, this is not enough isolation between different varieties of a crop species to minimize the amount of crossing that is possible under certain environmental or biological conditions that may be present (see "Environment," below).

Another serious flaw in the literature concerning isolation distances is that there is very little discussion of unusually promiscuous self-pollinated crops such as runner beans, favas, eggplants and peppers, which often cross at rates at or above 3 to 5%, with some types that can cross at rates above 20% or even 30% under certain environmental conditions. (See "Isolation Distances" under each of these respective crops for details.) These four crops, which are capable of outcrossing at such high rates, will be referred to as the "unusually promiscuous selfers" in this discussion.

Environment: Environmental factors such as heat and humidity can influence how easily the cleistogamous flower of a selfer will open when fertile. Under high temperatures, several crops of the Fabaceae may open slightly at pollen maturity, allowing any pollinator more access to pollen and nectar than usual. Also, the degree to which cleistogamous flowers will open when a probing insect in search of pollen or nectar attempts to pry the petals apart for access can vary with temperature and humidity. Crossing rates in field peas are higher under cool, clear conditions than when temperatures are excessively hot and dry.

Physical Barriers: The presence of physical barriers can play an important role in determining the isolation distance for a particular crop. Physical obstructions to the movement of insect pollinators (the main vectors of pollen in cross-pollinations among selfers) such as hills or forests can be very effective barriers that can cut the isolation distance between crops to half of what it should otherwise be in open terrain. However, when physical barriers are less than complete, the isolation distance needed should be somewhere in between the distances recommended for those with barriers and the standard distances given when no barriers are present (see "Isolation Distance" under each individual crop type).

Genetics: Different varieties of the same self-pollinated species may differ in their response to environmental challenges. In certain varieties of several crops, low relative humidity

levels may result in the failure of pollen to germinate on the stigma. This allows a window of opportunity for pollen from a less humidity-sensitive variety to fertilize these flowers during an excessively dry period. The take-home message for all seed growers to remember is that there are almost always genetic differences between varieties of any given crop in their susceptibility to environmental stresses. In other words, some varieties of a particular self-pollinated crop will outcross more readily than others, especially in the face of challenging environmental conditions. With the likelihood of environmental extremes increasing due to global warming and regional climate change, the precautions dictated by the isolation distances listed for self-pollinated species are warranted.

Recommendations: So what distance should you use as the minimum isolation distance for selfers? Because the potential influence of biological factors is quite variable and dependent on the environment of production, it is important to set a standard isolation distance that ensures a high degree of purity across most environments and production situations. Depending on whom you speak with in the seed industry, this can be a distance anywhere from 20 to 150 ft (6 to 46 m) to minimize possible crossing in most selfers for commercial purposes.

If we want to set a rigorous standard for isolation distances between selfers, then we must consider the standards set during the implementation of the Lend-Lease Act of 1941, when the United States greatly expanded its export of seed with the onset of World War II. Because Germany occupied many of the major seed production areas of Northern Europe during World War II, the United States was thrust into the role of a major supplier of seed, especially the seed of several important species of vegetable crops. In an attempt to ensure high-quality seed the isolation requirement for selfers was set at 150 ft (46 m) between crops. It must be remembered that this distance was established for large-scale productions where the risk of crossing is always greater due to the sheer numbers of plants. While this distance may seem excessive to many seed growers with experience in producing selfers, it will certainly eliminate most potential crossing events under most climatic conditions across the varied agricultural regions of the world. There are many climates where excesses of heat, humidity, or the presence of certain insects can cause a much higher rate of crossing in all selfers. Therefore you can confidently use this strict 150 ft (46 m) minimum isolation distance for most selfers in most cases (excluding the unusually promiscuous selfers, which require greater distances). If you are producing a particular self-pollinated crop in a climate where the isolation distance has proven to be reliably less than this recommendation, such as the production of common bean seed in the Treasure Valley of Idaho or lettuce seed in the Sacramento Valley of California, then you should use the locally established isolation recommendations. Always remember that seed production under organic conditions often favors a higher rate of crossing due to increased biological activity.

Cross-Pollinated Crops

Common Misinformation: While information on isolation distances for cross-pollinated crops can usually be found in the seed-growing literature, it is usually too general and doesn't make any distinction between insect-pollinated crossers and wind-pollinated crossers. There is a definite tendency for the isolation distance to be given as a "one size fits all" number for all crossers in many of these publications. In American publications the one number recommendation is typically either 0.5 mi (0.8 km) or 1 mi (1.6 km) for all crops, or sometimes 0.5 mi (0.8 km) is used for some of the crossers and 1 mi (1.6 km) is used for others, often in a seemingly arbitrary fashion. In a sample of the European literature published in English, the distances are usually either 1 km (0.6 mi) or 1.5 km (0.9 mi). While there is often a bit more reasoning behind the increased distance for some crossers in this group, there is still less-than-adequate information on the need for some crops in some situations to have upward of three to five times this much distance between them! (See specific recommendations as listed under "Isolation Distances" for each crop in chapters 4 through 12.)

Environment: Environmental factors such as temperature and relative humidity can have a marked effect on the extent of cross-pollination in crossers. Sometimes environmental extremes can diminish cross-pollination. If it's hot enough, the pollen may be damaged quickly at pollen shed or anytime during anthesis (the period of floral fertility and receptivity). If the relative humidity is too low, the pollen may desiccate and die before reaching another fertile flower. Cold weather can also diminish the viability of pollen and the receptivity of the female stigmas of a flowering crop. In contrast, some environmental conditions can increase the chances of pollinations occurring over time or space. High humidity can contribute to pollen's longevity, helping it remain viable longer and enhancing its chances of reaching a sexually compatible plant that is farther afield than it would normally travel, carried either by an insect or the wind. The wind, too, can contribute greatly to cross-pollinations, occurring far beyond the normal range of the wind-pollinated crops when the conditions are just right.

Physical Barriers: Physical barriers of the terrain, such as hills, vegetation, or even buildings, are just as important in reducing the flow of pollen between crops in crossers as they are with selfers. These obstructions will reduce or stop the insects or wind that can carry pollen. As with selfers, a solid obstruction such as a forest with dense vegetation between two crops will enable you to cut the isolation distance in half. Incomplete or partial barriers require you to make a judgment call as to the degree that you can reduce the isolation distance and avoid the vast majority of potential crosses. The isolation distance for partial physical obstruction must be gauged somewhere between the full recommended distance and half that distance.

Genetics: Genetic differences among the crossers include a number of traits that determine the quantity and possibly the quality of the pollen produced by a given cross-pollinated crop variety. Over the years many seed growers have recognized that some varietal populations are more robust in a number of floral characteristics than other varieties of the same crop: the number of flowers produced per plant, the average duration of flowering for the variety, and the amount of pollen produced by individual flowers at the time of pollen shed. All of these characters can vary greatly from variety to variety, with genetically healthy, robust populations producing more copious amounts of pollen due to the combination of the three. This increased quantity of pollen can have an impact on the potential for outcrosses, based on the idea that more pollen grains in a specific environment will increase the chances of wind or insects carrying it to another location.

Plant breeders have also long known that pollen from different genetic backgrounds within a particular crop species can be quite different in its duration of viability and in the vigor expressed in the growth of its pollen tube after landing on the stigma of another

flower (see chapter 2, Reproductive Biology of Crop Plants). The genetic variability of the male sexual cell or gamete's ability to reach an egg cell (some see it as a competition of the male gametes) can definitely influence the number of outcrosses between two varietal populations when all else is equal.

Wind- or Insect-Pollinated?: The two different types of pollination found in the cross-pollinated species—wind or insects being the primary mode by which pollen is transferred between plants—generally require two different sets of numbers for the isolation distances needed between crops of the same species. They also require a different mind-set and strategy when you're making decisions as to the placement of isolation plots. Most cross-pollinated crop species are primarily pollinated by insects. The notable wind-pollinated vegetable crops are sweet corn and the members of the Amaranthaceae (beets, orach, spinach, and Swiss chard). The wind-pollinated crops have pollen that is lightweight for its size and has a surface morphology that enables it to become airborne easily and potentially travel far distances. The classic story that is told to put one in awe of how airborne and travel-worthy this pollen can be is the tale of beet pollen being collected at 5,000 ft (1,524 m) in an airplane flying over a sugar beet seed production area. How high and how far this windborne pollen can go before desiccating and losing viability is the real issue in this case, but the main point—that windborne pollen can travel quite far—is well taken when considering the greater isolation distances that are recommended for wind-pollinated species. These are the questions that must be answered when placing wind-pollinated crops. Is there a predominant wind pattern in the cropping area under consideration? Is the crop directly downwind from another crop that it can cross with? Is it upwind of a crop it can cross with?

Also, the issue of crop type within a species comes to the fore. In commercial seed production two different crop morphological types are often separated by a greater crop isolation distance. For instance a smooth-leaved spinach and a savoy-leaved spinach require greater minimum isolation than two smooth-leaved spinach varieties. Why? Because if a cross does occur between a savoy and a smooth type, the savoy trait will become obvious in the out-crosses that appear in the next generation of the smooth-leaved variety when it is grown by the produce farmer!

The extent of crossing possible in insect-pollinated crops can increase depending on the type and number of insect pollinators present. There is very little information on the distances that wild pollinators will travel when foraging for pollen. We now know a good deal more about the range of honeybees and the other semi-domesticated bees, which are currently becoming even more important as pollinators as modern problems like colony collapse disorder keep mounting for the honeybee. Certainly the density of pollinators present on a per-acre basis is quite important when considering chances of crossing with nearby crops. If pollinators are numerous, then their incentive to travel farther to forage is increased, and this may increase the rate of crossing at what may otherwise be an adequate isolation distance.

Key to Isolation Distance Terminology Used in Each Crop Section

The distances given are the minimum distances that are necessary to eliminate most naturally occurring chances of crossing between two crops of the same species. Crosses may still occur at these distances due to insects or other animals, wind, and sometimes even human activity. Beet, Swiss chard, and spinach seed workers will need to change clothes between fields of the same crop if working in more than one field per day, since pollen clinging to cloth can be transmitted and cause crosses.

Barriers are physical obstructions to pollen flow on the terrain between two crops of the same species. These may include hills, mountains, rock formations, forests, or other forms of dense vegetation of some height and stature. They can also include buildings or other structures that have a significant height or stature. When these obstructions are prominent on the landscape and occur between two seed production fields, the seed grower can often reduce the minimum isolation requirement by half. If the barrier is partial, then the grower or seed company must decide if it will be significant in slowing the flow of pollen and insects. In any event, the reduction in isolation distance will be less than the one-half reduction that is possible when the barrier is prominent on the landscape.

– 14 –
MAINTAINING ADEQUATE POPULATION SIZE

One of the important things for seed growers to be mindful of when growing a seed crop is that the genetic diversity of the crop should be maintained. Each variety that you grow is essentially a biologically unique population. A **population** is defined as a group of potentially intermating individuals of a single species occupying a given place at a given time. For any population to remain adaptive and genetically healthy, it must contain adequate genetic diversity to respond to the varied challenges of the environment that it will encounter across generations.

In wild plant or animal species a genetically healthy population is usually considered a collection of individuals, each having a somewhat different combination of heritable traits that contribute to the adaptation of the species to its environment. This diversity of genetically controlled characteristics from individual to individual in the population is of great advantage because it enables the population to continue to adapt. In essence, it represents a breadth of raw material for natural selection to act upon, allowing the population to stretch and change with challenges from the environment. Conservation biologists concerned with maintaining this genetic diversity in wild plant or animal populations understand that this is largely a numbers game and, for a population to remain viable, it must be as large as possible in order to harbor a good representation of the genetic variation of the

population. When populations drop below a certain number of individuals, the danger of losing specific genes (hence genetic traits) that exist in the population at a fairly low frequency can increase dramatically. Shifts in gene frequencies when populations are small can cause a condition known as **genetic drift**, resulting in populations that are significantly different in any number of traits that may make them unique or, even more importantly, that may contribute to their ability to withstand environmental challenges.

The varieties of our crops face the same dilemma. They are populations that need to survive environmental pressures much like their wild relatives. Hence, it is critical that we maintain as much of the inherent diversity that gives them an adaptive advantage in our agricultural environments. Indeed, in low-input agricultural ecosystems there may be additional environmental and cultural challenges that face the crop. The more genetic elasticity and breadth within a population, the greater potential plants may have for nutrient scavenging, horizontal resistance to diseases and pests, or resistance to heat, cold, drought, or excessive moisture (or many other traits that will become increasingly important in the changing face of agriculture and the environment). Therefore, it is imperative that when we reproduce a crop we remain mindful of the minimum population sizes recommended in this chapter.

POPULATION SIZE IN COMMERCIAL SEED PRODUCTION

In commercial seed production considerations of population size are rarely of any concern. Even small-scale commercial seed growers who are producing unique, high-value specialty herb, flower, or vegetable varieties for small alternative seed companies usually grow well in excess of 200 plants whenever they produce seed and thus are in little danger of narrowing the inherent genetic diversity of these varieties. Indeed, the most pressing problem concerning narrow genetic breadth in commercial seed production of specialty varieties of this sort (whether they are heritage varieties or older commercial varieties that were saved from extinction by gardeners) is that they may have already gone through a genetic bottleneck by being reproduced in relatively small populations in backyard gardens in previous generations. In this instance the seed company should screen all available sources of the variety before production and identify the most robust and true-to-type version for their production.

WHEN POPULATION SIZE BECOMES AN ISSUE

There are at least two situations when population size becomes an issue of concern in seed production. Seed is often reproduced in relatively small quantities when commercial seed companies produce stockseed, foundation seed, or breeder seed for the genetic maintenance of all stocks.

The other important situation when smaller population size becomes an issue is for genetic preservation. When seed banks and other grassroots seed preservation organizations increase seed of any collections, they are usually faced with the need to grow far more varieties than they have the resources and time to accommodate. Therefore, they are often faced with having to grow as small a population as they believe is possible, and yet at the same time preserve each variety as faithfully as they can.

THE MINIMUM NUMBER OF PLANTS NEEDED

In order to maintain this genetic diversity, anyone producing seed will usually need to grow a minimum population of between 20 and 200 plants, depending on the crop and its mode of pollination, whenever they want to faithfully reproduce the crop and maintain the inherent genetic variation of a particular variety. Growing a self-pollinated crop may require at least 20 to 50 flowering plants that are all contributing to the population after selection is completed. This number recognizes that many selfers are more genetically diverse than what is usually recognized by many modern plant breeders or seed companies that only work with genetically narrow pure lines in these crops. Many older selfers have definitely accumulated genetic breadth over time that needs to be preserved. If the self-pollinated variety is newly released and bred in a strict pedigree fashion from a single selected plant, then you can faithfully reproduce the variety from less than 10 plants. But an older selfer variety often needs a population of at least 20 plants and as many as 50 plants to preserve the inherent variation of the variety.

Non-hybrid cross-pollinated crops are always harboring a wealth of genetic variation, especially if they are robust, healthy varieties that are suitable to organic agriculture. The minimum number of 200 plants comes from my own experience with crossers and some of the disasters I have witnessed when seed crop populations have dropped below this number. Corn breeders and corn preservationists often cite this number as well when talking about retaining genetic breadth in a corn population. Always count on losing at least half of the population that you started out with due to attrition, especially when storing roots or selecting diligently against many of the obvious flaws that are revealed in older, poorly maintained populations or varieties of crossers. Remember that the final selection isn't truly done until after the plant is well established in the flowering stage and reproductive maladies appear. The best seed growers are definitely the ones who view their crops on a regular basis through all of the stages of their life cycle in anticipation of eliminating any poor plants. In some well-maintained, highly selected cross-pollinated crop varieties it is possible to reproduce them successfully with a population of as few as 80 to 100 plants, but it is up to you as the seed grower to make sure that you are maintaining the valuable genetic variation of that variety when you are growing a seed crop of this type with less than 200 plants. It is also important to remember that all of the numbers given in this section refer to the number of fully fertile, healthy plants, for both selfers and crossers, that contribute to the final population.

15

SEED CROP CLIMATES

While there are several dozen plant families that contain species of crop plants that are commonly used by different agricultural societies around the world, there are only nine families that house the great majority of seed-propagated vegetables that are the most important across most cultures worldwide. Through learning a bit about the characteristics of these nine families of the most cultivated vegetable crops, it is possible to get a better feel for which crops are best suited to a particular climate, especially when growing them as a seed crop. Certainly by learning more about the crop's history in cultivation, its evolutionary past, and how its life cycle has been shaped by nature and by human hands, it is possible to understand the environment that it will thrive in under cultivation.

SEED PLANT CHARACTERISTICS

There are a number of prominent characteristics of cultivated plants that are quite similar within the nine plant families in which most of our vegetable crops are found. One of the first things someone researching our cultivated crop plants finds is that closely related crops within a particular family usually share a number of prominent features. We know that different crops within the same family often share certain phenotypic traits, such as structural or reproductive characteristics. Flower structure has

long been a principal way of categorizing plants into families. The type and structure of the fruit, which is indeed a fertilized ovary of the flower, has also classically been used to assign different plants of the angiosperms (the true flowering plants) to various species and genera. As to structural features, we all know that crop species in the same family usually share a common leaf type, arrangement of their leaves on the main stem, type of stem, and so forth.

Plant structure can also be a reflection of the function of a particular part of the plant. Certainly as you get to know the different crop members of a plant family you may begin to see more of the commonalities among these species. This way of viewing crops can prove quite useful when you consider growing unfamiliar seed crops for the first time and realize that it is possible to culturally handle them in a similar fashion to a seed crop with which you have experience. Here are a few categories in which crops within a particular family share traits that will help you decide whether the crop is suited to your environment:

1. **Evolutionary past**
   a. Center of origin. Is your climate similar to that of its evolutionary past?
   b. Climate. Is your climate similar to the climate where it's currently grown?
   c. Structure and flower parts of the family definitely relate to shared ancestry.
2. **Environment.** Characterize the climate that the crop thrives in.

a. Cool-season crops need cool weather to mature high-germination seed.

b. Intermediate crops will grow in cool or warm climes and mature seed in warm conditions.

c. Heat lovers need heat to thrive and produce high-germ seed.

3. **Life cycle.** While some patterns exist across families, there are clearly families that contain annual/biennial/perennial species.

a. Annuals complete their entire life cycle in one season.

b. Winter annuals are planted for fall growth and flowering early in the next growing season.

c. Biennials need most of two seasons to complete their life cycle, with vernalization between the first season of vegetative growth and the second season of reproductive growth.

d. Perennials. This includes very few seed-propagated vegetable crops.

4. **Daylength sensitivity.** Is the crop sensitive to daylength?

a. Daylength-sensitive crops only flower at certain daylengths.

b. Daylength-neutral crops flower at various daylengths.

5. **Reproductive biology.** Self-pollinated species versus cross-pollinated species.

a. Cross-pollinated species. Is on-farm isolation possible?

i. Wind-pollinated. Pollen travels far and doesn't require insects.

ii. Insect-pollinated. Are pollinating insects present?

b. Self-pollinated species. How many on-farm isolations are possible?

i. Faithful selfers are highly self-pollinated; several crops are possible.

ii. Promiscuous selfers—how many isolations are possible?

6. **Presence of disease.** Is disease a limiting factor in your environment?

a. Diseases of the vegetative stage—is it a limiting factor?

b. Seedborne diseases—are they endemic and economically limiting?

7. **Presence of insect pests.** Are insects a limiting factor in your environment?

a. Insects of the vegetative stage—are these a limiting factor?

b. Insects of the seed—are they endemic and economically limiting?

CLIMATIC ZONES

Here is a reference list of the four major climatic types in which vegetable seed crops are grown. The important climatic considerations that determine each zone's suitability are given, followed by the crops that are most well adapted to that particular zone. Note that some crops are suited to more than one climate and therefore have a wider adaptation to environmental conditions for producing high quality.

Cool-Season Dry-Seeded Crops

All dry-seeded crops are formed in dry pods or in clusters along the stem of the plant and are essentially harvested like grains. They produce the best quality seed when they mature and are harvested in seasonally dry, low-humidity regions; the so-called Mediterranean climate. These cool-season, dry-seeded crops are best grown in the cooler reaches of the Mediterranean climate, where cool, often wet weather predominates during prolonged springs, and summers are mild and dry with little or no rainfall through harvest. Cool-season crops do not handle hot weather, especially through the earliest stages of their reproductive cycle. These crops form the highest quality seed when temperatures are generally somewhere between 60 and 75°F (16 to 24°C) during pollination, fertilization, and the earliest stages of embryo and endosperm development in late spring and early summer. After this initial formation and development of the seed they are able to tolerate average summer daytime high temperatures between 75 and 85°F (24 to 29°C) but thrive in relatively cool summers, especially where daytime high temperatures rarely exceed 80°F (27°C) to produce the highest-quality seed.

Seed Crops That Excel under These Conditions

spinach
beet
cilantro
Asian greens
cabbage
cauliflower
kohlrabi
Chinese cabbage
parsnip
mustards
Swiss chard

Warm-Season Dry-Seeded Crops

This climate is similar to the Cool-Season Dry-Seeded parameters above but with temperatures that are consistently warmer throughout all the months of the growing season. Warmer spring temperatures result in more rapid early growth and development for these crops over the cool-season dry-seeded crops. Daytime high temperatures during flowering and seed setting should generally not exceed 78 to 85°F (26 to 29°C). But after this initial formation and development of the seed these crops are able to routinely tolerate summer daytime average high temperatures between 85 and 92°F (29 to 33°C) when producing high-quality seed.

Seed Crops That Excel under These Conditions

broccoli
kale
collards

celery
radish
turnip
lettuce
Swiss chard
favas
peas
runner beans
parsley
endive
escarole
chicories

Hot-Season Dry-Seeded Crops

All dry-seeded crops do best when there is little or no rainfall during seed maturation and harvest. This lessens the incidence of diseases of all kinds, especially seedborne diseases, and it lowers the threat of excessive rainfall shattering the seedheads that form with all dry-seeded crops. While summer highs do regularly exceed 92°F (33°C), a number of these crops must complete their early reproductive stages of pollination and anthesis to mature a high-germinating, high-quality seed crop, while early season daytime temperatures are between 80 and 92°F (27 and 33°C).

Crops That Excel under These Conditions

garden beans
lima beans
edamame
carrot
onion
sweet corn

Hot-Season Wet-Seeded Crops

The wet-seeded moniker refers both to the fact that most of the fruit of these crops is wet but also to the method used to extract the fruit, which is extracted through a wet fermentation or a series of water rinses (see Seed Harvest for each individual crop).

These crops are all heat lovers from the moment they are planted. They depend on warm spring temperatures that average above 65°F (18°C), to establish good early growth and need warm nighttime temperatures to realize a decent yield and mature a high-germinating, high-quality seed crop. Temperatures may routinely exceed 90°F (32°C) during flowering and early fruit and seed set,* and unlike the dry-seeded crops, some humidity is tolerated; in fact, the presence of humidity often is responsible for holding the heat into the evening and nighttime hours.

Crops That Excel under These Conditions

cucumbers
melons
watermelons
summer squash
winter squash
bitter melon
eggplant
peppers
tomatoes

* The exception for this group is cucumber, which does prefer slightly cooler temperatures (see Cucumber: Climatic and Geographic Suitability).

16

SEEDBORNE DISEASES

A number of risks are inherent in all agricultural endeavors. One of these, when growing crops, is the threat of disease. For all farmers, diseases of their crops are something that should be identified, monitored, and possibly controlled through various organically acceptable methods. If you're producing seed crops you need to be especially attentive to any diseases that might appear, as these diseases can affect the quality of your current crop—and in some cases can cause disease in crops grown from your seed.

There are a number of plant pathogens of our cultivated crop species that can be carried from one generation of a crop to the next via the seed. If a seed crop develops a given disease during the production cycle and the causal pathogen infects the seed-bearing parts of the plant, either internally or externally, then certain species of pathogens may remain virulent on or within the seed and can be passed on to the next generation of plants produced from this seed. In this way certain pathogenic microorganisms are said to be seedborne. Fortunately, not all of the microorganisms that can cause crop diseases will infect seed or are routinely transmitted via seed. However, the pathogens that are routinely transmitted via seed greatly increase disease incidence.

If you intend to successfully produce high-quality organic seed, you should learn as much as possible about the important seedborne diseases of the crops you plan to grow. To understand the relationship of seedborne diseases to your seed crops, it is first important to understand the nature of plant disease in the broad sense.

PLANT DISEASE BASICS

Plant disease can be defined as any changes from the normal growth, structure, or physiological processes in a plant "that is sufficiently pronounced and permanent to produce visible symptoms or to impair quality and economic value."

The majority of plant diseases are the result of a parasitic microorganism interacting with a host under certain environmental conditions that are conducive to the development of disease. These biotic diseases are the type that will be addressed in this chapter, as all of the seedborne maladies that will be considered are a result of pathogenic microorganisms. Alternatively, abiotic diseases can result from inanimate conditions of the environment, such as adverse weather, deficiencies or excesses of certain minerals, or pollutants in the soil, air, or water. While agriculturalists don't usually think of damage to plants from frost or drought as disease, there is a long history of considering the effects of mineral nutrient deficiencies of the soil as disease. Examples of this in vegetables would include heart rot in table beets or crack-stem in celery, both of which are caused by a boron deficiency in the soil. Copper deficiency in onions results in pale, thin, and brittle skin development, which leads to poor storability of the bulbs.

THE DISEASE TRIANGLE

In order for plants to become infected with a phytopathogen, three major elements must be present under the right set of circumstances. A particular disease will grow on a specific crop species when these three factors are combined: (1) There must be a disease-causing agent, or pathogen; (2) the pathogen must be in the presence of a suitable host; and (3) this interaction of pathogen and host must occur under environmental conditions that favor disease development.

These elements of pathogen, host, and environment make up what is commonly referred to as the Disease Triangle, which emphasizes the interdependent nature of the three elements. The concept of the triangle is used as an illustration to emphasize the fact that these three elements are equally important for the formation and development of disease in plants.

The environmental conditions under discussion here must be present long enough for the pathogen to commence growth, establish infection, and ultimately spread through the crop. The pathogen–host interaction can be very exacting. Most pathogens are quite specific in their ability to infect a certain genus or a particular species—in fact, the virulence of a specific pathogen can even affect different varieties of a particular crop species to varying degrees.

The virulence and genetic makeup of the pathogen can vary. The genetic makeup of the host as it relates to the susceptibility or resistance to a specific disease can also vary greatly. Plant pathologists frequently characterize different races of a pathogen based on the ability of a particular strain to infect specific varieties

of a crop. When the variables of these two elements interact, there can be quite a range in the resulting degree of disease and amount of crop damage that occurs.

When this combination of factors is further influenced by environmental conditions conducive to the spread of disease, it is easy to see that the occurrence of disease can be quite variable based on the region and local weather conditions. If any of these three elements of the disease triangle is altered, then the extent of disease formation can be increased, lessened, or eliminated entirely.

THE IMPORTANCE OF SEEDBORNE DISEASES IN SEED CROPS

All good farmers are conscious of the diseases that can affect the crops in their particular environment. No matter what type of crop you're producing, a disease outbreak can affect the quality or yield of your harvest. If the disease is caused by a pathogen that can be transmitted to the next generation via the seed, then that pathogen can have far-reaching and deleterious effects beyond your fields.

To help you gain an understanding of the range of seedborne pathogens capable of infecting your crops, this chapter includes "Seedborne Microorganisms of Common Vegetable Crops." This table lists the most serious North American seedborne diseases of vegetable crops, many of which are ubiquitous throughout most agricultural regions of the world. This table is intended to supply vegetable seed growers with the most basic knowledge needed to further research a potential seedborne disease problem. By first learning which pathogen is infecting your crop, it is then possible to do further research and gain a working knowledge of the pathogen, its life cycle, and the scope of its possible damaging effects. By becoming intimately familiar with any disease-causing agent your plants may come into contact with, you can minimize its potential consequences.

Seedborne pathogens can have a wide range of effects, whether you're producing the seed for your own use or are growing a commercial crop. This range of effects includes: (1) producing infected seed that will contaminate your own seedstock; (2) producing infected seed as a commercial crop that will potentially spread the disease to any and all users, sometimes with devastating effects; and (3) infecting the current production of the seed crop to a point where there is diminished seed quality and yield, or even to the point that you can't produce a harvestable seed crop, or the seed is unusable.

Understanding the potential risks that are inherent in selling seed that may have a seedborne pathogen is a very serious consideration for anyone in the seed trade. There have been instances where seed companies have been sued in civil court for selling infected seed that has led to disease epidemics, with devastating financial losses suffered by the farmers planting the seed. In recent memory this was most dramatically demonstrated by epidemics of bacterial fruit blotch (caused by *Acidovorox avenae* subsp. *citrulli*) in watermelon crops. This bacterial pathogen is definitely in the severe class of seedborne microorganisms, and it is suspected that an epidemic can be started from one seed in 10,000 being infected with the bacteria. There are several cases where watermelon farmers who had epidemics of this disease were able to trace the cause of the outbreak to infected seed and brought suits against the seed companies involved. Many companies now require all watermelon seed customers to sign a waiver of any liability before they will sell watermelon seed to them.

PREVENTING SEEDBORNE DISEASES IN YOUR SEED CROPS

Many of the basic agricultural practices that can contribute to the control of seedborne diseases on the farm are basically the same methods that are universally used by knowledgeable farmers to minimize the chances of either getting a disease outbreak or having it spread on the farm. By identifying the disease-causing agent, gaining insight into its life cycle, and understanding the environmental conditions that allow the disease to spread, it is possible to minimize the incidence of disease if it should appear. These practices can make the difference between a bountiful seed harvest and a seed crop that is unusable.

Your first line of defense against seedborne disease involves integrating a number of cultural practices when producing a seed crop. These cultural practices should always be used when growing seed under any conditions, but they are especially important in situations where there is a specific threat of a severe seedborne disease (as defined in the table "Seedborne Microorganisms of Common Vegetable Crops").

1. Always start with disease-free seed, especially for commercial seed production. If you don't specifically know the history of the seedstock you are planting, then test the lot before planting for any seedborne pathogens that can be especially troublesome in your area.
2. Avoid fields or districts that have a history of seedborne diseases, especially those that can be severe in their effects.
3. Isolate seed crops from any commercial production of the same crop species.

4. Use crop rotation and eliminate crop residues efficiently and effectively.

5. Clean all equipment used in fields that have or are suspected of having seedborne pathogens before using it in other fields.

6. When any symptoms of seedborne diseases appear, rogue or eliminate the affected plants from the field. In many cases it is important to remove the infected plants from the field and dispose of them properly.

7. If environmental conditions occur that are conducive to the spread of a severe or intermediate disease, especially if you've already identified symptoms of the disease in the crop, it may be important to use certified organic crop protection chemicals or other cultural practices to avoid or lessen the effects of the disease.

8. Have phytosanitary field inspections or laboratory testing done on seed lots. Many state or provincial agricultural institutions will offer field inspections or laboratory testing to determine if seed is infected with seedborne pathogens.

9. Perform hot-water treatment on seed lots (of appropriate species) where there is a known or suspected presence of seedborne pathogens.

The goal of these integrated steps to avoid seedborne diseases can be summarized as: Start with disease-free seed; avoid planting into or near fields with any seedborne pathogens; and make sure during the formation, maturation, harvest, and cleaning of the seed that it is disease-free. This is the ideal, one that may not always be possible in the real world, but once you become familiar with the execution of the above steps and understand the seedborne diseases that you are likely to encounter, then it is possible to minimize the effects of these potentially devastating diseases.

GUIDE TO USING THE SEEDBORNE DISEASE TABLE

The accompanying chart, "Seedborne Microorganisms of Common Vegetable Crops," offers an extensive list of the seedborne pathogens that can pose a serious risk of being transmitted via vegetable seed crops. In this section I have categorized the different types of seedborne pathogens based on the potential severity of the effects if a disease develops on a subsequent crop grown from a particular seed lot. Some of the pathogens listed are geographically specific to certain climatic regions, while others are nearly ubiquitous in their geographic range across most agricultural regions. It is incumbent on you to know which of these diseases are endemic to your region and to be fully aware of the most serious class of diseases (listed as severe in this chart) for any crop you are growing. This chart should help you easily learn which diseases are caused by which seedborne pathogens and the level of potential risk that each of these disease-causing organisms poses to subsequent crops.

I have categorized the seedborne pathogens that are most commonly recognized as leading to disease outbreaks via infected seed. This is based on the relative severity of their potential effects. In practical language, I have adopted the terms severe, intermediate, and minor as general descriptors for the classes of seedborne pathogens. These classes represent the relative degree of disease that will usually occur when inoculum of one of these seedborne pathogens is the source of infection in a crop planted from a particular lot of seed. While these classes can give you a frame of reference for the potential risk of these microorganisms, it is important to realize that there are no clear-cut lines of demarcation between classes of pathogens and the potential severity of the diseases they represent. In other words, a disease that may be intermediate or even minor in its effects under most circumstances may be severe under a different set of circumstances.

As with so many aspects of biology, the severity of the diseases that may develop from any given parasitic organism can vary considerably depending on the environmental factors present at the time of infection and the genetic factors of the pathogen and host. A number of other factors can also increase or decrease the relative severity of any potential seedborne disease outbreak, including: (1) the environment in which the seed is grown; (2) the amount of inoculum present at the time of harvest; (3) the relative virulence of the particular strain of the pathogen; (4) the degree of resistance to the disease of the infected host; and (5) the environment where the infected seed is being planted.

Combined, these factors create a very dynamic relationship of potential disease that all seed growers and seed companies need to consider. Please note: These classes of seedborne organisms are not definitive!

Classes of Seedborne Microorganisms

#1 Severe: These are the seedborne pathogens that pose the greatest risk to subsequent crops. Infected seed is usually the primary source of inoculum for these pathogens. Pathologist Jodie Lew-Smith of High Mowing Seeds calls these "seed-specific diseases." The resultant diseases are often among the most destructive for a number of crops. If seed infection is controlled, then the disease is typically controlled.

#2 Intermediate: These seedborne pathogens pose a significant risk to subsequent crops. While infected seed may not always be the primary source of inoculum for disease outbreak, there are many cases in which seed infected with these pathogens can lead to severe outbreaks when planted under environmental conditions that favor the growth and spread of a particular disease.

#3 Minor: These are important crop pathogens that can be transmitted via seed, but infected seed is not usually the primary source of inoculum for subsequent outbreaks of disease.

Seedborne Microorganisms of Common Vegetable Crops

Crops	Crop Species	Disease	Pathogen	Microorganism Type	Disease Severity Classes	Seed Treatment	Notes
Alliaceae							
Onion, Leek	*Allium sativa, A. ampeloprasum*	Neck rot	*Botrytis aclada*	Fungus	1		
		Black mold	*Aspergillus niger*	Fungus	2		
		Smudge	*Colletotrichum circinans*	Fungus	3		
		Botrytis blight	*Botrytis cinerea*	Fungus	3		
		Basal rot	*Fusarium oxysporum f. sp. cepae*	Fungus	3		
		Stemphylium leaf blight	*Stemphylium vesicarium*	Fungus	3		
		Purple blotch	*Alternaria porri*	Fungus	3		
Amaranthaceae							
Beet, Swiss Chard	*Beta vulgaris*	Phoma/Canker	*Phoma betae*	Fungus	2	20 min. at 125°F/52°C	
		Downy mildew	*Peronospora farinosa f. sp. betae*	Fungus	2	20 min. at 125°F/52°C	
		Bacterial leaf spot	*Pseudomonas syringae pv. aptata*	Bacteria	2		
		Cercospora leaf spot	*Cercospora beticola*	Fungus	2	20 min. at 125°F/52°C	

Crops	Crop Species	Disease	Pathogen	Microorganism Type	Disease Severity Classes	Seed Treatment	Notes
		Fusarium wilt	Fusarium oxysporum f. sp. betae	Fungus	3		
		Ramularia leaf spot	Ramularia betae	Fungus	3		
Spinach	Spinacia oleracea	Downy mildew	Peronospora farinosa f. sp. spinaciae	Fungus	2	25 min. at 122°F/50°C	
		Fusarium wilt	Fusarium oxysporum f. sp. spinaciae	Fungus	2	25 min. at 122°F/50°C	
		Verticillium wilt	Verticillium dahliae	Fungus	2		
		Cucumber mosaic virus (CMV) subgroup II		Virus	2	25 min. at 122°F/50°C	
		Anthracnose	Colletotrichum dematium	Fungus	2		
		Cladosporium leaf spot	Cladosporium variabile	Fungus	2		
		Stemphylium leaf spot	Stemphylium botryosum	Fungus	2		
		White rust	Albugo occidentalis	Fungus	2	25 min. at 122°F/50°C	
		Pseudomonas leaf spot	Pseudomonas syringae pv. spinaciae	Bacteria	2		
Apiaceae							
Carrot	Daucus carota	Bacterial leaf spot	Xanthomonas campestris pv. carotae	Bacteria	1	20 min. at 122°F/50°C	
		Black rot/Black crown	Alternaria radicina	Fungus	1		

Crop	Disease	Pathogen	Type		Treatment
	Alternaria leaf blight	*Alternaria dauci*	Fungus	1	20 min. at 122°F/50°C
	Cercospora leaf blight	*Cercospora carotae*	Fungus	2	20 min. at 122°F/50°C
	Itersonilia canker	*Itersonilia perplexans*	Fungus	3	
Celery, Celeriac *Apium graveolens*	Northern bacterial blight	*Pseudomonas syringae pv. apii*	Bacteria	1	
	Cercospora (early) blight	*Cercospora apii*	Fungus	2	
	Septoria (late) blight	*Septoria apiicola*	Fungus	1	
	Black rot/Black crown	*Alternaria radicina*	Fungus	1	
Cilantro *Coriandrum sativum*	Bacterial leaf spot	*Pseudomonas syringae pv. coriandricola*	Bacteria	2	
Parsley *Petroselinum crispum*	Alternaria leaf blight	*Alternaria petroselini*	Fungus	2	
	Black rot/Black crown	*Alternaria radicina*	Fungus	2	
	Cercosporoid leaf blight	*Passalora punctum*	Fungus	3	
	Septoria blight	*Septoria petroselini*	Fungus	1	
	Itersonilia canker	*Itersonilia perplexans*	Fungus	3	
Parsnip *Pastinaca sativa*	Itersonilia canker	*Itersonilia pastinaca*	Fungus	1	
	Black rot/Black crown	*Alternaria radicina*	Fungus	1	
	Phoma canker	*Phoma complanata*	Fungus	2	
	Cercosporoid leaf blight	*Passalora pastinacae*	Fungus	3	

Crops	Crop Species	Disease	Pathogen	Microorganism Type	Disease Severity Classes	Seed Treatment	Notes
Asteraceae							
Lettuce	*Latuca sativa*	Lettuce drop	*Sclerotinia sclerotiorum, S. minor*	Fungus	2	30 min. at 118°F/48°C	
		Lettuce mosaic virus (LMV)		Virus	1	30 min. at 118°F/48°C	
		Septoria leaf spot	*Septoria lactucae*	Fungus	1	30 min. at 118°F/48°C	
		Bacterial leaf spot	*Xanthomonas campestris* pv. *vitians*	Bacteria	2		
		Gray mold	*Botrytis cinerea*	Fungus	3		
Endive, Escarole, Chicories	*Cichorium endivia, C. intybus*	Black leaf spot	*Alternaria cichorii*	Fungus	2		
		Gray mold	*Botrytis cinerea*	Fungus	3		
Brassicaceae							
Various	*Brassicacea* spp., *Raphanus sativus*	Black rot	*Xanthomonas campestris* pv. *campestris*	Bacteria	1	20 or 25* min. at 122°F/50°C	
		Xanthomonas leaf spot	*Xanthomonas campestris* pv. *armoraciae*	Bacteria	1		
		Xanthomonas leaf spot	*Xanthomonas campestris* pv. *raphani*	Bacteria	1		
		Black leg	*Phoma lingam*	Fungus	1	20 or 25* min. at 122°F/50°C	

Crop/Family	Host	Disease	Pathogen	Type		Hot water treatment	Notes
		Black spot	*Alternaria brassicicola*	Fungus	1		
		Alternaria diseases	*Alternaria brassicae*	Fungus	1	20 or 25* min. at 122°F/50°C	
		Downy mildew	*Peronospora parasitica*	Fungus	3		
		Fusarium wilt	*Fusarium oxysporum* f. sp. *conglutinans*	Fungus	3		
		Peppery leaf spot	*Pseudomonas syringae* pv. *maculicola*	Bacteria	3		
		Verticillium wilt	*Verticillium dahliae*	Fungus	3		
		White rust/White blister	*Albugo candida*	Fungus	2		

*20 min. for broccoli, cauliflower, kale, collards, kohlrabi, rutabaga, turnip; 25 min. for Brussels sprouts and cabbage

Cucurbitaceae

Crop/Family	Host	Disease	Pathogen	Type		Notes
Cucumber, Squash, Melon	*Cucumis sativus*, *Cucurbita* spp., *Cucumis melo*	Fusarium crown and foot rot	*Fusarium solani* f. sp. *cucurbitae*	Fungus	2	
		Cucumber green mottle mosaic virus (CGMMV)		Virus	2	Cucumber only
		Scab	*Cladosporium cucumerinum*	Fungus	3	
		Fusarium wilt	*Fusarium oxysporum* f. sp. *cucurbitacearum*	Fungus	3	
		Squash mosaic virus (SqMV)		Virus	1	

Crops	Crop Species	Disease	Pathogen	Microorganism Type	Disease Severity Classes	Seed Treatment	Notes
Watermelon	Citrullus lanatus	Gummy stem blight	Didymella bryoniae	Fungus	1		
		Bacterial leaf spot	Xanthomonas campestris pv. cucurbitae	Bacteria	2		
		Charcoal rot	Macrophomina phaseolina	Fungus	3		Melon only
		Melon necrotic spot virus (MNSV)		Virus	2		Melon only
		Anthracnose	Colletotrichum orbiculare	Fungus	2		
		Bacterial leaf spot	Xanthomonas campestris pv. cucurbitae	Bacteria	2		
		Bacterial fruit blotch	Acidovorax avenae subsp. citrulli	Bacteria	1		
		Fusarium wilt	Fusarium oxysporum f. sp. niveum	Fungus	2		
Fabaceae							
Bean	Phaseolus spp.	Bean common mosaic virus (BCMV)		Virus	1		
		Gray mold	Botrytis cinerea	Fungus	3		
		Alfalfa mosaic virus (AMV)		Virus	2		
		Southern bean mosaic virus (SBMV)		Virus	2		
		Anthracnose	Colletotrichum spp.	Fungus	1		

Crop	Disease	Pathogen	Type	
	Cercospora leaf blotch	Cercospora canescens	Fungus	2
	Bacterial wilt	Curtobacterium flaccumfaciens pv. flaccumfaciens	Bacteria	1
	Charcoal rot/Ashy stem blight	Macrophomina phaseolina	Fungus	3
	Angular leaf spot	Phaeoisariopsis griseola	Fungus	2
	Bacterial brown spot	Pseudomonas syringae pv. syringae	Bacteria	1
	Common bacterial blight	Xanthomonas campestris pv. phaseoli	Bacteria	1
	Leaf and pod blight	Phoma exigua var. exigua	Fungus	2
	Halo blight	Pseudomonas syringae pv. phaseolicola	Bacteria	1
Edamame Glycine max	Anthracnose	Colletotrichum spp.	Fungus	2
	Bacterial blight	Pseudomonas savastanoi pv. glycinea	Bacteria	2
	Phomopsis seed decay	Phomopsis longicolla	Fungus	3
	Septoria leaf spot	Septoria glycines	Fungus	3
	Pod and stem blight	Diaporthe phaseolorum var. sojae	Fungus	3
	Soybean mosaic virus (SMV)		Virus	1
	Northern stem canker	Diaporthe phaseolorum var. caulivora	Fungus	3

Crops	Crop Species	Disease	Pathogen	Microorganism Type	Disease Severity Classes	Seed Treatment	Notes
Garden Pea	Pisum sativum	Tobacco ringspot virus (TRSV)		Virus	1		
		Ascochyta blights	Ascochyta spp., Mycosphaerella pinodes, Phoma medicaginis var. pinodella	Fungus	1		
		Gray mold	Botrytis cinerea	Fungus	3		
		Anthracnose	Colletotrichum spp.	Fungus	2		
		Fusarium wilt and root rot diseases	Fusarium spp.	Fungus	2		
		Pea early browning virus (PEBV)		Virus	2		
		Pea seedborne mosaic virus (PSbMV)		Virus	1		
		Downy mildew	Peronospora viciae	Fungus	3		
		Bacterial blight	Pseudomonas syringae pv. pisi	Bacteria	1		
Poaceae							
Sweet Corn	Zea mays	Black bundle disease	Acremonium strictum	Fungus	2		
		Southern leaf blight	Helminthosporium maydis	Fungus	3		
		Black kernel rot	Botryodiplodia theobromae	Fungus	3		

		Disease	Pathogen	Type	
		Goss's bacterial blight and wilt	*Clavibacter michiganensis subsp. nebraskensis*	Bacteria	3
		Anthracnose	*Colletotrichum graminicola*	Fungus	3
		Fusarium diseases	*Fusarium spp.*	Fungus	2
		Northern corn leaf spot	*Helminthosporium carbonum*	Fungus	3
		Charcoal rot	*Macrophomina phaseolina*	Fungus	3
		Maize dwarf mosaic virus (MDMV)		Virus	3
		Seedling blight	*Penicillium spp.*	Fungus	3
		Downy mildew	*Peronosclerospora sorghi*	Fungus	3
		Head smut	*Sporisorium holci-sorghi*	Fungus	3
		Ear rot; stalk rot; seedling blight	*Stenocarpella macrospora, S. maydis*	Fungus	2
Solanaceae					
Eggplant	*Solanum melongena*	Charcoal rot/Ashy stem blight	*Macrophomina phaseolina*	Fungus	3
		Anthracnose	*Colletotrichum spp.*	Fungus	2
		Eggplant mosaic virus (EMV)		Virus	2
		Phomopsis fruit rot; blight	*Phomopsis vexans*	Fungus	2
		Phytophthora diseases	*Phytophthora capsici*	Fungus	3

Crops	Crop Species	Disease	Pathogen	Microorganism Type	Disease Severity Classes	Seed Treatment	Notes
Pepper	Capsicum annuum	Bacterial wilt	Pseudomonas solanacearum	Bacteria	3		
		Tomato mosaic virus (ToMV)		Virus	2		
		Anthracnose	Colletotrichum spp.	Fungus	2	25 min. at 122°F/50°C	
		Alfalfa mosaic virus (AMV)		Virus	2		
		Charcoal rot/Ashy stem blight	Macrophomina phaseolina	Fungus	3		
		Phytopthora diseases	Phytophthora capsici	Fungus	3		
		Bacterial wilt	Pseudomonas solanacearum	Bacteria	3		
		Tomato mosaic virus (ToMV)		Virus	2		
		Pepper mild mottle virus (PMMoV)		Virus	1		
		Bacterial spot	Xanthomonas campestris pv. vesicatoria	Bacteria	1	25 min. at 122°F/50°C	
Tomato	Solanum lycopersicum	Anthracnose	Colletotrichum spp.	Fungus	3	25 min. at 122°F/50°C	
		Alfalfa mosaic virus (AMV)		Virus	2		
		Bacterial spot	Xanthomonas campestris pv. vesicatoria	Bacteria	1	25 min. at 122°F/50°C	

Disease	Pathogen	Type		Treatment
Early blight	Alternaria solani	Fungus	2	25 min. at 122°F/50°C
Bacterial canker	Clavibacter michiganensis subsp. michiganensis	Bacteria	1	25 min. at 122°F/50°C
Stem rot; stem canker	Didymella lycopersici	Fungus	3	
Leaf mold	Fulvia fulva	Fungus	3	25 min. at 122°F/50°C
Fusarium wilt	Fusarium oxysporum f. sp. lycopersici	Fungus	2	25 min. at 122°F/50°C
Phytopthora diseases	Phytophthora capsici	Fungus	3	
Bacterial wilt	Pseudomonas solanacearum	Bacteria	3	
Bacterial speck	Pseudomonas syringae pv. tomato	Bacteria	1	25 min. at 122°F/50°C
Syringae leaf spot	Pseudomonas syringae pv. syringae	Bacteria	2	
Tobacco mosaic virus (TMV)		Virus	2	
Tomato mosaic virus (ToMV)		Virus	2	

– 17 –

STOCKSEED BASICS

Stockseed is the seed that is used to produce a high-quality, commercial seed crop. The term stockseed is a quality designation. Stockseed implies that the seed used by a reputable, commercial seed grower to produce a seed crop has in fact had more scrutiny and care in its development than the more common production seed that is sold to the farmer. As the "genetic source" of commercial seed lots, it is held to a higher quality standard in the scrutiny applied by the seed company. This includes the extent to which each plant in the population is evaluated for its phenotypic uniformity and trueness to type. The increased quality demands of producing high-quality stockseed also dictate that it is grown at a greater isolation distance than a commercial seed lot would be produced.

The true goal of developing and using stockseed is to maintain the specific genetic identity and purity of a particular variety across generations of usage. The ability to maintain varietal integrity and trueness to type within any variety has become the hallmark of the seed companies that earn the respect of farmers and in turn earns their longtime loyalty as customers. Indeed, the reliability of a favored variety for a farmer's specific needs of climate, maturity, quality, market acceptance, or resistance to environmental challenges can provide a stability that is one of the economic pillars of agriculture. Certainly, every grower with experience knows the economic consequences of growing a variety that has worked well in

the past and has somehow changed—perhaps with the overall phenotype being different, or perhaps with the appearance of an inordinate percentage of off-types (plants that vary from the normal ideotype) when a new seed lot of the variety was purchased.

When a farmer buys seed that is corrupted and is not true to type or does not have the varietal quality that is expected, the problem can often be traced back to the stockseed that produced that particular problematic lot of the variety. Worse yet, some seed companies will produce a seed crop from last year's production lot, with all of the pitfalls that can occur during production (see Avoiding Genetic Changes, "The Basics," below). A seed company that doesn't maintain high-quality stockseed won't stay in business for long. Indeed, a company that doesn't use stockseed is courting disaster.

AVOIDING GENETIC CHANGES

The Basics

Each time a crop variety is reproduced sexually to increase its seed, its varietal integrity can only be changed by the altering of the genetic makeup of the population that constitutes the "variety." There are essentially four ways in which the genetics of a population can change, producing undesirable variants that are not within the agreed-upon description of the variety. The first three—off-types, outcrosses, and seed mixes—can usually be seen by an experienced eye that checks the crop regularly throughout the different stages of the crop's development. This must be done by someone who not only knows the phenotype of the variety but who also has a good general knowledge of the form and developmental stages of the particular species. In this way the stockseed person performing the genetic maintenance by roguing (identifying and removing individuals from the field that aren't true to type) will be able to recognize subtle differences at different stages of the crop's life. The fourth way that genetic changes occur is through genetic drift. This often happens when unintended selection occurs, either by the environment conditions in which the crop is being produced or because of the cultural methodology used by the grower.

Here are the four major circumstances that can change the genetic constitution of a crop when it is grown for seed:

1. Off-Types. A true off-type is usually considered to be part of the original makeup of the varietal population, but is outside the defined phenotype. Through the normal workings of genetic recombination and minor variation that is slowly revealed in all sexually reproducing species, changes can occur that are different enough from the regularly defined phenotype to be taken out or rogued when evaluating the variety in a stockseed production plot. These new variants were often attributed to genetic mutation in the past.

2. Outcrosses. As the name implies, this is when the variety you are producing has unwanted matings with individuals outside the varietal population. The unwanted parents from outside the varietal population can be from another population of the same crop species or a wild version of the same species (for example, Queen Anne's lace is thought to be a feral form of carrot), or they might be from a closely related wild species (for instance, wild lettuce or wild radish). Ultimately there are three categories of outcrosses.

a. The first case is the most common type of outcross. The variety that you are producing can cross to another variety of the same crop due to the fact that the two populations are growing too close to each other (see chapter 13, Isolation Distances for Maintaining Varietal Integrity). Also, exceptional environmental events can occur, such as an insect pollinator foraging farther afield than usual or the pollen of a species with windborne pollen traveling farther than normal (viable beet and corn pollen has been retrieved by researchers more than 5 mi/8 km from flowering fields of these crops). People working with seed crops can also carry pollen from one isolated field to another on their clothes when inspecting or roguing fields, and stockseed people at the larger production research companies will change clothes between roguing different stockseed fields of the same crop species.

b. The second type of outcross occurs between different crops of the same species. Seed growers must also learn all of the cases of distinctly different crops that are of the same species and therefore will cross readily. For instance, Swiss chard and beets are both *Beta vulgaris* (as are sugar beets and mangels) and are fully sexually compatible. The Brassicaceae family has perhaps the most cases of this type of sexual compatibility between different crops. Almost everyone that works with vegetable seed knows that cabbage, cauliflower, broccoli, kale, collards, and Brussels sprouts are all *Brassica oleracea* and that cross-pollinations will readily occur between them. But how many people know that this species also includes kohlrabi? How about seemingly quite different crops like turnips, Chinese cabbage, and Siberian kale? These three are all *B. rapa* and cross-pollinate readily! There are numerous other examples that are referred to in each of the crop family chapters.

3. Seed Mixes. The threat of physical seed mixing can occur at almost all of the myriad steps involved in seed production. When handling seed at each stage of the processes listed below, smart seedspeople should always be mindful of the importance of their work (no daydreaming) and the potential consequences of allowing a seed mix, and they should always give every seed lot they handle two identification tags: one for the outside of the bin, box, or bag, and another that's placed inside the container, nestled into the seed in case the outer tag is torn off or falls off during handling. Both tags should list the species, the variety name, and the lot number, and they should be written in pencil or indelible ink. The stories that are told of unmarked or poorly marked seed would make your hair stand on end, as it can cause chaos and confusion that frequently takes years to rectify, at great expense of time, money, and land. Here are the processes in seed production during which seed mixes most often occur:

a. Planting. There are at least two ways that seed can be mixed when planting. Seed companies or growers will often plant more than one variety of the same crop on the same day or within the same week. If you're using seed-planting equipment, all seed boxes or hoppers must be meticulously cleaned between varieties to avoid seed mixing. Also, all seed that is taken to the field should be in clearly marked bags and the identification information double-checked before you pour any seed into the planter. Sometimes a portion of stockseed is transferred from clearly marked bags in storage to poorly marked field bags, and mistakes are made. The second thing that can happen at planting is there may be volunteers of the same crop that you are planting growing in the vicinity. If you are in a seed production area for the crop of interest, or if you have recently grown the same seed crop in an adjacent field, volunteers of last year's crop could be growing in an adjacent field (or, if proper rotation was ignored, could even be in the very same field). This is a more common cause of mixing than many realize.

b. Harvesting. Because threshing equipment is frequently used on multiple seed crops of the same species, the risk of contaminating subsequent seed lots with seed from a previous threshing is great. There are many places for seed to get stuck in most threshers, and it is a real act of patience to thoroughly clean the thresher out after each lot, but it is of critical importance that this is done. Threshing smaller lots by hand can also pose a problem, as many growers will bring the bundles of several different varieties of the same crop to thresh in the same place: a shed, greenhouse, or other dry, protected space. Then the seed, which is all in close proximity by this time, is often threshed onto the same tarp or other surface. During threshing, seed is flying everywhere and could land in the next pile waiting to be threshed, thus creating a seed mix. Of course, most people are very conscious of the inherent dangers of this, but it is important to remember the pitfalls of both machine and hand threshing and make sure mixing doesn't occur.

c. Cleaning. Seed can be mixed during seed cleaning much as physical mixing can occur during threshing. Seed-cleaning equipment must be meticulously cleaned between lots, and if you're cleaning seed in small lots by hand use great caution in performing the steps by making sure that all lots are physically separated during the entire process and that all surfaces and seed-cleaning implements are thoroughly cleaned between lots. Remember to keep all lots marked with two tags at every step of the cleaning.

d. Packaging. Stockseed for any particular variety is usually produced only once every 4 to 7 years. Depending on the size of the seed operation, the projected seed production for that period, and which species is being produced, you may need anywhere from several ounces of seed (as with a fine-seeded flower) to several hundred pounds for a large-seeded crop such as beans or corn. (Certainly many of the transnational seed companies produce much greater quantities of stockseed.) But for most smaller regional seed companies, stockseed occupies a fairly small seed room that is isolated from the regular production seed room. Stockseed is often packaged by hand in close quarters, and there is always the risk of people grabbing the wrong bag for packaging or putting two different lots into the same bag (which everyone involved in seed has heard of happening). Also, stockseed is economically precious when considering the time necessary to produce it, and when some of it is spilled the people packaging it tend to scoop it up and put it back into the bag. But this is a very risky practice—there may be seed of another variety of the same species on the floor. Hence the old seed-company adage, "If it hits the ground, leave it down!"

4. Genetic Drift. Genetic drift refers to the genetic changes that can happen by chance, especially in any smaller population (remember that all varieties are populations), as we frequently have when we produce stockseed. Random matings between plants in the population can lead to some genes increasing or decreasing in their frequency in the population. When this happens, genes for certain traits that are an important part of the variety may decrease, and genes that confer traits that are not representative of the population (or are downright undesirable) may increase, thus changing the varietal integrity of the variety. Alternatively, some unintended selection can occur due to the environment in which the crop was grown, or the cultural techniques used. For instance, if you have a lettuce variety that is bolt-resistant or long standing (it remains vegetative for a long time during the season before bolting), and you then attempt to produce seed of it in an

area with a shorter growing season than usual, you may inadvertently select it to be a faster-bolting variety, especially if it has some variability in the bolting trait. If grown in a shorter-season area where the last plants to bolt do not mature, you will only harvest seed from the earliest bolters. Over several generations you will inadvertently select for an earlier-bolting lettuce, and thus the variety will no longer be as long standing as it was when you started this process.

CONCLUSIONS

Producing high-quality commercial seed requires a commitment to quality. All commercial plantings must come from highly selected, well-maintained stockseed. If stockseed is well maintained, the chances of off-types, out-crosses, seed mixes, or appreciable genetic drift appearing are minimized, helping to ensure a seed crop with a high level of varietal purity.

Because production plots rarely get the level of scrutiny necessary to catch all of these potential variants, it is important that both growers and seed companies know that the planting stock they are using is true to type for the variety and requires only minimum checking at a couple of pivotal times in its life cycle to ensure that there aren't any variants in the field. The more familiar they are with the specific crop variety and the particular stockseed that is used for planting, the better prepared both growers and seed companies will be to watch for and recognize any plants that don't fit the ideotype of the variety. This is where important partnerships between farmers and seed companies can be built.

GLOSSARY

Annual: The type of plant that normally starts from seed, produces flowers, sets seed, and then dies within one growing season.

Anther: The sac-like structure of the male part (stamen) of a flower in which the pollen is formed. There are normally two lobes, which dehisce at anthesis and allow the pollen to disperse.

Anthesis: The flowering stage when the anthers burst, pollen is shed, and the stigma is ready to receive the dispersed pollen.

Apomixis: The development of seed without the sexual fusion of an egg and a sperm cell.

Asexual reproduction: Reproduction by vegetative means without fusion of two sexual cells.

Biennial: The type of plant that normally produces only vegetative growth the first growing season, overwinters, and then produces a seed crop, after which the plant dies. The plant requires two growing seasons to complete its life cycle.

Carpel: The individual structure of the pistil composed of the ovary, style, and stigma. It is equivalent to the pistil if there is only one carpel, but in most vegetable crops there are multiple, fused carpels that comprise the pistil and subsequent fruit.

Cell: The basic structural unit of living organisms. The cell is made up of protoplasm enclosed, in plants, in a cell wall. The protoplasm consists of a nucleus and cytoplasm, which contains plastids and other small bodies. Cells may contain a cavity filled with starches, salts, sugars, or other substances.

Character, characteristic: An identifiable hereditary property of a variety, such as the specific component for flower color, a morphological detail, or resistance to a specific disease.

Composite variety: A plant population in which at least 70% of progeny result from cross of the parent lines.

Conditioning of seed: A term used to describe the cleaning of seed, usually to improve mechanical purity.

Cotyledon: Seed leaves of the embryo, which are usually thickened for storage of food reserves and may serve as true foliage leaves.

Crop rotation: Growing of crops in a regularly scheduled sequence on the same land area, as contrasted with continuous culture of one crop or the growing of different crops in haphazard order.

Cross-pollinate: To fertilize with pollen from another plant.

Cross-pollinated crops: The group of plant species that bear seeds that are largely the result of cross-fertilization between the pollen and ovaries of different plants of that species, though there is always some percentage of seed produced by self-pollination.

Cultivar: A variety of a cultivated crop. Short for "cultivated variety." See Variety.

Cytoplasm: The contents of a cell between the nucleus and the cell wall. In reproduction, the cytoplasmic constituents from the female parent become part of the cytoplasm of the offspring. There may be a transfer of traits determined by factors contained in the cytoplasm not associated with chromosomes.

Damping-off: A disease syndrome of seed and young seedlings that can be caused by several different fungi and/or bacteria, sometimes occurring due to a complex of more than one pathogen at a time.

Daylength: The number of hours of daylight in each 24-hour cycle.

Dehiscence: The act or process by which a structure splits or opens when mature, such as with a fruit or anthers. Can commonly refer to either an anther splitting to release pollen or to a dried fruit splitting to release seed. (The latter case is also referred to as *shattering* by seed growers.)

Detassel: To remove the tassel or pollen-producing organ at the top of a corn plant before pollen is released, usually in hybrid corn production.

Dioecious: A type of plant that has stamens and pistils on different plants. The plants are unisexual, so plants of both types must be present to produce seed. An example is spinach.

Diploid: An organism with two sets of chromosomes.

Disease: Any changes from the normal growth, structure, or physiological processes in a plant that are sufficiently pronounced and permanent to produce visible symptoms or to impair quality and economic value.

Domesticate: To convert a wild plant species into a cultivated crop by selection and adaptation.

Emasculation: The removal of anthers from a flower before pollen is shed to prevent self-pollination.

Embryo: The rudimentary plant contained within the seed.

Endosperm: The storage tissue in most angiosperm seed, containing nourishment for the developing embryo.

Feral: A domesticated species that has reverted to a wild or untamed state.

Fertilization: Fusion of a sperm nucleus from the pollen tube with the egg nucleus from an ovary.

Foundation seed: The approved progeny of Breeder or Select seed produced by seed growers. Foundation is the highest official pedigreed class of commercial seed.

Frame: The basic vegetative structure of a plant before flower initiation and growth. In many vegetable crops this is the basal rosette of vegetation that develops before the growth of a flower stalk.

Fungi: Microscopic plants consisting of a vegetative structure called a mycelium, lacking chlorophyll and conductive tissue and reproduced by spores.

Gene: The unit of inheritance composed of DNA (deoxyribonucleic acid) forming part of a chromosome, which controls the transmission and development of inherited characteristics. Its effect is generally conditioned by its interaction with other genes, the cytoplasm, and environmental factors.

Genetic code: The means of storing genetic information as sequences of nucleotide bases in the chromosomal DNA.

Genetic modification (GM): The deliberate modification of an organism's characteristics by manipulation of DNA that occurs without normal sexual recombination.

Genotype: The genetic composition of the plant.

Germination: The resumption of growth by the embryo and development of a young plant from seed.

Germplasm: Plant genetic resources that serve as a basis of crop improvement or a reservoir of heritable traits for breeding or research. This term is often used to refer to the total hereditary makeup of a particular species.

Haploid: A term indicating one-half the normal diploid complement of chromosomes.

Hard seed: A seed that is dormant due to the nature of its seed coat, which is impervious to either water or oxygen, or both.

Head: An inflorescence in which the floral units on the peduncle are tightly clustered and surrounded by a group of flower-like bracts called an involucre. An example is sunflower.

Heterosis: See Hybrid vigor.

Heterozygous: A term that refers to not breeding true for a specific hereditary characteristic, usually determined by both dominant and recessive alleles. Plants may be heterozygous for some characteristics and homozygous for others.

Hilum: The scar remaining on the seed (ovule) at the place of its detachment from the seed stalk (funiculus).

Homozygous: Refers to breeding true for a specific hereditary characteristic, usually by identical alleles.

Host: A plant that is invaded or parasitized by a disease-producing agent and from which the parasite obtains its sustenance.

Hybrid: The first-generation progeny of a cross between two different plants of the same species, often resulting in a plant that is more vigorous and productive than either parent. These are crop varieties that are created from the controlled crossing of genetically distinct, highly uniform parents. The term F1 designates that the resultant hybrid seed is of the first filial generation after the cross. This is the hybrid seed of commerce. Seed that is saved from the F1 hybrid plants will not breed true.

Hybrid vigor: The increase in vigor of hybrids over their parental inbred types, also known as heterosis.

Hypocotyl: The part of the embryo axis between the cotyledons and the primary root that gives rise to the stalk of the young plant.

Inbred: A relatively true breeding strain resulting from several successive generations of controlled self-fertilization or back-crossing to a recurrent parent through selection or its equivalent.

Increase: To multiply a quantity of parent seed through a generation of production.

Inflorescence: The arrangement of flowers of a plant, such as umbel, raceme, spike, tassel, and panicle.

Inoculum: Material from a pathogen, such as fungal spores, bacteria, etc., that is capable of spreading and infecting a plant with disease.

Intellectual property protection (IPP): The legal measures, such as patents, Plant Breeders' Rights, trademarks, contracts, and licenses, usually developed to ensure adequate returns on investment in the development of a new technology.

Isolation distance: The distance required to isolate pedigreed seed crops from other crops that may be a source of pollen or seed contamination. Used by most seed certification agencies as one of the requirements for maintaining varietal purity of pedigreed seed crops.

Kernel: The seed of a grain plant.

Legume: A plant that is a member of the Fabaceae family, having the characteristic of forming nitrogen-fixing nodules on roots and dry, dehiscent, multiseeded pods.

Lodging: The displacement of the stems of crops from an upright position.

Male sterility: An inherited factor, useful in hybrid seed production; it prevents viable pollen from being produced.

Monocotyledon: A term referring to plants with a single seed leaf at the first node of the lead shoot or stem.

Monoculture: The production of a single species, often the same cultivar, over a wide geographic area.

Monoecious: A type of plant that has stamens and pistils in different floral structures on the same plant, as in maize and most cultivated cucurbit crop varieties.

Morphology: The form, structure, and development of plants.

Multiline: A composite (blend) population of several genetically related lines of a self-pollinated crop.

Mutation: A sudden heritable variation that results from changes in a gene or genes.

Nicking: The synchronization of the receptivity of the male sterile plant to the maximum pollen load of the pollinator for cross-pollination in hybrid seed production.

Off-type: Plants in a variety that deviate in one or more characteristics from the official description of the variety.

Open-pollinated (OP): Seed produced as a result of natural pollination, as opposed to hybrid seed produced as a result of a controlled pollination.

Open-pollinated (OP) variety: A heterogeneous cultivar resulting from a cross-pollinated crop allowed to interpollinate freely during seed production (as opposed to a controlled crossed pollination).

Outcross: A cross-pollination between two different crop varieties (usually unintended) of the same species. The term is also used by seed growers as a noun to refer to a plant resulting from such a cross.

Parasite: An organism that subsists in whole or in part on living tissue at the expense of the host.

Pathogen: Any organism capable of causing disease in a host or range of hosts.

Perennial: A plant that produces vegetative growth each year without replanting.

Perfect flower: A flower having both staminate (male) and pistillate (female) organs.

Petiole: The stalk of a leaf.

Phenotype: The appearance of an individual, as contrasted with its genetic makeup or genotype. A set of observable characteristics of an individual or group, usually determined by genotype and environment. Also used to describe a group of individuals with similar appearance—though not necessarily identical genotypes.

Pistil: The complete female organ within a flower. The pistil produces the ovule which contains a stigma, a style, and one or more carpels that house the developing seeds after fertilization of an ovary or multiple ovaries.

Plant breeding: An organized effort to produce progressively improved plants.

Pollen: The cells that are borne in the anthers of flowers and contain the male generative cells.

Pollen parent: The parent plant furnishing the pollen that fertilizes the ovules of the other parent in the production of seed.

Pollination: The process by which pollen is transferred from an anther to the stigmatic surface of the pistil of a flower.

Population: A community of individuals of a particular species that share a common gene pool.

Progeny: Offspring or plants grown from seed.

Propagule: Any type of plant or part of a plant that is used to propagate the plant. This can include a seedling, cutting, bulb, tuber, scion wood, steckling, etc.

Raceme: A type of flower cluster in which single-flowered pedicels are arranged along the sides of a flower shoot terminus. There is space along the shoot between the pedicels.

Radicle: A rudimentary root; the lower end of the hypocotyl of the embryo and the primary root of the seedling.

Resistance: The ability of a plant to remain relatively unaffected by the effects of disease due to inherited factors that it possesses. This inherited quality is often expressed by degrees, with a slight, moderate, or high degree of resistance possible.

Rogue (noun): A term used by seed growers and plant breeders to indicate an off-type plant in a population of plants representing a specific variety. Rogue plants may originate as a result of several circumstances: (1) an unwanted cross-pollination with a different variety of the same species in a previous generation; (2) an unwanted mixture of seed from two different varieties; (3) a plant within a population with one or more unique traits that is the result of a spontaneous genetic change or mutation; and (4) a volunteer plant from an earlier planting appearing in the same field.

Roguing (transitive verb): The act of identifying and removing rogue plants from seed fields at any stage of a seed crop's growth or development.

Saprophyte: An organism that subsists upon dead organic matter and inorganic materials.

Sclerotia (sclerotium): A compact mass of fungal hyphae, usually with a black outer surface and white inside. It may remain dormant for long periods and eventually gives rise to more fungus.

Seed: A mature ovule consisting of an embryonic plant together with a store of food (the endosperm) and surrounded by a protective coat. In angiosperms, the seed develops after the fertilization of an egg cell by a male generative cell from a pollen grain.

Seedborne: Carried on or in the seed, usually referring to a pathogen (as in seedborne disease).

Seed coat: The protective covering of a seed, usually composed of inner and outer integuments.

Seedling: A young plant grown from seed.

Selection: Identification of individual plants that contain traits that are more desirable than other plants in the population and are used to contribute to the next generation.

Self-fertilization: The fusion of the male and female gametes from a single individual plant, resulting from a self-pollination.

Self-incompatibility: The failure of pollen from an individual to fertilize the flower and successfully produce seed in the same plant.

Self-pollinate: The transfer of pollen to the stigma within the same flower. Self-pollinated crops always have perfect flowers.

Self-pollinated crops: The group of plant species that bear seeds that are largely the result of self-fertilization of each

plant's ovaries, though there is always some percentage of seed produced by cross-pollination.

Shatter: The dehiscence of the seed from a dry fruit (e.g. a pod or a silique) that occurs with minimal stimulation, often used when seed is released when the seed crop is still in the field before the grower has harvested the crop.

Silique: A fruit that is borne of a gynoecium with two carpels with seeds attached to a thickened central membrane. Siliques are only found in the Brassicaceae.

Single-cross hybrid: The first generation of a cross between two specified inbred lines.

Stamen: The complete male organ within the flower. The part of the flower, bearing the male reproductive cells, composed of the anther on a filament (stalk).

Steckling: The prepared root of a biennial crop like carrots, beets, or parsnips that are evaluated, trimmed, and prepared for replanting using the root-to-seed method.

Stigma: The upper part of the pistil that receives the pollen.

Stockseed: Seed used to produce a crop eligible for pedigreed status.

Strain: A term used to designate an improved or divergent selection of a crop variety.

Tassel: The flower cluster at the tip of monoecious plants, such as corn, comprising pollen-bearing flowers (staminate inflorescence).

Top-cross hybrid: The first generation of a cross between an inbred line and an open-pollinated variety.

Varietal purity: Trueness to type or variety.

Variety: A group of plants of a particular species that shares a set of characteristics or traits that differentiates it from other varieties of the same crop. These characteristics must be distinct and relatively uniform across all of the plants of the variety. A synonym of cultivar. A variety must be uniform, stable, and reproducible. Variety names are typically framed by single quotation marks: 'Scarlet Nantes' carrot.

Vernalization: The exposure of seed and young plants to certain conditions of cold temperature and photoperiod, which promotes floral induction without development of the plant.

Vigor: The vitality or strength of germination, especially under unfavorable conditions.

Volunteer plants: Unwanted plants growing from residual seed from the previous crop.

Weed: Any plant growing in a place where it is considered a nuisance. Usually denotes uncultivated plants growing in fields.

Weed seed: In commercial seed lots the weed seed component is given as the percentage by weight of the seed lot, which is composed of seed of plants considered to be weeds.

Winter annual: A plant that develops a seedling stage in the early fall, becomes vernalized over the winter, and then produces vegetative and reproductive growth the following season.

REFERENCES

Bates, D. M., R. W. Robinson, and C. Jeffrey, eds. 1990. *Biology and Utilization of the Cucurbitaceae.* Ithaca, NY: Cornell University Press.

Brewster, J. L. 1994. *Onions and Other Vegetable Alliums.* Vol. 3 of *Crop Production Science in Horticulture.* Oxon, UK: CABI.

Carle, B. 2011. Personal communication on cucumber, melon, and watermelon. Hollar Seeds, Rocky Ford, CO.

Chupp, C., and A. F. Sherf. 1960. *Vegetable Diseases and Their Control.* New York: Wiley and Sons.

du Toit, L. J. 2004. Personal communication on seedborne diseases. Washington State University, Mt. Vernon, WA.

Ferriol, M., and B. Pico. 2008. "Pumpkin and Winter Squash" in *Vegetables I: Asteraceae, Brassicaceae, Chenopodiaceae, and Cucurbitaceae.* Vol. 1. of *Handbook of Plant Breeding*, edited by J. Prohens and F. Nuez. New York: Springer.

Franklin, D. F. 1953. "Growing carrot seed in Idaho." *Agricultural Experiment Station Bulletin*, Moscow, ID: University of Idaho.

George, Raymond A. T. 1999. *Vegetable Seed Production*, Second edition. Oxon, UK: CABI.

Goldman, I. L. 2012. Personal communication on table beets. University of Wisconsin, Madison, WI.

Goldman, I. L., and J. P. Navazio. 2008. "Table Beets" in *Vegetables I: Asteraceae, Brassicaceae, Chenopodiaceae, and Cucurbitaceae.* Vol. 1. of *Handbook of Plant Breeding*, edited by J. Prohens and F. Nuez. New York: Springer.

Griffiths, A. E., W. W. Jones, and A. H. Finch. 1946. "Vegetable and Herb Seed Production in Arizona." *Agricultural Experiment Station Bulletin 204.* Tucson, AZ: University of Arizona.

Hawthorn, L. R., and L. H. Pollard. 1954. *Vegetable and Flower Seed Production.* New York: Blakiston Co.

Kakizaki, Y. 1924. "The Flowering Habit and Natural Crossing in the Eggplant." *Japanese Journal of Genetics* 3:29–38.

Kreitlow, K. W., C. L. Lefebrve, J. T. Presley, and W. J. Zaumeyer. 1961. "Diseases that Seeds Can Spread." In *Seeds: The Yearbook of Agriculture.* Washington, D.C.: US Government Printing Office.

Leukel, R. W. 1953. "Treating Seeds to Prevent Diseases." In *Plant Diseases: The Yearbook of Agriculture.* Washington, D.C.: US Government Printing Office.

McCormack, J. H. 2004–2010. *Isolation Distances: Principles and Practices of Isolation Distances for Seed Crops: An Organic Seed Production Manual for Seed Growers in the Mid-Atlantic and Southern U.S.* Saving Our Seeds. http://www.savingourseeds.org /pubs/isolation_distances_ver_1pt8.pdf.

———. 2005–2010. *Pepper Seed Production: An Organic Seed Production Manual for Seed Growers in the Mid-Atlantic and South.* Saving Our Seeds. http://www .savingourseeds.org/pubs/pepper_seed _production_ver_1pt4.pdf.

McGregor, Samuel E. 1976. "Pumpkin and Squash." In *Insect Pollination of Cultivated Crop Plants.* Agriculture Handbook No. 496. Washington, D.C.: US Government Printing Office.

Meader, E. M. 1988. Personnel communication on gynoecious cucumbers. University of New Hampshire, Durham, NH.

Merrick, L. C. 1990. "Systematics and Evolution of a Domesticated Squash, Cucurbita argyrosperma, and Its Wild and Weedy Relatives." In *Biology and Utilization of the Cucurbitaceae*. edited by D. M. Bates, R. W. Robinson, and C. Jeffrey. Ithaca, NY: Cornell University Press.

———. 2012. Personal communication on winter squash. Iowa State University, Ames, IA.

Morton, F. 2006–2012. Personal communication on Asteraceae and Brassicaceae. Wild Garden Seed, Philomath, OR.

Myers, J. R. 2011. Personal communication on runner beans, common beans, and garden peas. Oregon State University, Corvallis, OR.

Nabhan, Gary Paul. 2011. Personal communication on chilis. University of Arizona, Tucson, AZ.

Paris, H. S. 2008. "Summer Squash." In *Vegetables I: Asteraceae, Brassicaceae, Chenopodiaceae, and Cucurbitaceae*. Vol. 1. of *Handbook of Plant Breeding*, edited by J. Prohens and F. Nuez. New York: Springer.

Robinson, R. W., and D. S. Decker-Walters. 1997. *Cucurbits*. Vol. 6 of *Crop Production Science in Horticulture*. Oxon, UK: CABI.

Rubatzky, Vincent E., C. F. Quiros, and P. W. Simon. 1999. *Carrots and Related Umbelliferae*. Vol. 10 of *Crop Production Science in Horticulture*. Oxon, UK: CABI.

Rubatzky, Vincent E. and Mas Yamaguchi. 1997. *World Vegetables: Principles, Production, and Nutrition Values*. Second edition. New York: Chapman and Hall.

Rupp, R. 1987. *Blue Corn and Square Tomatoes: Unusual Facts About Common Vegetables*. Pownal, VT: Storey Communications.

Ryder E. J. 1998. *Lettuce, Endive, and Chicory*. Vol. 9 of *Crop Production Science in Horticulture*. Oxon, UK: CABI.

Schudel, H. L. 1952. "Vegetable Seed Production in Oregon." *Agricultural Experiment Station Bulletin* 512. Corvallis, OR: Oregon State College.

Shepherd, L. M. 2012. Personal communication on seedborne diseases. Iowa State University, Ames, IA.

Simon, P. W. 2012. Personal communication on carrots. University of Wisconsin, Madison, WI.

Stackman, E. C., and J. G. Harrar. 1957. *Principles of Plant Pathology*. New York: Ronald Press Company.

Staub, J. E., M. D. Robbins, and T. C. Wehner. 2008. "Cucumber." In *Vegetables I: Asteraceae, Brassicaceae, Chenopodiaceae, and Cucurbitaceae*. Vol. 1. of *Handbook of Plant Breeding*, edited by J. Prohens and F. Nuez. New York: Springer.

Tracy, W. F. 2012. Personal communication on sweet corn. University of Wisconsin, Madison, WI.

Walker, J. C. 1952. *Diseases of Vegetable Crops*. New York: McGraw Hill Book Co.

———. 1957. *Plant Pathology*. Second edition. New York: McGraw Hill Book Co.

Watts, Ralph L., and Gilbert S. Watts. 1940. *The Vegetable Growing Business*. War Department Education Manual EM 885. Madison, WI: United States Armed Forces Institute.

Wehner, T. C. 2008. "Watermelon." In *Vegetables I: Asteraceae, Brassicaceae, Chenopodiaceae, and Cucurbitaceae*. Vol. 1. of *Handbook of Plant Breeding*, edited by J. Prohens and F. Nuez. New York: Springer.

Whitaker, T. W., and G. N. Davis. 1962. *Cucurbits: Botany, Cultivation, and Utilization*. New York: Interscience.

Whitaker, T. W., and R. W. Robinson. 1986. "Squash Breeding." In *Breeding Vegetable Crops*, edited by M. J. Bassett. Westport, CT: AVI Publishing.

INDEX

ABOUT THE AUTHOR

Emily Flan

John Navazio, PhD is the senior scientist and a plant breeder with the Organic Seed Alliance. He also serves as the organic seed research and extension specialist for Washington State University. John lives in Port Townsend, Washington, on the Salish Sea. *The Organic Seed Grower* is his first book.

Dr. Navazio trains farmers and students in organic seed production practices and the techniques used in on-farm plant breeding. His own breeding work has resulted in a number of new vegetable varieties with improved quality and flavor, as well as a greater ability to compete with weeds and resist disease, cold temperatures, drought, and other challenges common to organic farming systems. He also works closely with farmers across North America to develop crop varieties for regional seed independence through participatory plant-breeding projects.

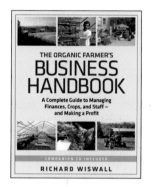

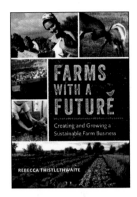

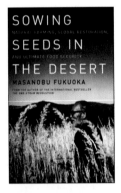
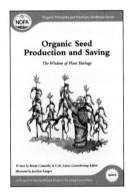